D0503178
LIBRARY
LEWIS - CLARK STATE COLLEGE
WITHDRAWN

R_s - resistivity of adjacent shale
R_{SFL} - resistivity of Spherically Focused Log*
R_t - resistivity of uninvaded zone
R_w - resistivity of formation water
R_{wa} - apparent formation water resistivity
R_{xo} - resistivity of flushed zone
S_h - hydrocarbon saturation $(1.0 - S_w)$
S.N. - short normal log
SNP - sidewall neutron porosity
SP - spontaneous potential
SPI - secondary porosity index
SSP - static spontaneous potential $SSP = -K \log (R_{mf}/R_w)$
S_{wirr} - irreducible water saturation
S_{wa} - water saturation of uninvaded zone (Archie method)
S_{wr} - water saturation of uninvaded zone (Ratio Method)
S_w/S_{xo} - moveable hydrocarbon index
S_{xo} - water saturation of flushed zone
T_f - formation temperature
V_{sh} - volume of shale

Log Names Used In Text

bulk density log
caliper log
Compensated Density Log
Combination Gamma Ray Neutron-Density Log
Combination Neutron-Density Log
Compensated Neutron Log
density log
Dual Induction Log
Dual Induction Focused Log
Dual Laterolog*
electrode logs
gamma ray log
Induction Electric Log
induction log
Lateral Log
Laterolog*
Laterolog-8*
microcaliper
Microlaterolog* (MLL)*
Microlog* (ML)*
Microspherically Focused Log* or (MSFL)*
neutron log
normal logs
nuclear logs
porosity log
Proximity Log* or (PL)*
resistivity logs
Resistivity Spherically Focused Log* or (R_{SFL})*
short normal log
Sidewall Neutron Log
sonic log
Spectralog**
Spherically Focused Log* (SFL)*
spontaneous potential log (SP)

Basic Well Log Analysis for Geologists

By
George B. Asquith
Pioneer Production Corporation

with
Charles R. Gibson
Alpar Resources Inc.

Methods in Exploration Series
Published by
The American Association of Petroleum Geologists
Tulsa, Oklahoma USA

Published by
The American Association of Petroleum Geologists
Tulsa, Oklahoma 74101, USA

Copyright • 1982 by
The American Association of Petroleum Geologists
All Rights Reserved

Published October 1982
Second Printing (revised), August 1983
Library of Congress Catalog Card No. 82-73052
ISBN: 0-89181-652-6

For AAPG:
Editor: M. K. Horn
Science Director: E. A. Beaumont
Project Editors: A. L. Asquith, R. L. Hart

Table of Contents:

Acknowledgements:

The construct of this book would have been entirely different had it not been for the creative contributions of Edward A. Beaumont and Ronald L. Hart. As Science Director of the American Association of Petroleum Geologists, Mr. Beaumont early-on recognized the need for a logging course designed especially for geologists. His efforts and encouragement led to the development of the AAPG school on basic logging from which the text was derived. The considerable editorial talents of Ronald L. Hart, Manager of the AAPG Projects Department and his assistance with formating and writing figure captions, helped ensure the book would meet its goal of introducing the reader to fundamental concepts of well logging.

Perhaps the most significant contribution of all, however, was by Ann L. Asquith. She helped her husband with both the writing and editing of the manuscript. Her unflagging efforts to improve readability, her assistance with writing, and her suggestions concerning content were an incalculable asset. She was assisted in her editing tasks by Robert V. Brown, who critically read the manuscript and offered many useful suggestions; the text's introduction owes much to his insight. Robert J. Mitchell also lent his technical expertise to a review of the manuscript, as did Edith C. Campbell and Leon Williams. Their assistance is recognized and gratefully acknowledged.

Many charts and figures used in the text were provided by Dresser Industries and by Schlumberger Well Services. Their cooperation in allowing reproduction of these items, and their unwavering courtesy eased the task of authorship.

By assisting her husband with some of the drafting and graphic layout work, Pearl Gibson helped ensure that the text's complex figures would be legible and easily understood. The quality of the graphics work was also enhanced by Rick Blackburn's efforts on behalf of the photographic reproduction of various charts and figures. Bette Haimes typed the finished manuscript copy; her commitment to accuracy, with a difficult and often tedious task, does not pass unnoticed.

Preface:

This book is a basic introduction to open hole logging.

Study of the properties of rocks by petrophysical techniques using electric, nuclear, and acoustical sources is as important to a geologist as the study of rock properties by more conventional means using optical, x-ray, and chemical methods. Nevertheless, despite the importance of petrophysics, it is frequently underutilized by many geologists who are either intimidated by logging terminology and mathematics, or who accept the premise that an in-depth knowledge of logging is only marginally useful to their science because, they feel, it more properly belongs in the province of the log analyst or engineer.

The enormous importance of logging dictates that as geologists, we put aside old notions and apply ourselves diligently to learning log interpretation. The rewards are obvious; in fact, no less than achieving an understanding of the ancient record hangs in the balance. And, it is likely that the success or failure of an exploration program may hinge on a geologist's logging expertise.

In the interest of conciseness, and so that logs used most often in petroleum exploration are thoroughly discussed, the text is restricted to open hole logs. I hope that the reader initiates his or her own study of other log types which are beyond the scope of this book.

Unfortunately, learning about open hole logging requires more of the reader than a light skimming of the text's material. The plain truth is that a great deal of hard work, including memorizing log terminology, awaits the serious student; and even then, a facility with logs develops only after plenty of real-life experience. The intent here is simply to provide a foundation of knowledge which can be built upon later. Consequently, many exceptions to rules are left to more advanced books.

It is quite possible that some colleagues will raise objections about the lack of time devoted to tool theory; they may also comment on the paucity of qualifying statements in the text. These objections are understood and indeed there may be disagreements about what constitutes over-simplification. In defense of brevity, it should be pointed out that the surfeit of information available on petrophysics often discourages all but the most ardent beginner. Certainly, many of the difficult decisions which had to be faced in preparing the manuscript dealt with selecting information judged indispensable at an elementary level.

Many in the audience will note frequent references to a book by Douglas Hilchie, Golden, Colorado entitled *Applied Open Hole Log Interpretation* (1978). For those who are interested in expanding their knowledged of logs, his book will be a great help. Another helpful book is *The Glossary of Terms and Expressions Used in Well Logging,* The Society of Professional Well Log Analysts (1975), which explains the meaning of logging terms by extended definitions.

Finally, a last word - a substantial effort was expended to ensure that a minimum number of errors would appear in the text. However, given the nature of the subject and the almost infinite possibility for mistakes, there may be slip-ups, regardless; hopefully they will not be too serious.

George B. Asquith
Pioneer Production Corporation
Amarillo, Texas
October, 1982

Biographical Sketches:

George B. Asquith: George B. Asquith received his Ph.D. degree in geology from the University of Wisconsin/Madison and has some 15 years experience throughout North America involving geological consulting, prospect development and evaluation, research, and teaching. In addition to independent consulting work, he has held various positions with Humble Oil and Refining Co.; Atlantic Richfield Co.; Alpar Resources Inc., Search Drilling Company, and with the University of Wisconsin, West Texas State University, and Killgore Research Center. He is presently Exploration Coordinator, for the Pioneer Production Corporation, Amarillo, Texas.

Dr. Asquith has authored two books, *Subsurface Carbonate Depositional Models,* and *Log Analysis by Microcomputer.* He has also written numerous articles and abstracts in the fields of carbonate petrology, sandstone petrology, and computer geology, and has served as a reviewer for the AAPG Bulletin, Texas Journal of Science, and Journal of Sedimentary Petrology.

His areas of specialization in petroleum exploration include: subsurface carbonate and clastic depositional models, identification of lithologies from logs, and computer applications to log interpretation. He has applied his areas of specialization to a number of different basins including the Anadarko, San Juan, Permian, and Williston basins, and also in the Gulf Coast (onshore and offshore), Central Texas, the Rocky Mountains, and Canada.

During 1979-1981, Dr. Asquith presented short course lectures for the American Association of Petroleum Geologists throughout the United States and in Canada and Brazil. He is currently a lecturer and Science Advisor with the American Association of Petroleum Geologists' Continuing Education Program and serves as an AAPG Visiting Petroleum Geologist.

Charles R. Gibson: Charles Richard Gibson is exploration manager and vice-president for Alpar Resources, Inc., Perryton, Texas. As a geologic undergraduate in 1965, he was employed by the Colorado Fuel and Iron Corporation in their Arizona project for geologic field mapping of iron ore deposits and base metal geochemistry exploration, collecting soil samples and running wet chemical analyses. Education was postponed for military duty where, at the termination of his advanced military training, he was selected and qualified to serve with the 3rd Infantry (The Old Guard), Fort Myer, Virginia. Returning to the University of Southern Colorado, Gibson received his B.S. degree in geology in 1970, and was granted a Graduate Teaching Assistantship to continue advanced studies at West Texas State University.

Since joining Alpar Resources, Inc., in 1972, Gibson has been involved in a diverse range of subsurface clastic and carbonate exploration and development studies from the Gulf Coast to the Williston basin and has co-authored several published technical papers.

He has a special interest in applying computerized log analyses to solve complex lithologic and production problems. Gibson obtained his M.S. degree from West Texas State University in 1977. He is certified by the A.I.P.G, A.A.P.G., and is a member of the Society of Professional Well Log Analysts.

Publisher's Note:

Because most new geologists come out of college with little understanding of the industry's primary tool, and because many experienced geologists use logs only as a means to correlate productive zones (unaware of the many other applications of logging), we have published this book.

As with the other titles in the AAPG Methods in Geology Series, we intend for this book to become a training standard in both industry and academia. The book is oriented toward geologists rather than engineers, and can be used in a class environment or as a self-help program. A set of six case histories in Chapter 8 provides the reader with diverse, yet typical, log-based decisions founded in both geology and economics.

As a special note of thanks, AAPG acknowledges the logging companies and engineers who cooperated with their advice and examples. Because it is important to offer examples in a book of this nature, specific Schlumberger and Dresser log types are mentioned by name; this is in no way an endorsement of these two companies, nor does it reflect on the fine logging service companies whose examples were not used. In the text and examples, the single asterisk (*) indicates a mark of Schlumberger; a double asterisk (**) indicates a trademark of Dresser Industries, Inc.

Note also, that many service company charts are overprinted with colored ink to highlight an example. This selection is the author's and the associated service company is not responsible for its accuracy.

AAPG Publications
Tulsa, Oklahoma

BASIC RELATIONSHIPS OF WELL LOG INTERPRETATION

Introduction

As logging tools and interpretive methods are developing in accuracy and sophistication, they are playing an expanded role in the geological decision-making process. Today, petrophysical log interpretation is one of the most useful and important tools available to a petroleum geologist.

Besides their traditional use in exploration to correlate zones and to assist with structure and isopach mapping, logs help define physical rock characteristics such as lithology, porosity, pore geometry, and permeability. Logging data is used to identify productive zones, to determine depth and thickness of zones, to distinguish between oil, gas, or water in a reservoir, and to estimate hydrocarbon reserves. Also, geologic maps developed from log interpretation help with determining facies relationships and drilling locations.

Of the various types of logs, the ones used most frequently in hydrocarbon exploration are called *open hole* logs. The name open hole is applied because these logs are recorded in the uncased portion of the well bore. All the different types of logs and their curves discussed in the text are this type.

A geologist's first exposure to log interpretation can be a frustrating experience. This is not only because of its lengthy and unfamiliar terminology, but also because knowledge of many parameters, concepts, and measurements is needed before an understanding of the logging process is possible.

Perhaps the best way to begin a study of logging is by introducing the reader to some of the basic concepts of well log analysis. Remember that a borehole represents a dynamic system; that fluid used in the drilling of a well affects the rock surrounding the borehole, and therefore, also log measurements. In addition, the rock surrounding the borehole has certain properties which affect the movement of fluids into and out of it.

The two primary parameters determined from well log measurements are porosity, and the fraction of pore space filled with hydrocarbons. The parameters of log interpretation are determined both directly or inferred indirectly, and are measured by one of three general types of logs: (1) electrical, (2) nuclear, and (3) acoustic or sonic. The names refer to the sources used to obtain the measurements. The different sources create records (logs) which contain one or more curves related to some property in the rock surrounding the well bore (see Society of Professional Well Log Analysts, 1975). For the reader unfamiliar with petrophysical logging, some confusion may develop over the use of the word *log*. In common usage, the word *log* may refer to a particular curve, a suite or group of curves, a logging tool (sonde), or the process of logging.

Rock properties or characteristics which affect logging measurements are: *porosity, permeability, water saturation,* and *resistivity*. It is essential that the reader understand these properties and the concepts they represent before proceeding with a study of log interpretation.

Porosity—can be defined as the percentage of voids to the total volume of rock. It is measured as a percent and has the symbol ϕ.

$$\text{Porosity } (\phi) = \frac{\text{volume of pores}}{\text{total volume of rock}}$$

The amount of internal space or voids in a given volume of rock is a measure of the amount of fluids a rock will hold. The amount of void space that is interconnected, and so able to transmit fluids, is called *effective porosity*. Isolated pores and pore volume occupied by adsorbed water are excluded from a definition of effective porosity.

Permeability—is the property a rock has to transmit fluids. It is related to porosity but is not always dependent upon it. Permeability is controlled by the size of the connecting passages (pore throats or capillaries) between pores. It is measured in darcies or millidarcies and is represented by the symbol K_a. The ability of a rock to transmit a single fluid when it is 100% saturated with that fluid is called *absolute* permeability. *Effective* permeability refers to the presence of two fluids in a rock, and is the ability of the rock to transmit a fluid in the presence of another fluid when the two fluids are immiscible.

Formation water (connate water in the formation) held by capillary pressure in the pores of a rock serves to inhibit the transmission of hydrocarbons. Stated differently, formation water takes up space both in pores and in the connecting passages between pores. As a consequence, it may block or otherwise reduce the ability of other fluids to move through the rock.

Relative permeability is the ratio between effective permeability of a fluid at partial saturation, and the permeability at 100% saturation (absolute permeability). When relative permeability of a formation's water is zero, then the formation will produce water-free hydrocarbons (i.e. the relative permeability to hydrocarbons is 100%). *With increasing relative permeabilities to water, the formation will produce increasing amounts of water relative to hydrocarbons.*

Water saturation—is the percentage of pore volume in a rock which is occupied by formation water. Water saturation is measured in percent and has the symbol S_w.

$$\text{water saturation } (S_w) = \frac{\text{formation water occupying pores}}{\text{total pore space in the rock}}$$

Water saturation represents an important log interpretation concept because you can determine the hydrocarbon saturation of a reservoir by subtracting water saturation from the value, one (where 1.0 = 100% water saturation).

Irreducible water saturation or $S_{w\,irr}$ is the term used to describe the water saturation at which all the water is adsorbed on the grains in a rock, or is held in the capillaries by capillary pressure. At irreducible water saturation, water will not move, and the relative permeability to water equals zero.

Resistivity—is the rock property on which the entire science of logging first developed. Resistance is the inherent property of all materials, regardless of their shape and size, to resist the flow of an electric current. Different materials have different abilities to resist the flow of electricity.

Resistivity is the measurement of resistance; the reciprocal of resistivity is *conductivity.* In log interpretation, hydrocarbons, the rock, and freshwater all act as insulators and are, therefore, non-conductive and highly resistive to electric flow. Saltwater, however, is a conductor and has a low resistivity. The unit of measure used for the conductor is a cube of the formation one meter on each edge. The measured units are ohm-meter^2/meter, and are called ohm-meters.

$$R = \frac{r \times A}{L}$$

Where:

R = resistivity (ohm-meters)
r = resistance (ohms)
A = cross sectional area of substance being measured (meters^2)
L = length of substance being measured (meters)

Resistivity is a basic measurement of a reservoir's fluid saturation and is a function of porosity, type of fluid (i.e. hydrocarbons, salt or fresh water), and type of rock. Because both the rock and hydrocarbons act as insulators but saltwater is conductive, resistivity measurements made by logging tools can be used to detect hydrocarbons and estimate the porosity of a reservoir. Because during the drilling of a well fluids move into porous and permeable formations surrounding a borehole, resistivity measurements recorded at different depths into a formation often have different values. Resistivity is measured by electric logs.

Conrad Schlumberger in 1912 began the first experiments which led, eventually, to the development of modern day petrophysical logs. The first electric log was run September 5, 1927 by H. G. Doll in Alsace-Lorraine, France. In 1941, G. E. Archie with Shell Oil Company presented a paper to the AIME in Dallas, Texas, which set forth the concepts used as a basis for modern quantitative log interpretation (Archie, 1942).

Archie's experiments showed that the resistivity of a water-filled formation (R_o), filled with water having a resistivity of R_w can be related by means of a formation resistivity factor (F):

$$R_o = F \times R_w$$

where the formation resistivity factor (F) is equal to the resistivity of the formation 100% water saturated (R_o) divided by the resistivity of the formation water (R_w).

Archie's experiments also revealed that formation factors can be related to porosity by the following formula:

$$F = \frac{1.0}{\phi^m}$$

where m is a cementation exponent whose value varies with grain size, grain size distribution, and the complexity of the paths between pores (tortuosity). The higher the value for tortuosity the higher the m value.

Water saturation (S_w) is determined from the water filled resistivity (R_o) and the formation resistivity (R_t) by the following relationship:

$$S_w = \left(\frac{R_o}{R_t}\right)^{1/n}$$

where n is the saturation exponent whose value varies from 1.8 to 2.5 but is most commonly 2.

By combining the formulas: $R_o = F \times R_w$ and $S_w = (R_o/R_t)^{1/n}$ the water saturation formula can be rewritten in the following form:

$$S_w = \left(\frac{F \times R_w}{R_t}\right)^{1/n}$$

This is the formula which is most commonly referred to as the Archie equation for water saturation (S_w). And, all present methods of interpretation involving resistivity curves are derived from this equation.

Now that the reader is introduced to some of the basic concepts of well log interpretation, our discussion can be continued in more detail about the factors which affect logging measurements.

Borehole Environment

Where a hole is drilled into a formation, the rock plus the

fluids in it (rock-fluid system) are altered in the vicinity of the borehole. A well's borehole and the rock surrounding it are contaminated by the drilling mud, which affects logging measurements. Figure 1 is a schematic illustration of a porous and permeable formation which is penetrated by a borehole filled with drilling mud.

The definitions of each of the symbols used in Figure 1 are listed as follows:

- d_h - hole diameter
- d_i - diameter of invaded zone (inner boundary; flushed zone)
- d_j - diameter of invaded zone (outer boundary; invaded zone)
- Δ_{rj} - radius of invaded zone (outer boundary)
- h_{mc} - thickness of mudcake
- R_m - resistivity† of the drilling mud
- R_{mc} - resistivity of the mudcake
- R_{mf} - resistivity of mud filtrate
- R_s - resistivity of shale
- R_t - resistivity of uninvaded zone (true resistivity)
- R_w - resistivity of formation water
- R_{xo} - resistivity of flushed zone
- S_w - water saturation of uninvaded zone
- S_{xo} - water saturation flushed zone

Some of the more important symbols shown in Figure 1 are:

Hole Diameter (d_h)—A well's borehole size is described by the outside diameter of the drill bit. But, the diameter of the borehole may be larger or smaller than the bit diameter because of (1) wash out and/or collapse of shale and poorly cemented porous rocks, or (2) build-up of mudcake on porous and permeable formations (Fig. 1). Borehole sizes normally vary from 7 7/8 inches to 12 inches, and modern logging tools are designed to operate within these size ranges. The size of the borehole is measured by a *caliper log*.

Drilling Mud (R_m)—Today, most wells are drilled with rotary bits and use special mud as a circulating fluid. The mud helps remove cuttings from the well bore, lubricate and cool the drill bit, and maintain an excess of borehole pressure over formation pressure. The excess of borehole pressure over formation pressure prevents blow-outs.

The density of the mud is kept high enough so that hydrostatic pressure in the mud column is always greater than formation pressure. This pressure difference forces some of the drilling fluid to invade porous and permeable formations. As invasion occurs, many of the solid particles (i.e. clay minerals from the drilling mud) are trapped on the side of the borehole and form *mudcake* (R_{mc}; Fig. 1). Fluid that filters into the formation during invasion is called *mud filtrate* (R_{mf}; Fig. 1). The resistivity values for drilling mud, mudcake, and mud filtrate are recorded on a log's header (Fig. 2).

Invaded Zone–The zone which is invaded by mud filtrate is called the invaded zone. It consists of a flushed zone (R_{xo}) and a transition or *annulus* (R_i) *zone*. The flushed zone (R_{xo}) occurs close to the borehole (Fig. 1) where the mud filtrate has almost completely flushed out a formation's hydrocarbons and/or water (R_w). The transition or annulus (R_i) zone, where a formation's fluids and mud filtrate are mixed, occurs between the flushed (R_{xo}) zone and the uninvaded (R_t) zone. The *uninvaded zone is defined as the area beyond the invaded zone where a formation's fluids are uncontaminated by mud filtrate*.

The depth of mud filtrate invasion into the invaded zone is referred to as the diameter of invasion (d_i and d_j; Fig. 1). The diameter of invasion is measured in inches or expressed as a ratio: d_j/d_h (where d_h represents the borehole diameter). The amount of invasion which takes place is dependent upon the permeability of the mudcake and not upon the porosity of the rock. In general, *an equal volume of mud filtrate can invade low porosity and high porosity rocks if the drilling muds have equal amounts of solid particles*. The solid particles in the drilling muds coalesce and form an impermeable mudcake. The mudcake then acts as a barrier to further invasion. Because an equal volume of fluid can be invaded before an impermeable mudcake barrier forms, the diameter of invasion will be greatest in low porosity rocks. This occurs because low porosity rocks have less storage capacity or pore volume to fill with the invading fluid, and, as a result, pores throughout a greater volume of rock will be affected. General invasion diameters are:

$d_j/d_h = 2$ for high porosity rocks;
$d_j/d_h = 5$ for intermediate porosity rocks;
and $d_j/d_h = 10$ for low porosity rocks.

Flushed Zone (R_{xo})—The flushed zone extends only a few inches from the well bore and is part of the invaded zone. If invasion is deep or moderate, most often the flushed zone is *completely* cleared of its formation water (R_w) by mud filtrate (R_{mf}). When oil is present in the flushed zone, you can determine the degree of flushing by mud filtrate from the difference between water saturations in the flushed (S_{xo}) zone and the uninvaded (S_w) zone (Fig. 1). Usually, about 70 to 95% of the oil is flushed out; the remaining oil is called *residual oil* ($S_{ro} = [1.0 - S_{xo}]$ where S_{ro} equals residual oil saturation [ROS]).

Uninvaded Zone (R_t)—The uninvaded zone is located beyond the invaded zone (Fig. 1). Pores in the uninvaded

---

†Resistivity $(R) = \frac{r \times A}{L}$

R - resistivity in ohm-meters²/meters (ohm-meter)
r - resistance (ohms)
A - cross sectional area (meters²)
L - length (meter)

zone are uncontaminated by mud filtrate; instead, they are saturated with formation water (R_w), oil, or gas.

Even in hydrocarbon-bearing reservoirs, there is always a layer of formation water on grain surfaces. Water saturation (S_w; Fig. 1) of the uninvaded zone is an important factor in reservoir evaluation because, *by using water saturation data, a geologist can determine a reservoir's hydrocarbon saturation*. The formula for calculating hydrocarbon saturation is:

$$S_h = 1.0 - S_w$$

S_h = hydrocarbon saturation (i.e. the fraction of pore volume filled with hydrocarbons).

S_w = water saturation uninvaded zone (i.e. fraction of pore volume filled with water)

The ratio between the uninvaded zone's water saturation (S_w) and the flushed zone's water saturation (S_{xo}) is an index of *hydrocarbon moveability*.

Invasion and Resistivity Profiles

Invasion and resistivity profiles are diagrammatic, theoretical, cross sectional views moving away from the borehole and into a formation. They illustrate the horizontal distributions of the invaded and uninvaded zones and their corresponding relative resistivities. There are three commonly recognized invasion profiles: (1) step, (2) transition, and (3) annulus. These three invasion profiles are illustrated in Figure 3.

The step profile has a cylindrical geometry with an invasion diameter equal to d_j. Shallow reading, resistivity logging tools read the resistivity of the invaded zone (R_i), while deeper reading, resistivity logging tools read true resistivity of the uninvaded zone (R_t).

The transition profile also has a cylindrical geometry with two invasion diameters: d_i (flushed zone) and d_j (transition zone). It is probably a more realistic model for true borehole conditions than the step profile. Three resistivity devices are needed to measure a transitional profile; these three devices measure resistivities of the flushed, transition, and uninvaded zones R_{xo}, R_i, and R_t; (see Fig. 3). By using these three resistivity measurements, the deep reading resistivity tool can be corrected to a more accurate value of true resistivity (R_t), and the depth of invasion can be determined. Two modern resistivity devices which use these three resistivity curves are: the Dual Induction Log with a Laterolog-8* or Spherically Focused Log (SFL)* and the Dual Laterolog* with a Microspherically Focused Log (MSFL)*.

An annulus profile is only sometimes recorded on a log because it rapidly dissipates in a well. The annulus profile is detected only by an induction log run soon after a well is drilled. However, it is very important to a geologist because the profile can only occur in zones which bear hydrocarbons. As the mud filtrate invades the hydrocarbon-bearing zone, hydrocarbons move out first. Next, formation water is pushed out in front of the mud filtrate forming an annular (circular) ring at the edge of the invaded zone (Fig. 3). The annulus effect is detected by a higher resistivity reading on a deep induction log than by one on a medium induction log.

Log resistivity profiles illustrate the resistivity values of the invaded and uninvaded zones in the formation being investigated. They are of particular interest because, by using them, a geologist can quickly scan a log and look for potential zones of interest such as hydrocarbon zones. Because of their importance, resistivity profiles for both water-bearing and hydrocarbon-bearing zones are discussed here. These profiles vary, depending on the relative resistivity values of R_w and R_{mf}. All the variations and their associated profiles are illustrated in Figures 4 and 5.

Water-Bearing Zones—Figure 4 illustrates the borehole and resistivity profiles for water-bearing zones where the resistivity of the mud filtrate (R_{mf}) is much greater than the resistivity of the formation water (R_w) in freshwater muds, and where resistivity of the mud filtrate (R_{mf}) is approximately equal to the resistivity of the formation water (R_w) in saltwater muds. A freshwater mud (i.e. $R_{mf} > 3\ R_w$) results in a "wet" log profile where the shallow (R_{xo}), medium (R_i), and deep (R_t) resistivity tools separate and record high (R_{xo}), intermediate (R_i), and low (R_t) resistivities (Fig. 4). A saltwater mud (i.e. $R_w \simeq R_{mf}$) results in a wet profile where the shallow (R_{xo}), medium (R_i), and deep (R_t) resistivity tools all read low resistivity (Fig. 4). Figures 6a and 6b illustrate the resistivity curves for wet zones invaded with both freshwater and saltwater muds.

Hydrocarbon-Bearing Zones—Figure 5 illustrates the borehole and resistivity profiles for hydrocarbon-bearing zones where the resistivity of the mud filtrate (R_{mf}) is much greater than the resistivity of the formation water (R_w) for freshwater muds, and where R_{mf} is approximately equal to R_w for saltwater muds. A hydrocarbon zone invaded with freshwater mud results in a resistivity profile where the shallow (R_{xo}), medium (R_i), and deep (R_t) resistivity tools all record high resistivities (Fig. 5). In some instances, the deep resistivity will be higher than the medium resistivity. When this happens, it is called the annulus effect. A hydrocarbon zone invaded with saltwater mud results in a resistivity profile where the shallow (R_{xo}), medium (R_i), and deep (R_t) resistivity tools separate and record low (R_{xo}), intermediate (R_i) and high (R_t) resistivities (Fig. 5). Figures 7a and 7b illustrate the resistivity curves for hydrocarbon zones invaded with both freshwater and saltwater muds.

Basic Information Needed in Log Interpretation

Lithology—In quantitative log analysis, there are several reasons why it is important to know the lithology of a zone (i.e. sandstone, limestone, or dolomite). Porosity logs require a lithology or a matrix constant before a zone's porosity (ϕ) can be calculated. And the formation factor (F), a variable used in the Archie water saturation equation ($S_w = \sqrt{F \times R_w/R_t}$), varies with lithology. As a consequence, water saturations change as F changes. Table 1 is a list of the different methods for calculating formation factor, and illustrates how lithology affects the formation factor.

Temperature of Formation—Formation temperature (T_f) is also important in log analysis because the resistivities of the drilling mud (R_m), the mud filtrate (R_{mf}), and the formation water (R_w) vary with temperature. The temperature of a formation is determined by knowing: (1) formation depth; (2) bottom hole temperature (BHT); (3) total depth of the well (TD); and (4) surface temperature. You can determine a reasonable value for the formation temperature by using these data and by assuming a linear geothermal gradient (Fig. 8).

Table 1. Different Coefficients and Exponents Used to Calculate Formation Factor (F). (Modified after Asquith, 1980).

$F = a/\phi^m$	general relationship Where: a = tortuosity factor[†] m = cementation exponent ϕ = porosity
[††]$F = 1/\phi^2$	for carbonates
[††]$F = 0.81/\phi^2$	for consolidated sandstones
[††]$F = 0.62/\phi^{2.15}$	Humble formula for unconsolidated sands
$F = 1.45/\phi^{1.54}$	for average sands (after Carothers, 1958)
$F = 1.65/\phi^{1.33}$	for shaly sands (after Carothers, 1958)
$F = 1.45/\phi^{1.70}$	for calcareous sands (after Carothers, 1958)
$F = 0.85/\phi^{2.14}$	for carbonates (after Carothers, 1958)
$F = 2.45/\phi^{1.08}$	for Pliocene sands, Southern California (after Carothers and Porter, 1970)
$F = 1.97/\phi^{1.29}$	for Miocene sands, Texas-Louisiana Gulf Coast (after Carothers and Porter, 1970)
$F = 1.0/\phi^{(2.05-\phi)}$	for clean granular formations (after Sethi, 1979)

[†]Tortuosity is a function of the complexity of the path the fluid must travel through the rock.
[††]Most commonly used.

The formation temperature is also calculated (Asquith, 1980) by using the linear regression equation:

$$y = mx + c$$

Where:
x = depth
y = temperature
m = slope—in this example it is the geothermal gradient
c = a constant—in this example it is the surface temperature

An example of how to calculate formation temperature is illustrated here:

Temperature Gradient Calculation

Assume:
y = bottom hole temperature (BHT) = 250°F
x = total depth (TD) = 15,000 ft
c = surface temperature = 70°F

Solve for m (i.e. slope or temperature gradient)

$$m = \frac{y - c}{x}$$

Therefore:

$$m = \frac{250° - 70°}{15{,}000 \text{ ft}}$$

m = 0.012°/ft or 1.2°/100 ft

Formation Temperature Calculation

Assume:
m = temperature gradient = 0.012°/ft
x = formation depth = 8,000 ft
c = surface temperature = 70°

Remember:
y = mx + c

Therefore:
y = (0.012) × (8,000) + 70°
y = 166° formation temperature at 8,000 ft

After a formation's temperature is determined either by chart (Fig. 8) or by calculation, the resistivities of the different fluids (R_m, R_{mf}, or R_w) can be corrected to formation temperature. Figure 9 is a chart that is used for correcting fluid resistivities to formation temperature. This chart is closely approximated by the Arp's formula:

$$R_{Tf} = R_{temp} \times (\text{Temp} + 6.77)/(T_f + 6.77)$$

Where:
R_{Tf} = resistivity at formation temperature
R_{temp} = resistivity at a temperature other than formation temperature

Temp = temperature at which resistivity was measured
T_f = formation temperature

Using a formation temperature of 166° and assuming an R_w of 0.04 measured at 70°, the R_w at 166° will be:

$R_{w166} = 0.04 \times (70 + 6.77)/(166 + 6.77)$
$R_{w166} = 0.018$

Resistivity values of the drilling mud (R_m), mud filtrate (R_{mf}), mudcake (R_{mc}), and the temperatures at which they are measured, are recorded on a log's header (Fig. 2). The resistivity of a formation's water (R_w) is obtained by analysis of water samples from a drill stem test, a water producing well, or from a catalog of water resistivity values. Formation water resistivity (R_w) is also determined from the spontaneous potential log (discussed in Chapter II) or can be calculated in water zones (i.e., $S_w = 100\%$) by the apparent water resistivity (R_{wa}) method (see Chapter VI).

Fundamental Equations

Table 2 is a list of fundamental equations that are used for the log evaluation of potential hydrocarbon reservoirs. These formulas are discussed in detail in subsequent chapters.

Table 2. Fundamental Equations of Well Log Interpretation.

Porosity:

Sonic Log $$\phi_{SONIC} = \frac{\Delta t - \Delta t_{ma}}{\Delta tf - \Delta t_{ma}}$$

Density Log $$\phi_{DEN} = \frac{\rho_{ma} - \rho_b}{\rho_{ma} - \rho_f}$$

Neutron-Density Log $$\phi_{N\text{-}D} = \sqrt{\frac{\phi_N^2 + \phi_D^2}{2}}$$

Formation Factor:

$F = a/\phi^m$ General
$F = 1.0/\phi^2$ Carbonates
$F = 0.81/\phi^2$ Consolidated Sandstones
$F = 0.62/\phi^{2.15}$ Unconsolidated Sands

Formation Water Resistivity:

$SSP = -K \times \log(R_{mf}/R_w)$
$R_{we} \rightarrow R_w$
$$R_w = \frac{R_o}{F}$$

Water Saturations:

$S_w^{n\dagger} = F \times (R_w/R_t)$ water saturation uninvaded zone

$S_{xo}^n = F \times (R_{mf}/R_{xo})$ water saturation flushed zone

$$S_w = \left(\frac{R_{xo}/R_t}{R_{mf}/R_w}\right)^{0.625}$$ water saturation ratio method

Bulk Volume Water:

$BVW = \phi \times S_w$

Permeability

$K_e = [250 \times (\phi^3/S_{w\,irr})]^2$ oil
K_e = permeability in millidarcies
$K_e = [79 \times (\phi^3/S_{w\,irr})]^2$ gas
$S_{w\,irr}$ = irreducible water saturation

†n = saturation exponent which varies from 1.8 to 2.5 but most often equals 2.0

Review - Chapter I

1. The four most fundamental rock properties used in petrophysical logging are (1) porosity; (2) permeability; (3) water saturation; and (4) resistivity.

2. The Archie equation for water saturation is:

$$S_w = \left(\frac{F \times R_w}{R_t}\right)^{1/n}$$

Where:

S_w = water saturation of uninvaded zone
F = formation factor
R_w = formation water resistivity
R_t = formation resistivity (uninvaded zone)

3. Where a porous and permeable formation is penetrated by the drill bit, the drilling mud invades the formation as mud filtrate (R_{mf}).

4. The invasion of the porous and permeable formation by mud filtrate creates invasion zones (R_{xo} and R_i) and an uninvaded zone (R_t). Shallow, medium, and deep reading resistivity logging tools provide information about the invaded and uninvaded zones and about the depth of invasion.

5. The lithology of a formation must be known because: (1) porosity logs require a matrix value—sandstone, limestone, or dolomite—in order to determine porosity; (2) the formation factor varies with lithology; (3) the variation in formation factor causes changes in water saturation values.

6. The four fluids that affect logging measurements are: (1) drilling mud, R_m; (2) mud filtrate, R_{mf}; (3) formation water, R_w; and (4) hydrocarbons.

7. The resistivities of the drilling mud (R_m), mudcake (R_{mc}), mud filtrate (R_{mf}) and formation water (R_w) all vary with changes in temperature. Consequently, a formation's temperature (T_f) must be determined and all resistivities corrected to T_f.

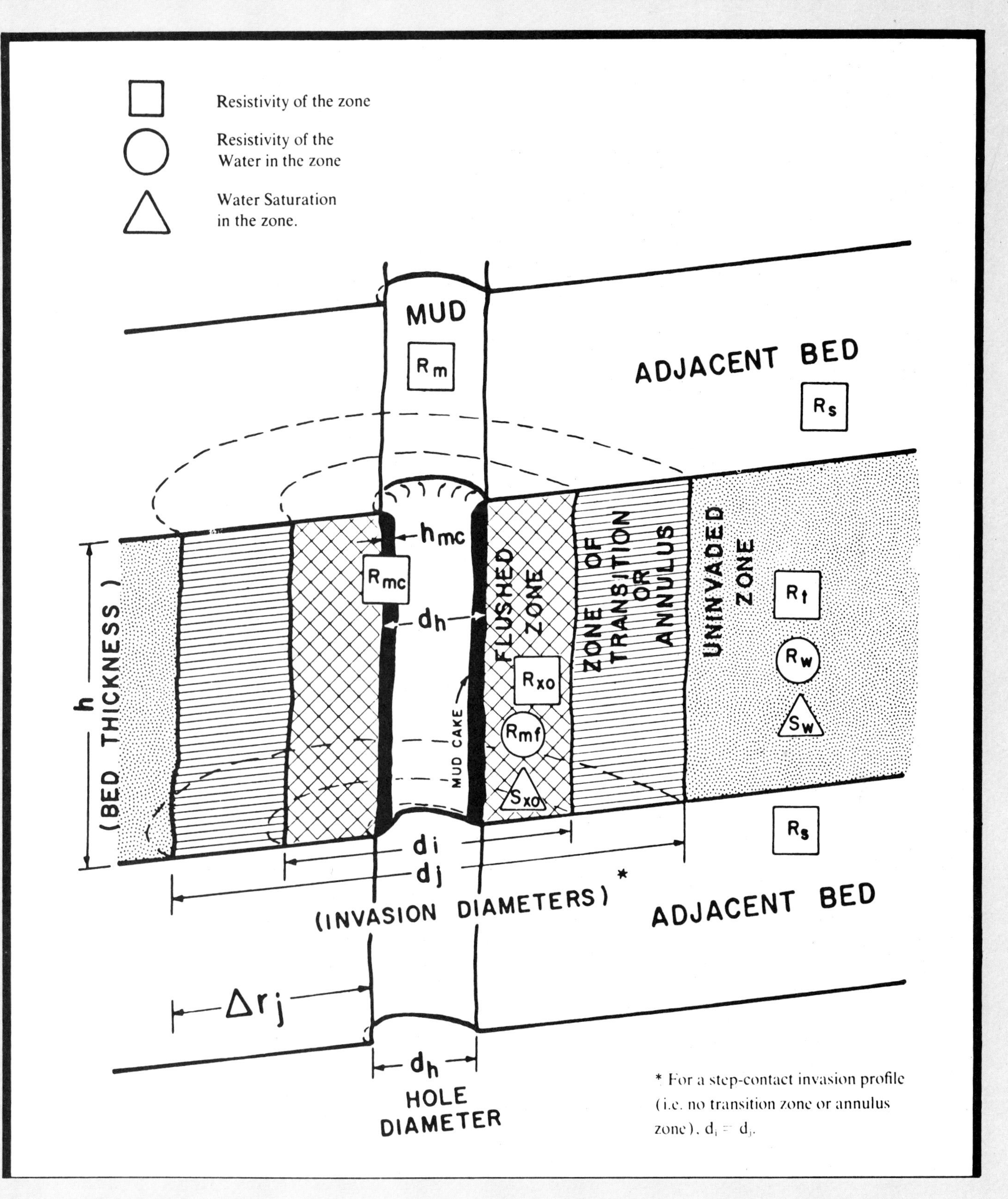

Figure 1. The borehole environment and symbols used in log interpretation. This schematic diagram illustrates an idealized version of what happens when fluids from the borehole invade the surrounding rock. Dotted lines indicate the cylindrical nature of the invasion.

Courtesy, Schlumberger Well Services.
Copyright 1977, Schlumberger.

Schlumberger

DUAL INDUCTION-SFL
WITH LINEAR CORRELATION LOG

COUNTY ___ FIELD ___ LOCATION ___ WELL ___ COMPANY ___

COMPANY ___

WELL ___

FIELD ___

COUNTY ___ STATE ___

LOCATION ___

API SERIAL NO.	SEC.	TWP	RANGE

Other Services:
FDC/CNL/GR
HDT

Permanent Datum: GROUND LEVEL ; Elev.: 3731
Log Measured From KB 11 Ft. Above Perm. Datum
Drilling Measured From KB

Elev.: K.B. 3742
D.F. ----
G.L. 3731

Date	6-11-79			
Run No.	ONE			
Depth—Driller	5000			
Depth—Logger	4990			
Btm. Log Interval	4984			
Top Log Interval	1601			
Casing—Driller	8 5/8 @ 1601	@	@	@
Casing—Logger	1601			
Bit Size	7 7/8			
Type Fluid in Hole	DRISPAC			
Dens. | Visc.	9.2 | 44	|	|	|
pH | Fluid Loss	9.0 | 6.8 ml	| ml	| ml	| ml
Source of Sample	FLOWLINE			
R_m @ Meas. Temp.	2.44 @ 81 °F	1.72 @ 115 °F	@ °F	@ °F
R_{mf} @ Meas. Temp.	2.04 @ 63 °F	1.12 @ 115 °F	@ °F	@ °F
R_{mc} @ Meas. Temp.	---- @ -- °F	@ °F	@ °F	@ °F
Source: R_{mf} | R_{mc}	M | --	|	|	|
R_m @ BHT	1.72 @ 115 °F	@ °F	@ °F	@ °F
TIME: Circulation Stopped	2000/6-10			
TIME: Logger on Bottom	0000/6-11			
Max. Rec. Temp.	115 °F	°F	°F	°F
Equip. | Location	7688 | LIBERAL	|	|	|
Recorded By				
Witnessed By	MR.			

TIGHT HOLE

*BHT = Bottom Hole Temperature

Figure 2. Reproduction of a typical log heading. Information on the header about the resistivity values for drilling mud (R_m) and mud filtrate (R_{mf}) are especially useful in log interpretation and are used in calculations.

NOTE: Sometimes, as in this example, a value for the resistivity of mudcake (R_{mc}) is not recorded on the heading.

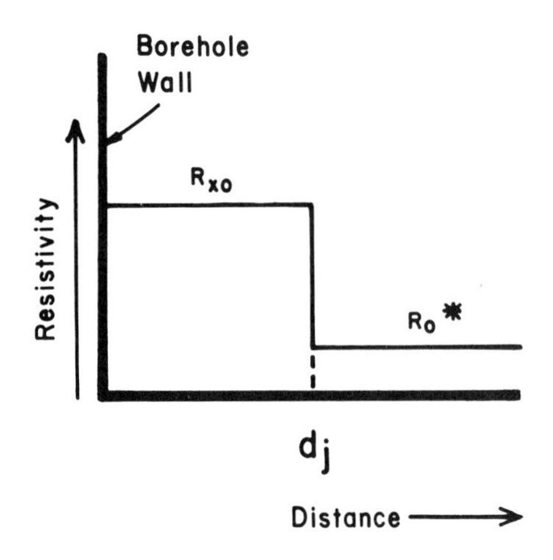

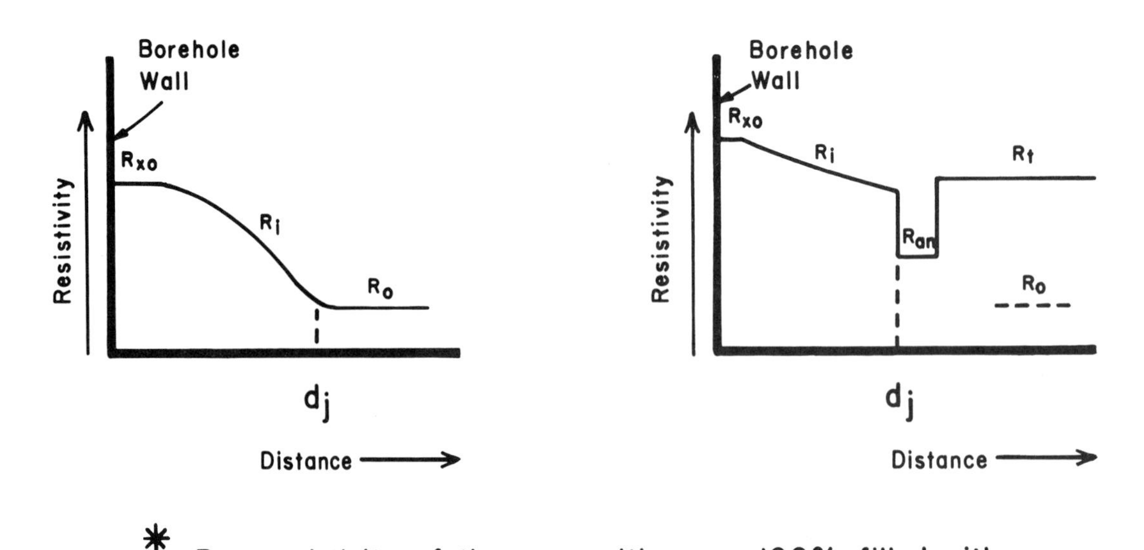

* R_o = resistivity of the zone with pores 100% filled with formation water (R_w). Also called wet resistivity.

Figure 3. Typical invasion profiles for three idealized versions of fluid distributions in the vicinity of the borehole.

As mud filtrate (R_{mf}) moves into a porous and permeable formation, it can invade the formation in several different ways. Various fluid distributions are represented by the step, transition, or annulus profiles.

A. Step Profile—Mud filtrate is distributed with a cylindrical shape around the borehole and creates an invaded zone. The cylindrically shaped invaded zone is characterized by its abrupt contact with the uninvaded zone. The diameter of the cylinder is represented as d_j. In the invaded zone, pores are filled with mud filtrate (R_{mf}); pores in the uninvaded zone are filled with formation water (R_w) or hydrocarbons. In this example the uninvaded zone is wet (100% water and no hydrocarbons), thus the resistivity beyond the invaded zone is low. The resistivity of the invaded zone is R_{xo}, and the resistivity of the uninvaded zone is either R_o if a formation is water-bearing, or R_t if a formation is hydrocarbon-bearing.

B. Transition Profile—This is the most realistic model of true borehole conditions. Here again invasion is cylindrical, but in this profile, the invasion of the mud filtrate (R_{mf}) diminishes gradually, rather than abruptly, through a transition zone toward the outer boundary of the invaded zone (see d_j on diagram for location of outer boundary).

In the flushed part (R_{xo}) of the invaded zone, pores are filled with mud filtrate (R_{mf}), giving a high resistivity reading. In the transition part of the invaded zone, pores are filled with mud filtrate (R_{mf}), formation water (R_w), and, if present, residual hydrocarbons (RH). Beyond the outer boundary of the invaded zone (d_j on diagram), pores are filled with either formation water, or (if present) hydrocarbons. In this diagram, hydrocarbons are not present, so resistivity of the uninvaded zone is low. The resistivity of the invaded zone's flushed part is R_{xo}, and the resistivity of the transition part is R_i. Resistivity of the uninvaded zone is R_t if hydrocarbon-bearing or R_o if water-bearing.

C. Annulus Profile—This reflects a temporary fluid distribution, and is a condition which should disappear with time (if the logging operation is delayed, it may not be recorded on the logs at all). The annulus profile represents a fluid distribution which occurs between the invaded zone and the uninvaded zone and denotes the presence of hydrocarbons.

In the flushed part (R_{xo}) of the invaded zone, pores are filled with both mud filtrate (R_{mf}) and residual hydrocarbons (RH). Thus the resistivity reads high. Pores beyond the flushed part of the invaded zone (R_i) are filled with a mixture of mud filtrate (R_{mf}), formation water (R_w), and residual hydrocarbons (RH).

Beyond the outer boundary of the invaded zone is the annulus zone where pores are filled with residual hydrocarbons (RH) and formation water (R_w). When an annulus profile is present, there is an abrupt drop in measured resistivity at the outer boundary of the invaded zone. The abrupt resistivity drop is due to the high concentration of formation water (R_w) in the annulus zone. Formation water has been pushed ahead by the invading mud filtrate into the annulus zone. This causes a temporary absence of hydrocarbons which, in their turn, have been pushed ahead of formation water.

Beyond the annulus is the uninvaded zone where pores are filled with formation water (R_w) and hydrocarbons. Remember that true resitivity of a formation can be measured in the uninvaded zone because of its virgin nature. True resistivity (R_t) will be higher than the wet resistivity (R_o) because hydrocarbons have a higher resistivity than saltwater.

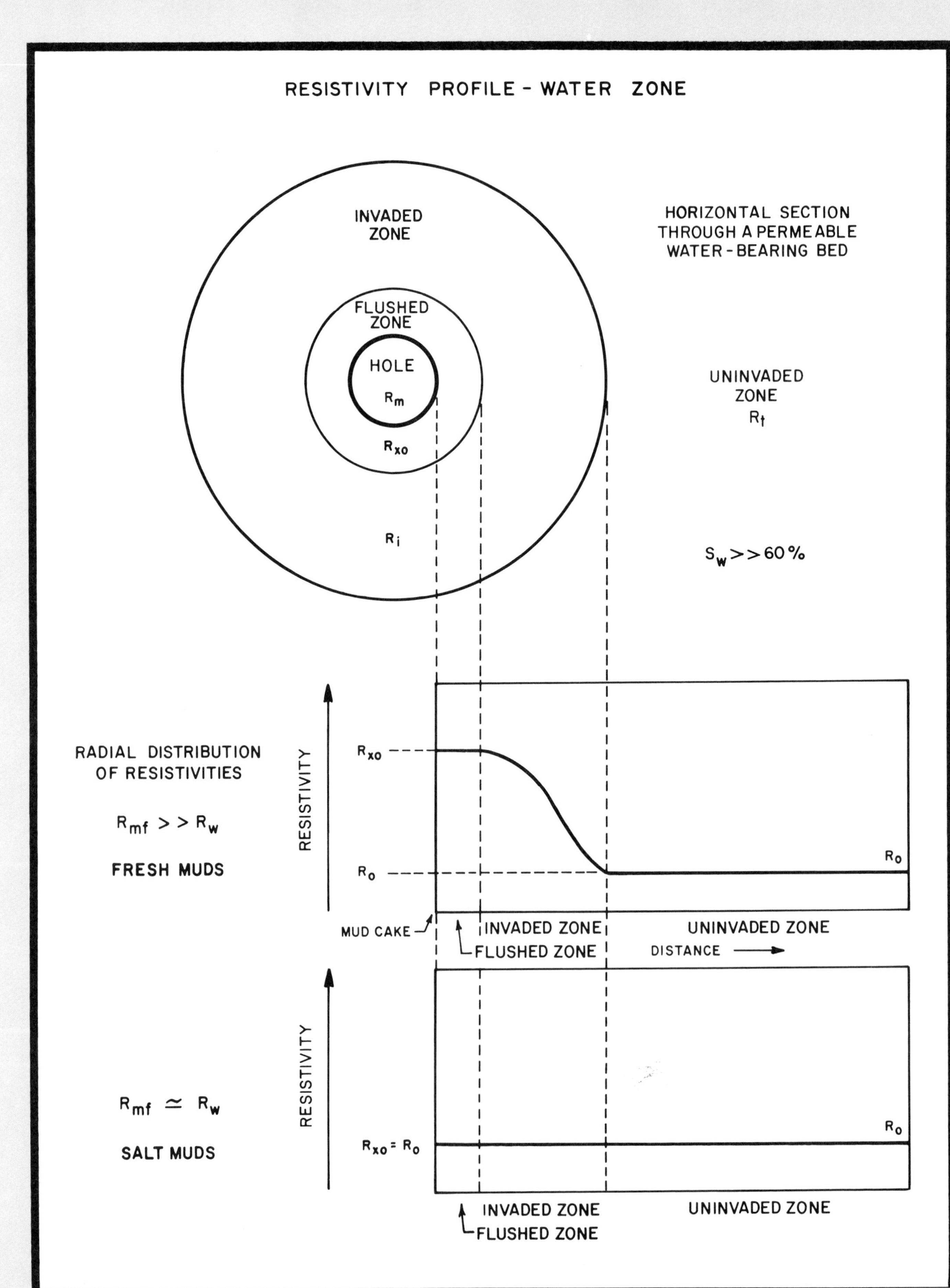
RESISTIVITY PROFILE - WATER ZONE
INVADED ZONE
FLUSHED ZONE
HOLE
R_m
R_{xo}
R_i
HORIZONTAL SECTION THROUGH A PERMEABLE WATER-BEARING BED
UNINVADED ZONE
R_t
$S_w >> 60\%$
RADIAL DISTRIBUTION OF RESISTIVITIES
$R_{mf} >> R_w$
FRESH MUDS
RESISTIVITY
R_{xo}
R_o
R_o
MUD CAKE
INVADED ZONE
FLUSHED ZONE
UNINVADED ZONE
DISTANCE
RESISTIVITY
$R_{mf} \simeq R_w$
SALT MUDS
$R_{xo} = R_o$
R_o
INVADED ZONE
FLUSHED ZONE
UNINVADED ZONE

Figure 4. Horizontal section through a permeable water-bearing formation and the concomitant resistivity profiles which occur when there is invasion by either freshwater- or saltwater-based drilling muds (see Fig. 5 for resistivity profiles in a hydrocarbon-bearing formation).

Note: These examples are shown because freshwater muds and saltwater muds are used in different geographic regions, usually exclusively. The geologist needs to be aware that a difference exists. To find out which mud is used in your area, ask your drilling engineer. The type of mud used affects the log package selected, as we will see later.

Freshwater Muds—The resistivity of the mud filtrate (R_{mf}) is greater than the resistivity of the formation water (R_w) because of the varying salt content (remember, saltwater is conductive). A general rule when freshwater muds are used is: $R_{mf} > 3R_w$. The flushed zone (R_{xo}), which has a greater amount of mud filtrate, will have higher resistivities. Away from the borehole, the resistivity of the invaded zone (R_i) will decrease due to the decreasing amount of mud filtrate (R_{mf}) and the increasing amount of formation water (R_w).

With a water-bearing formation, the resistivity of the uninvaded zone will be low because the pores are filled with formation water (R_w). In the uninvaded zone, true resistivity (R_t) will be equal to wet resistivity (R_o) because the formation is 100% saturated with formation water ($R_t = R_o$ where the formation is 100% saturation with formation water).

To summarize: in a water-bearing zone, the resistivity of the flushed zone (R_{xo}) is greater than the resistivity of the invaded zone (R_i) which in turn has a greater resistivity than the uninvaded zone (R_t). Therefore: $R_{xo} > R_i >> R_t$ in water-bearing zones.

Saltwater Muds—Because the resistivity of mud filtrate (R_{mf}) is approximately equal to the resistivity of formation water ($R_{mf} \approx R_w$), there is no appreciable difference in the resistivity from the flushed (R_{xo}) to the invaded zone (R_i) to the uninvaded zone ($R_{xo} = R_i = R_t$); all have low resistivities.

Both the above examples assume that the water saturation of the uninvaded zone is much greater than 60%.

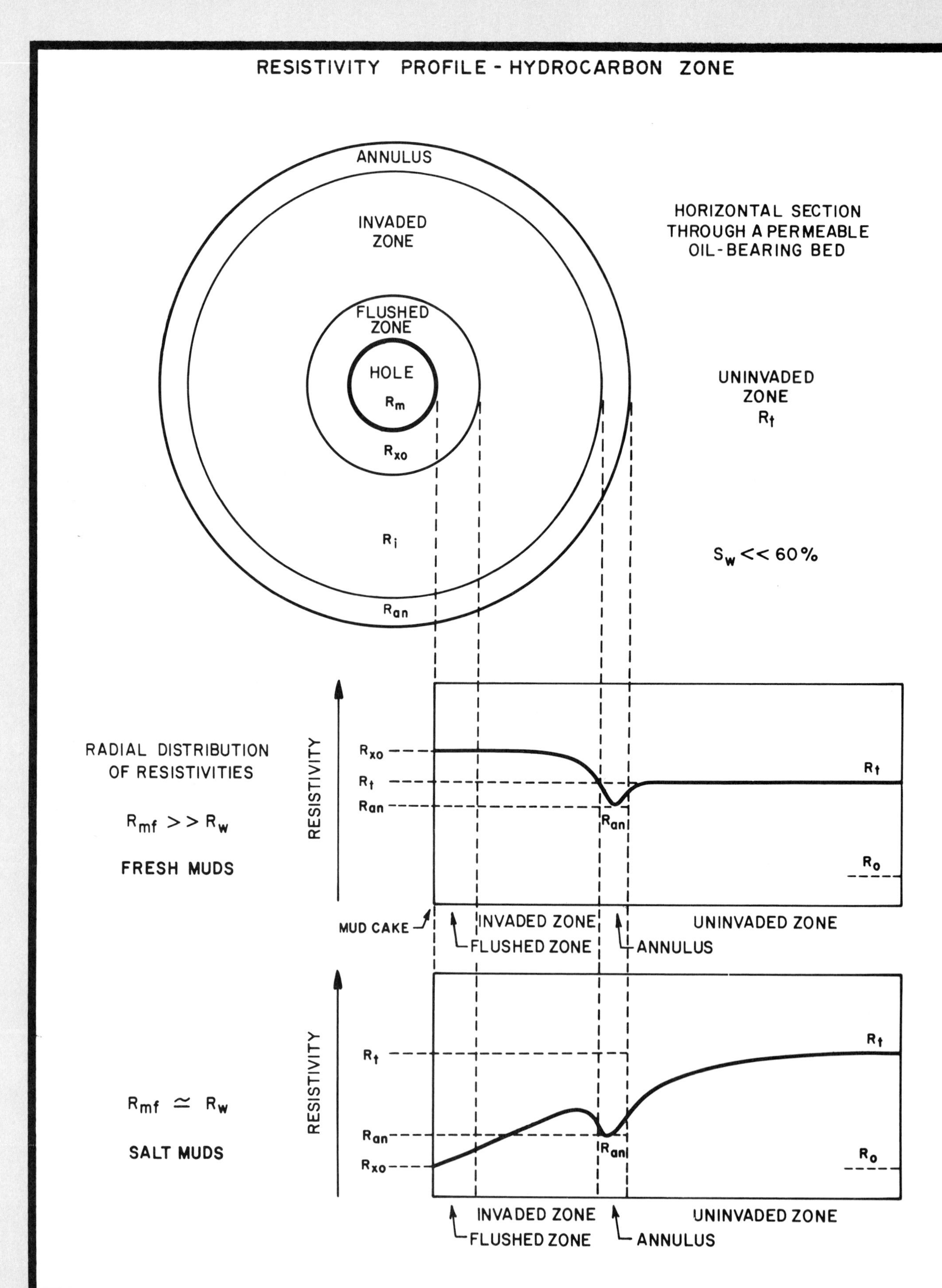
RESISTIVITY PROFILE - HYDROCARBON ZONE
ANNULUS
INVADED ZONE
FLUSHED ZONE
HOLE
Rm
Rxo
Ri
Ran
HORIZONTAL SECTION THROUGH A PERMEABLE OIL-BEARING BED
UNINVADED ZONE
Rt
Sw << 60%
RADIAL DISTRIBUTION OF RESISTIVITIES
Rmf >> Rw
FRESH MUDS
RESISTIVITY
Rxo
Rt
Ran
Ran
Rt
Ro
MUD CAKE
INVADED ZONE
FLUSHED ZONE
ANNULUS
UNINVADED ZONE
Rmf ≃ Rw
SALT MUDS
RESISTIVITY
Rt
Ran
Rxo
Ran
Rt
Ro
INVADED ZONE
FLUSHED ZONE
ANNULUS
UNINVADED ZONE

Figure 5. Horizontal section through a permeable hydrocarbon-bearing formation and the concomitant resistivity profiles which occur when there is invasion by either freshwater- or saltwater-based drilling muds (see Fig. 4 for resistivity profiles in a water-bearing formation).

Freshwater Muds—Because the resistivity of both the mud filtrate (R_{mf}) and residual hydrocarbons (RH) is much greater than formation water (R_w), the resistivity of the flushed zone (R_{xo}) is comparatively high (remember that the flushed zone has mud filtrate and some residual hydrocarbons).

Beyond its flushed part (R_{xo}), the invaded zone (R_i) has a mixture of mud filtrate (R_{mf}), formation water (R_w), and some residual hydrocarbons (RH). Such a mixture causes high resistivities. In some cases, resistivity of the invaded zone (R_i) almost equals that of the flushed zone (R_{xo}).

The presence of hydrocarbons in the uninvaded zone causes higher resistivity than if the zone had only formation water (R_w), because hydrocarbons are more resistant than formation water. So, $R_t > R_o$. The resistivity of the uninvaded zone (R_t) is normally somewhat less than the resistivity of the flushed and invaded zones (R_{xo} and R_i). However, sometimes when an annulus profile is present, the invaded zone's resistivity (R_i) may be slightly lower than the uninvaded zone's resistivity (R_t).

To summarize: therefore, $R_{xo} > R_i \gtrless R_t$ in hydrocarbon-bearing zones.

Saltwater Muds—Because the resistivity of the mud filtrate (R_{mf}) is approximately equal to the resistivity of formation water ($R_{mf} \simeq R_w$), and the amount of residual hydrocarbons (RH) is low, the resistivity of the flushed zone (R_{xo}) is low.

Away from the borehole as more hydrocarbons mix with mud filtrate in the invaded zone, the resistivity of the invaded zone (R_i) begins to increase.

Resistivity of the uninvaded zone (R_t) is much greater than if the formation was at 100% water saturation (R_o) because hydrocarbons are more resistant than saltwater. Resistivity of the uninvaded zone is greater than the resistivity of the invaded (R_i) zone. So, $R_t > R_i > R_{xo}$.

Both the above examples assume that the water saturation of the uninvaded zone is much less than 60%.

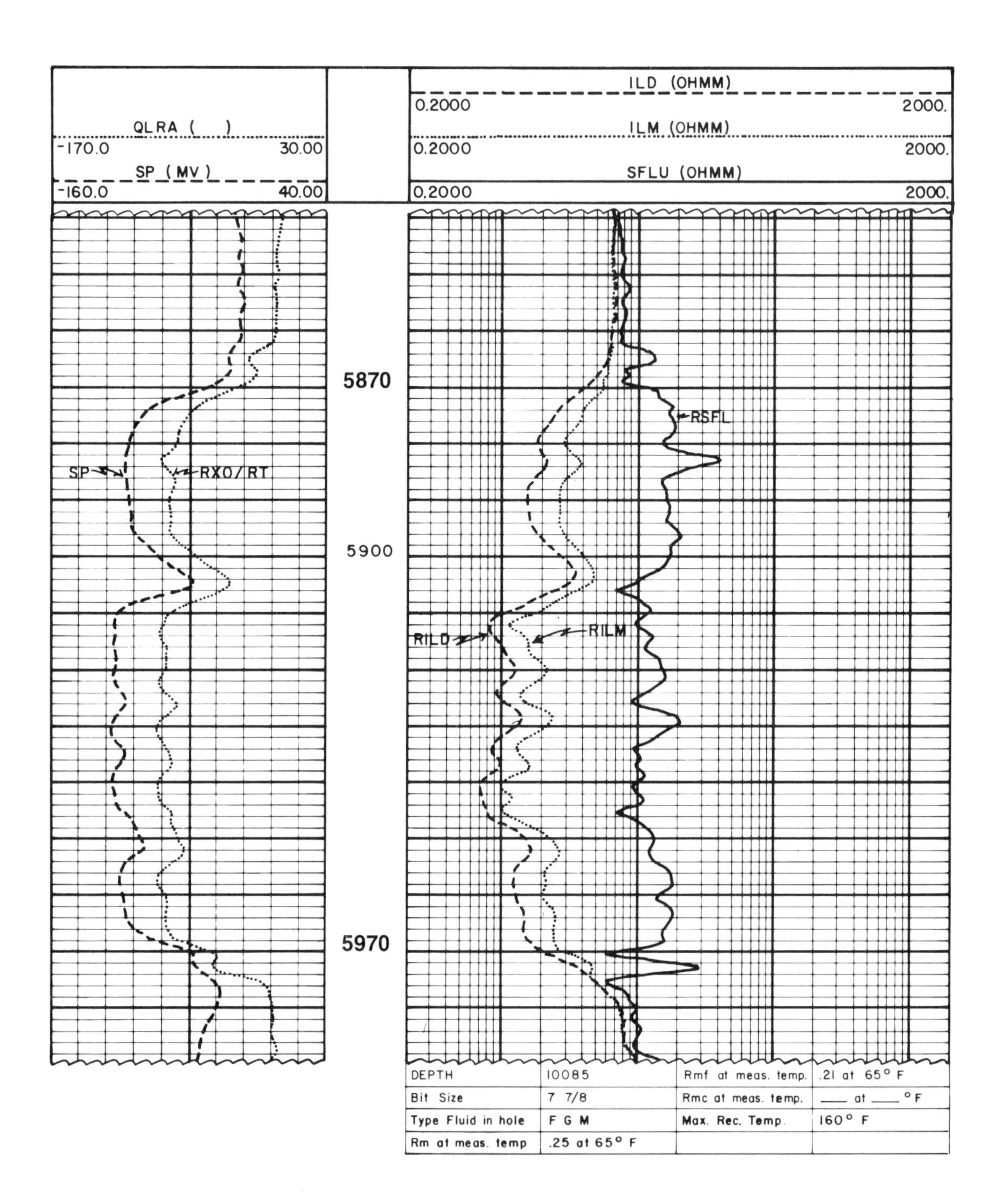

DEPTH	10085	Rmf at meas. temp.	.21 at 65° F
Bit Size	7 7/8	Rmc at meas. temp.	___ at ___ °F
Type Fluid in hole	F G M	Max. Rec. Temp.	160° F
Rm at meas. temp	.25 at 65° F		

Figure 6A. Example of Dual Induction Focused Log curves through a water-bearing zone. Given: the drilling mud is freshwater-based ($R_{mf} > 3R_w$).

We've seen that where freshwater drilling muds invade a water-bearing formation ($S_w >> 60\%$), there is high resistivity in the flushed zone (R_{xo}), a lesser resistivity in the invaded zone (R_i), and a low resistivity in the uninvaded zone (R_t). See Figure 4 for review.

Ignore the left side of the log on the opposite page, and compare the three curves on the right side of the log (tracks #2 and #3). Resistivity values are higher as distance increases from the left side of the log.

Log Curve R_{ILD}—Deep induction log resistivity curves measure true resistivity (R_t) or the resistivity of the formation, deep beyond the outer boundary of the invaded zone. This is a measure of the uninvaded zone. In water-bearing zones (in this case from 5,870 to 5,970 ft), the curve will read a low resistivity because the resistivity of the formation water (R_w) is less than the resistivity of the mud filtrate (R_{mf}).

Log Curve R_{ILM}—Medium induction log resistivity curves measure the resistivity of the invaded zone (R_i). In a water-bearing formation, the curve will read intermediate resistivity because of the mixture of formation water (R_w) and mud filtrate (R_{mf}).

Log Curve R_{SFL}—Spherically Focused Log* resistivity curves measure the resistivity of the flushed zone (R_{xo}). In a water-bearing zone, the curve will read high resistivity because freshwater mud filtrate (R_{mf}) has a high resistivity. The SFL* pictured here records a greater resistivity than either the deep (R_{ILd}) or medium (R_{ILm}) induction curves.

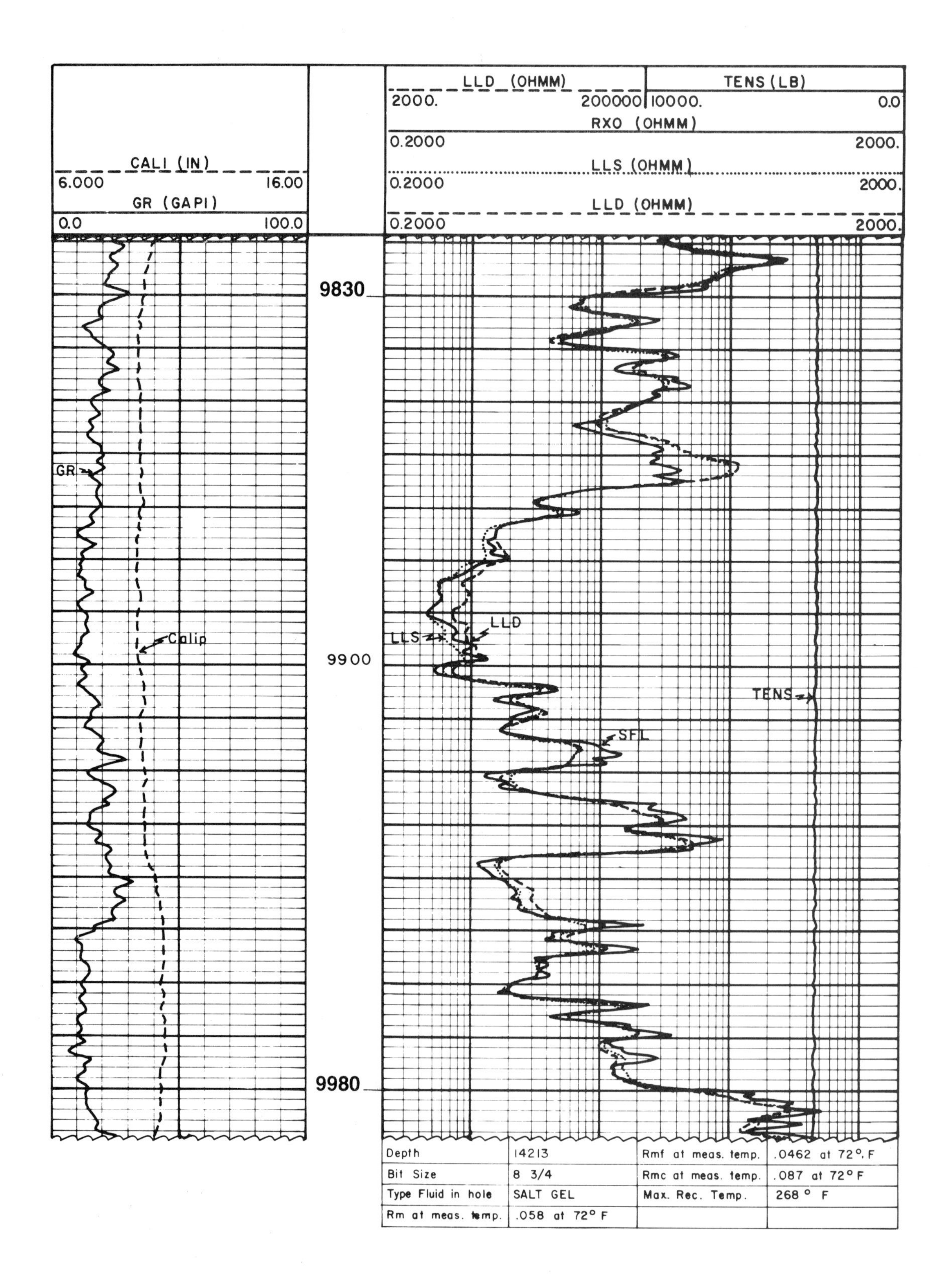

Depth	14213	Rmf at meas. temp.	.0462 at 72°F
Bit Size	8 3/4	Rmc at meas. temp.	.087 at 72°F
Type Fluid in hole	SALT GEL	Max. Rec. Temp.	268° F
Rm at meas. temp.	.058 at 72°F		

Figure 6B. Example of Dual Laterolog* - Microspherically Focused Log (MSFL)* curves through a water-bearing zone. Given: the drilling mud is saltwater-based ($R_{mf} \simeq R_w$).

We've seen that where saltwater drilling muds invade a water-bearing formation ($S_w >> 60\%$), there is low resistivity in the flushed zone (R_{xo}), a low resistivity in the invaded zone (R_i), and low resistivity in the uninvaded zone (R_t). Because R_{mf} is approximately equal to R_w, the pores in the flushed (R_{xo}), invaded (R_i), and uninvaded (R_t) zones are all filled with saline waters; the presence of salt results in low resistivity. See Figure 4 for review.

Ignore the left side of the log on the opposite page, and compare the three curves on the right side of the log (tracks #2 and #3). Resistivity values are higher as distance increases from the left side of the log.

Log Curve LLD—Deep Laterolog* resistivity curves measure true resistivity (R_t) or the resistivity of the formation deep beyond the outer boundary of the invaded zone. In water-bearing zones (in this case from 9,830 to 9,980 ft), the curve will read low resistivity because the pores of the formation are saturated with connate water (R_w).

Log Curve LLS—Shallow Laterolog* resistivity curves measure the resistivity in the invaded zone (R_i). In a water-bearing zone the shallow Laterolog* (LLS) will record a low resistivity because R_{mf} is approximately equal to R_w.

Log Curve SFL—Microspherically Focused Log* resistivity curves measure the resistivity of the flushed zone (R_{xo}). In water-bearing zones the curve will record low resistivity because saltwater mud filtrate has low resistivity. The resistivity recorded by the Microspherically Focused Log* will be low and approximately equal to the resistivities of the invaded and uninvaded zones.

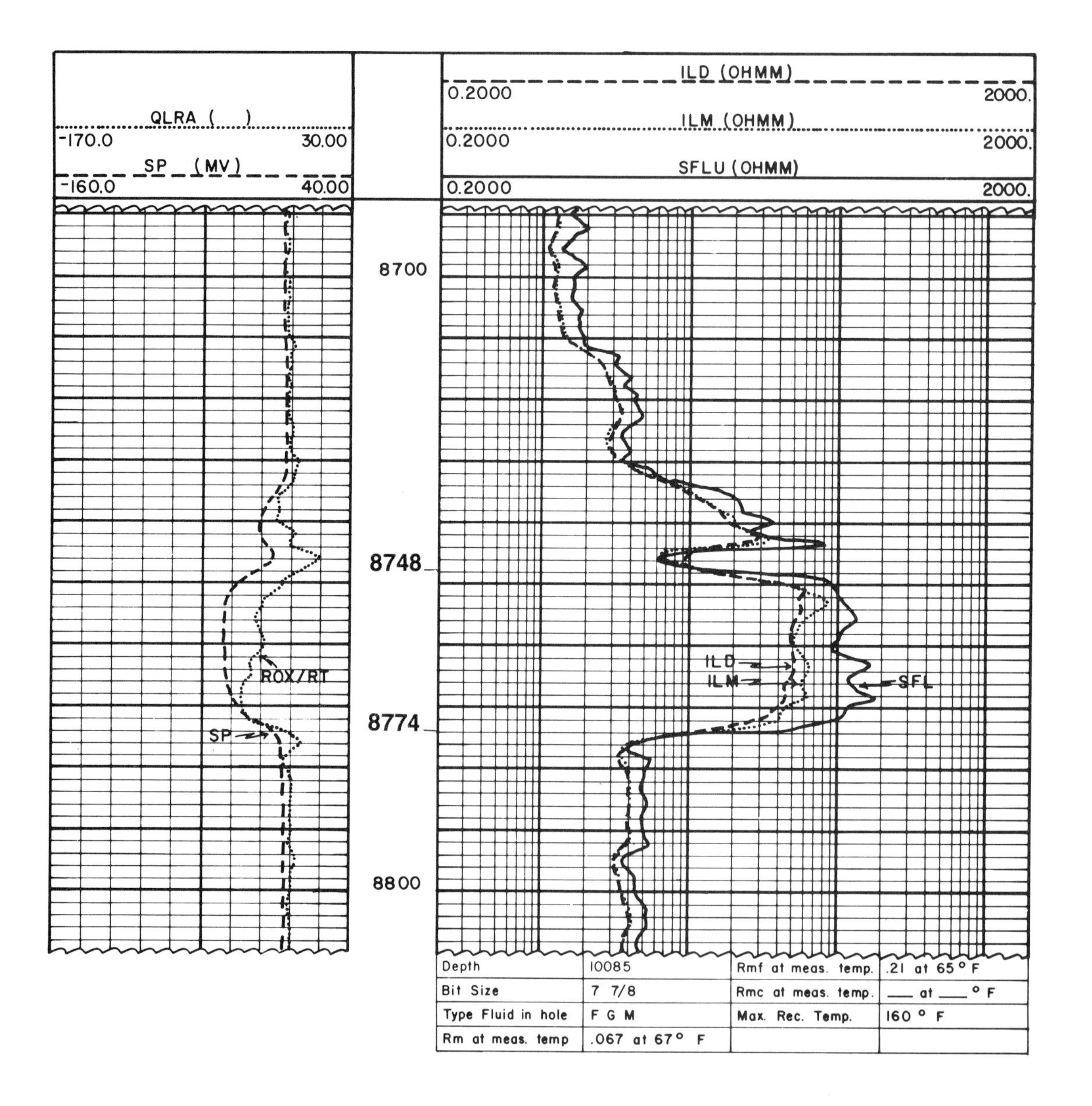
QLRA ( )
-170.0
30.00
SP (MV)
-160.0
40.00
ILD (OHMM)
0.2000
2000.
ILM (OHMM)
0.2000
2000.
SFLU (OHMM)
0.2000
2000.
8700
8748
8774
8800
ROX/RT
SP
ILD
ILM
SFL
Depth
10085
Rmf at meas. temp.
.21 at 65° F
Bit Size
7 7/8
Rmc at meas. temp.
___ at ___ ° F
Type Fluid in hole
F G M
Max. Rec. Temp.
160 ° F
Rm at meas. temp
.067 at 67° F

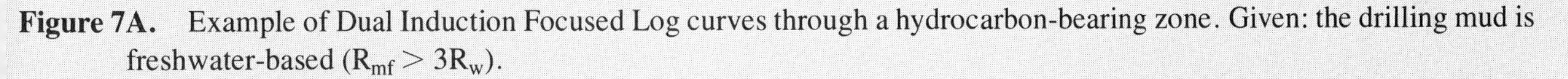

Figure 7A. Example of Dual Induction Focused Log curves through a hydrocarbon-bearing zone. Given: the drilling mud is freshwater-based ($R_{mf} > 3R_w$).

We've seen that where freshwater drilling muds invade a hydrocarbon-bearing formation ($S_w \ll 60\%$), there is high resistivity in the flushed zone (R_{xo}), high resistivity in the invaded zone (R_i), and high resistivity in the uninvaded zone (R_t). But, normally, beyond the flushed zone some diminishment of resistivity takes place. See Figure 5 for review.

Ignore the left side of the log on the opposite page, and compare the three curves on the right side of the log (tracks #2 and #3). Resistivity values are higher as distance increases from the left side of the log.

Log Curve ILD—Deep induction log resistivity curves measure the true resistivity (R_t) or the resistivity of the formation deep beyond the outer boundary of the invaded zone. This is a measure of the uninvaded zone. In hydrocarbon-bearing zones (in this case from 8,748 to 8,774 ft), the curve will read a high resistivity because hydrocarbons are more resistant than saltwater in the formation ($R_t > R_o$).

Log Curve ILM—Medium induction log resistivity curves measure the resistivity of the invaded zone (R_i). In a hydrocarbon-bearing zone, because of a mixture of mud filtrate (R_{mf}), formation water (R_w), and residual hydrocarbons (RH) in the pores, the curve will record a high resistivity. This resistivity is normally equal to or slightly more than the deep induction curve (ILD). But, in an annulus situation, the medium curve (ILM) may record a resistivity slightly less than the deep induction (ILD) curve.

Log Curve SFL—Spherically Focused Log* resistivity curves measure the resistivity of the flushed zone (R_{xo}). In a hydrocarbon-bearing zone, the curve will read a higher resistivity than the deep (ILD) or medium (ILM) induction curves because the flushed zone (R_{xo}) contains mud filtrate and residual hydrocarbons. The SFL* pictured here records a greater resistivity than either the deep (ILD) or medium (ILM) induction curves.

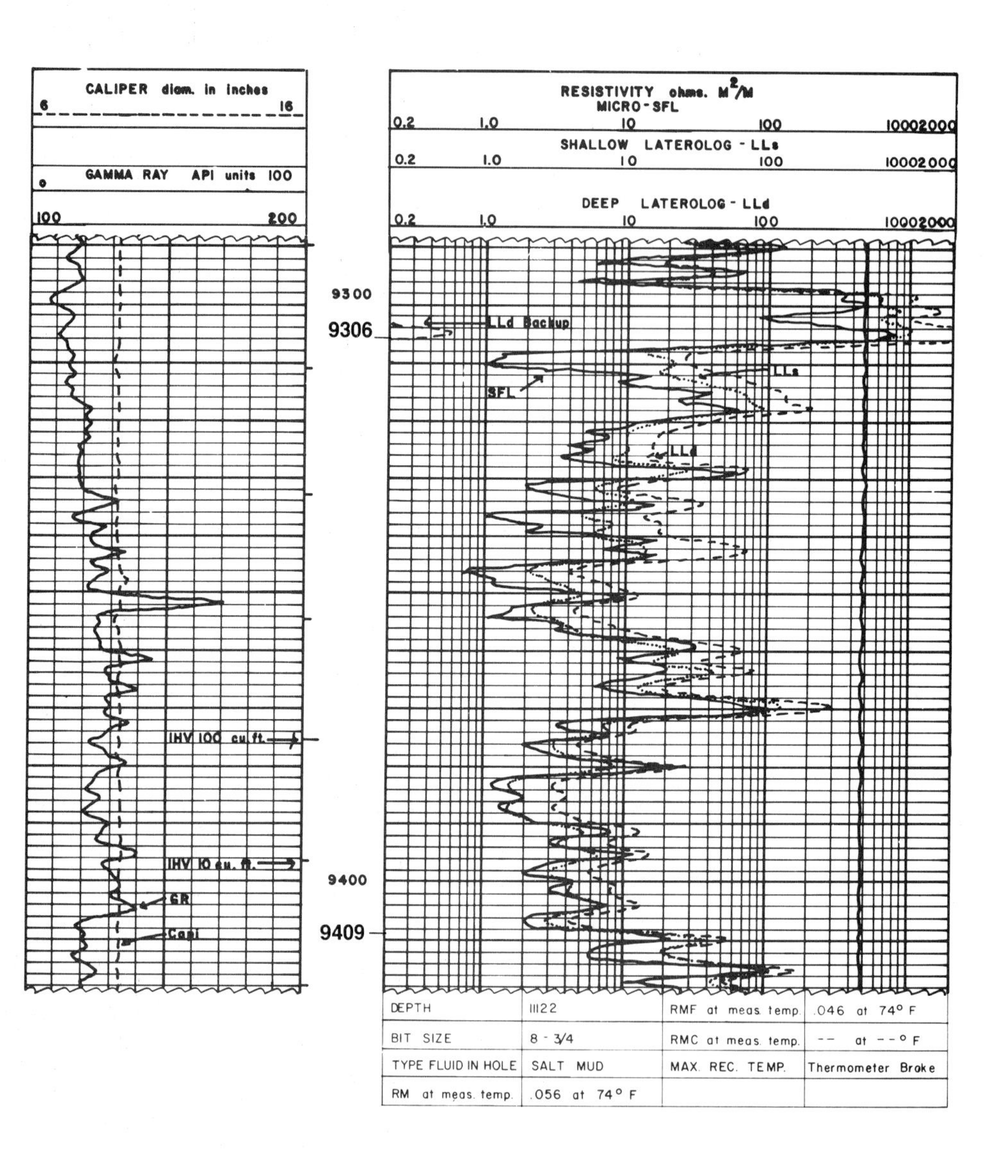

DEPTH	11122	RMF at meas. temp.	.046 at 74° F
BIT SIZE	8 - 3/4	RMC at meas. temp.	-- at --° F
TYPE FLUID IN HOLE	SALT MUD	MAX. REC. TEMP.	Thermometer Broke
RM at meas. temp.	.056 at 74° F		

Figure 7B. Example of Dual Laterolog* - Microspherically Focused Log (MSFL)* curves through a hydrocarbon-bearing zone. Given: the drilling mud is saltwater-based ($R_{mf} \simeq R_w$).

We've seen that where saltwater drilling muds invade a hydrocarbon-bearing zone ($S_w \ll 60\%$), there is low resistivity in the flushed zone (R_{xo}), an intermediate resistivity in the invaded zone (R_i), and high resistivity in the uninvaded zone (R_t). The reason for the increase in resistivities deeper into the formation is because of the increasing hydrocarbon saturation. See Figure 5 for review.

Ignore the left side of the log on the opposite page, and compare the three curves on the right side of the log (tracks #2 and #3). Resistivity values are higher as distance increases from the left side of the log.

Log Curve LLD—Deep Laterolog* resistivity curves measure true resistivity (R_t), or the resistivity of the formation deep beyond the outer boundary of the invaded zone. In hydrocarbon-bearing zones (in this case from 9,306 to 9,409 ft), the curve will read high resistivity because of high hydrocarbon saturation in the uninvaded zone (R_t).

Log Curve LLS—Shallow Laterolog* resistivity curves measure the resistivity in the invaded zone (R_i). In a hydrocarbon-bearing zone, the shallow Laterolog* (LLS) will record a lower resistivity than the deep Laterolog* (LLD) because the invaded zone (R_i) has a lower hydrocarbon saturation than the uninvaded zone (R_t).

Log Curve SFL—Microspherically Focused Log* resistivity curves measure the resistivity of the flushed zone (R_{xo}). In hydrocarbon-bearing zones, the curve will record low resistivity because saltwater mud filtrate has low resistivity and the residual hydrocarbon (RH) saturation in the flushed zone (R_{xo}) is low. Therefore, in a hydrocarbon-bearing zone with saltwater-based drilling mud, the uninvaded zone (R_t) has high resistivity, the invaded zone (R_i) has a lower resitivity, and the flushed zone (R_{xo}) has the lowest resistivity.

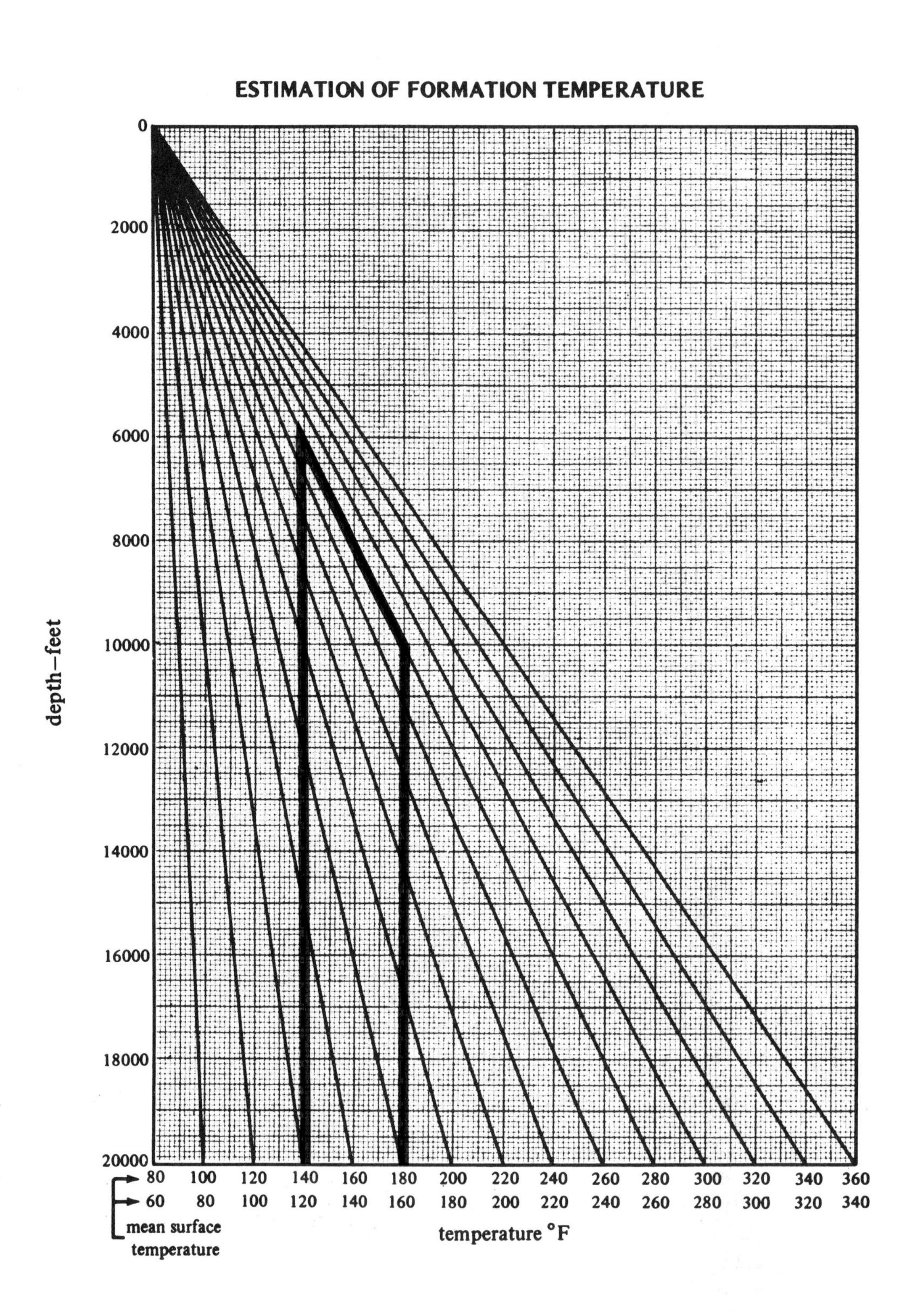
ESTIMATION OF FORMATION TEMPERATURE
depth–feet
0
2000
4000
6000
8000
10000
12000
14000
16000
18000
20000
80 100 120 140 160 180 200 220 240 260 280 300 320 340 360
60 80 100 120 140 160 180 200 220 240 260 280 300 320 340
mean surface temperature
temperature °F

igure 8. Chart for estimating formation temperature (T_f) with depth (linear gradient assumed).

Courtesy, Dresser Industries.
Copyright 1975, Dresser Atlas.

Given:

Surface temperature = 80°
Bottom hole temperature (BHT) = 180°
Total depth (TD) = 10,000 feet
Formation depth = 6,000 feet

Procedure:

1. Locate BHT (180°F) on the 80 scale (bottom of the chart; surface temperature = 80°F).
2. Follow BHT (180°) vertically up until it intersects 10,000 ft (TD) line. This intersection defines the temperature gradient.
3. Follow the temperature gradient line up to 6,000 ft (formation depth).
4. Formation temperature (140°) is read on the bottom scale vertically down from the point where the 6,000 ft line intersects the temperature gradient.

NOTE: In the United States (as an example) 80° is used commonly as the mean surface temperature in the Southern States, and 60° is used commonly in the Northern States. However, a person can calculate his own mean surface temperature if such precision is desired.

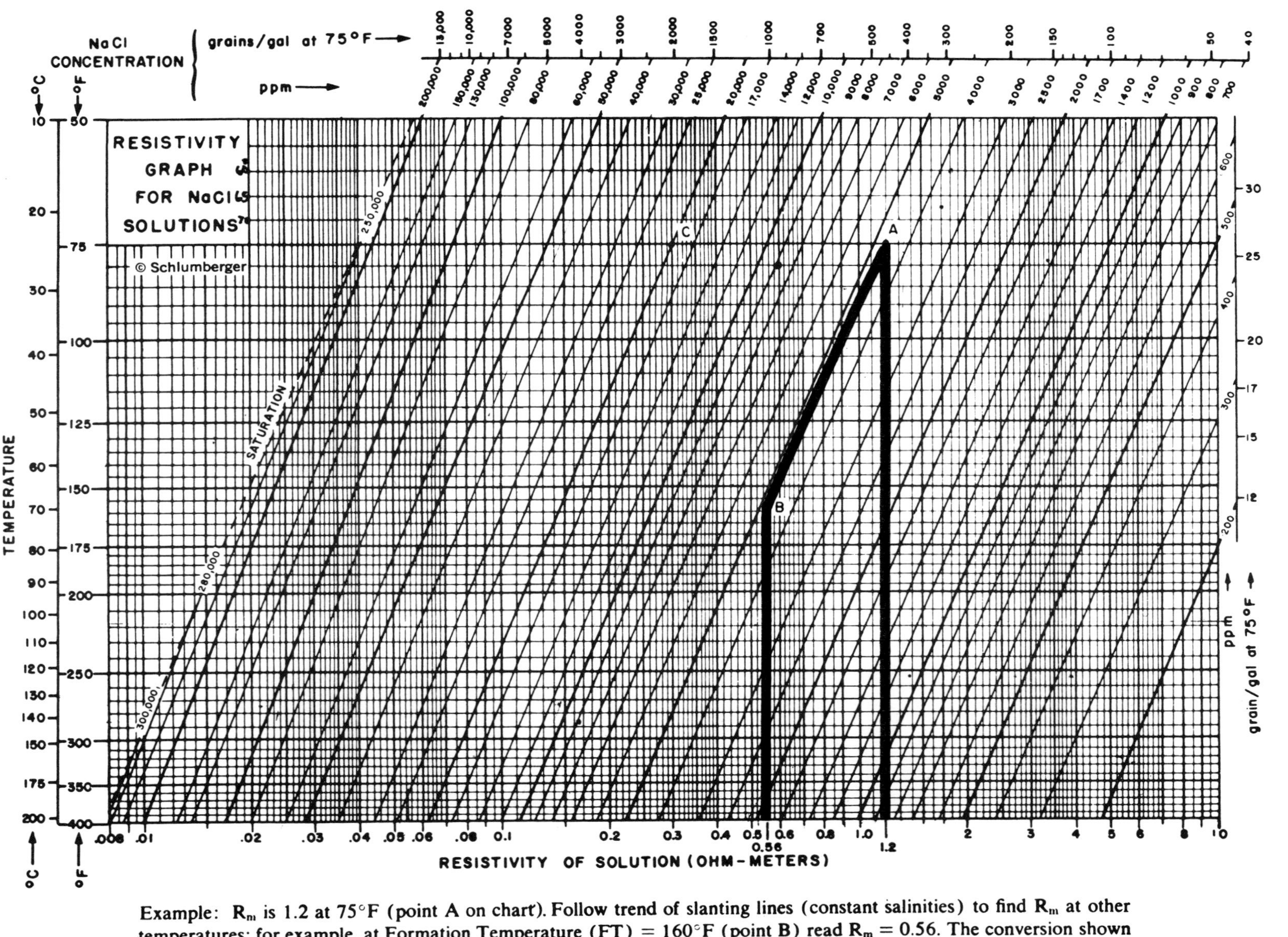

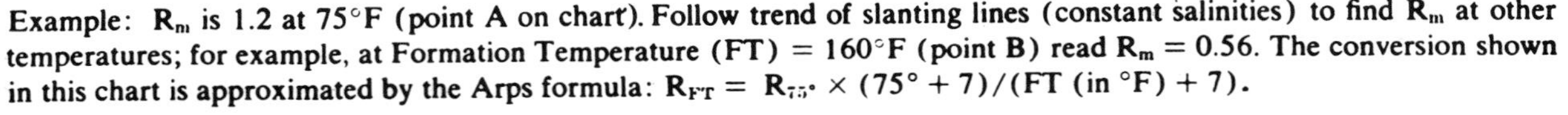

Example: R_m is 1.2 at 75°F (point A on chart). Follow trend of slanting lines (constant salinities) to find R_m at other temperatures; for example, at Formation Temperature (FT) = 160°F (point B) read $R_m = 0.56$. The conversion shown in this chart is approximated by the Arps formula: $R_{FT} = R_{75°} \times (75° + 7)/(FT\ (\text{in } °F) + 7)$.

'igure 9. Because resistivity varies with changes in temperature, you must adjust before calculation. Use the chart on the opposite page.

Courtesy, Schlumberger Well Services.
Copyright 1972, Schlumberger.

Given: Resistivity of drilling mud (R_m) equals 1.2 at 75°F. Formation temperature (T_f) = 160°.

Procedure:

1. Locate the resistivity value, 1.2, on the scale at the bottom of the chart.
2. Follow the vertical line up to a temperature value of 75°F (point A on the chart).
3. Follow the diagonal line (constant salinity) to where it intersects a temperature value of 160°F (point B on the chart).
4. From point B, follow the vertical line to the scale at the bottom, and find a resistivity value of 0.56.

CHAPTER II

THE SPONTANEOUS POTENTIAL LOG

General

This chapter and succeeding chapters (III through V) introduce the reader to specific log types such as SP, resistivity, porosity, and gamma ray logs. The text discusses how different log types measure various properties in the well bore and surrounding formations, what factors affect these measurements, where a particular curve is recorded, and how data are obtained from the log using both charts and mathematical formulas.

The spontaneous potential (SP) log was one of the earliest electric logs used in the petroleum industry, and has continued to play a significant role in well log interpretation. By far the largest number of wells today have this type of log included in their log suites. Primarily the spontaneous potential log is used to identify impermeable zones such as shale, and permeable zones such as sand. However, as will be discussed later, the SP log has several other uses perhaps equally important.

The spontaneous potential log is a record of direct current (DC) voltage differences between the naturally occurring potential of a moveable electrode in the well bore, and the potential of a fixed electrode located at the surface (Doll, 1948). It is measured in millivolts.

Electric currents arising primarily from electrochemical factors within the borehole create the SP log response. These electrochemical factors are brought about by differences in salinities between mud filtrate (R_{mf}) and formation water resistivity (R_w) within permeable beds. *Because a conductive fluid is needed in the borehole for the SP log to operate, it cannot be used in non-conductive (i.e. oil-based) drilling muds.*

The SP log is recorded on the left hand track of the log in track #1 and is used to: (1) detect permeable beds, (2) detect boundaries of permeable beds, (3) determine formation water resistivity (R_w), and (4) determine the volume of shale in permeable beds. An auxiliary use of the SP curve is in the detection of hydrocarbons by the suppression of the SP response.

The concept of static spontaneous potential (SSP) is important because SSP represents the *maximum* SP that a thick, shale-free, porous and permeable formation can have for a given ratio between R_{mf}/R_w. SSP is determined by formula or chart and is a necessary element for determining accurate values of R_w and volume of shale. The SP value that is measured in the borehole is influenced by bed thickness, bed resistivity, invasion, borehole diameter, shale content, and most important—the ratio of R_{mf}/R_w (Fig.10a).

Bed thickness—As a formation thins (i.e. < 10 feet thick) the SP measured in the borehole will record an SP value less than SSP (Fig. 10b). However, the SP curve can be corrected by chart for the effects of bed thickness. As a general rule whenever the SP curve is narrow and pointed in shape, the SP should be corrected for bed thickness.

Bed resistivity—Higher resistivities reduce the deflection of the SP curves.

Borehole and invasion—Hilchie (1978) indicates that the effects of borehole diameter and invasion on the SP log are very small and, in general, can be ignored.

Shale content—The presence of shale in a permeable formation reduces the SP deflection (Fig. 10b). In water-bearing zones the amount of SP reduction is proportional to the amount of shale in the formation. In hydrocarbon-bearing zones the amount of SP reduction is greater than the volume of shale and is called "hydrocarbon suppression" (Hilchie, 1978).

The SP response of shales is relatively constant and follows a straight line called a shale baseline. SP curve deflections are measured from this shale baseline. *Permeable zones are indicated where there is SP deflection from the shale baseline*. For example, if the SP curve moves either to the left (negative deflection; $R_{mf} > R_w$) or to the right (positive deflection; $R_{mf} < R_w$) of the shale baseline, permeable zones are present. *Permeable bed boundaries are detected by the point of inflection from the shale baseline*.

But, take note, when recording non-permeable zones or permeable zones where R_{mf} is equal to R_w, the SP curve will not deflect from the shale baseline. The magnitude of SP deflection is due to the difference in resistivity between mud filtrate (R_{mf}) and formation water (R_w) and not to the amount of permeability.

Resistivity of Formation Water (R_w) Calculated from the SP Curve

Figure 11 is an electric induction log with an SP curve from a Pennsylvanian upper Morrow sandstone in Beaver County, Oklahoma. In this example, the SP curve is used to find a value for R_w by the following procedure: After you determine the formation temperature, you correct the resistivities (obtained from the log heading) of the mud filtrate (R_{mf}) and drilling mud (R_m) to formation temperature (see Chapter I).

Next, to minimize for the effect of bed thickness, the SP

is corrected to static SP (SSP). SSP represents the maximum SP a formation can have if unaffected by bed thickness. Figure 12 is a chart used to correct SP to SSP. The data necessary to use this chart are: (1) bed thickness, (2) resistivity from the shallow-reading resistivity tool (R_i), and (3) the resistivity of the drilling mud (R_m) at formation temperature.

Once the value of SSP is determined, it is used on the chart illustrated in Figure 13 to obtain a value for the R_{mf}/R_{we} ratio. *Equivalent resistivity* (R_{we}) is obtained by dividing R_{mf} by the R_{mf}/R_{we} value from the chart (Fig. 13).

The value of R_{we} is then corrected to R_w, using the chart illustrated in Figure 14, for average deviation from sodium chloride solutions, and for the influence of formation temperature. A careful examination of Figures 11-14 should help you gain an understanding of the R_w from SP procedure. But, rather than using charts in the procedure, you might prefer using the mathematical formulas listed in Table 3.

It is important to remember that normally the SP curve has less deflection in hydrocarbon-bearing zones; this is called *hydrocarbon suppression*, and results in too high a

Table 3. Mathematical Calculation of R_w from SSP (modified after Bateman & Konen, 1977).

Instead of charts, some individuals may prefer using these formulas, especially if they want to computerize the procedure.

R_{mf} at 75°F = $R_{mf\ temp}{}^{\dagger} \times (temp + 6.77)/81.77$
Correction of R_{mf} to 75°

$K = 60 + (0.133 \times T_f)$

$R_{mfe}/R_{we} = 10^{-SSP/K}$ ††

$R_{mfe} = (146 \times R_{mf} - 5)/(337 \times R_{mf} + 77)$

R_{mfe} formula if R_{mf} at 75°F < 0.1

$R_{mfe} = 0.85 \times R_{mf}$

R_{mfe} formula if R_{mf} at 75° > 0.1

$R_{we} = R_{mfe}/(R_{mfe}/R_{we})$

R_w at 75°F = $(77 \times R_{we} + 5)/(146 - 377 \times R_{we})$

R_w at 75° formula if $R_{we} < 0.12$

R_w at 75°F = $-\ [0.58 - 10^{(0.69xRwe-0.24)}]$

R_w at 75° formula if $R_{we} > 0.12$

R_w at formation temperature = R_w at 75° $\times\ 81.77/(T_f + 6.77)$

†$R_{mf\ temp}$ = R_{mf} at a temperature other than 75°F
††The e subscript (i.e. R_{mfe}) stands for *equivalent resistivity*.

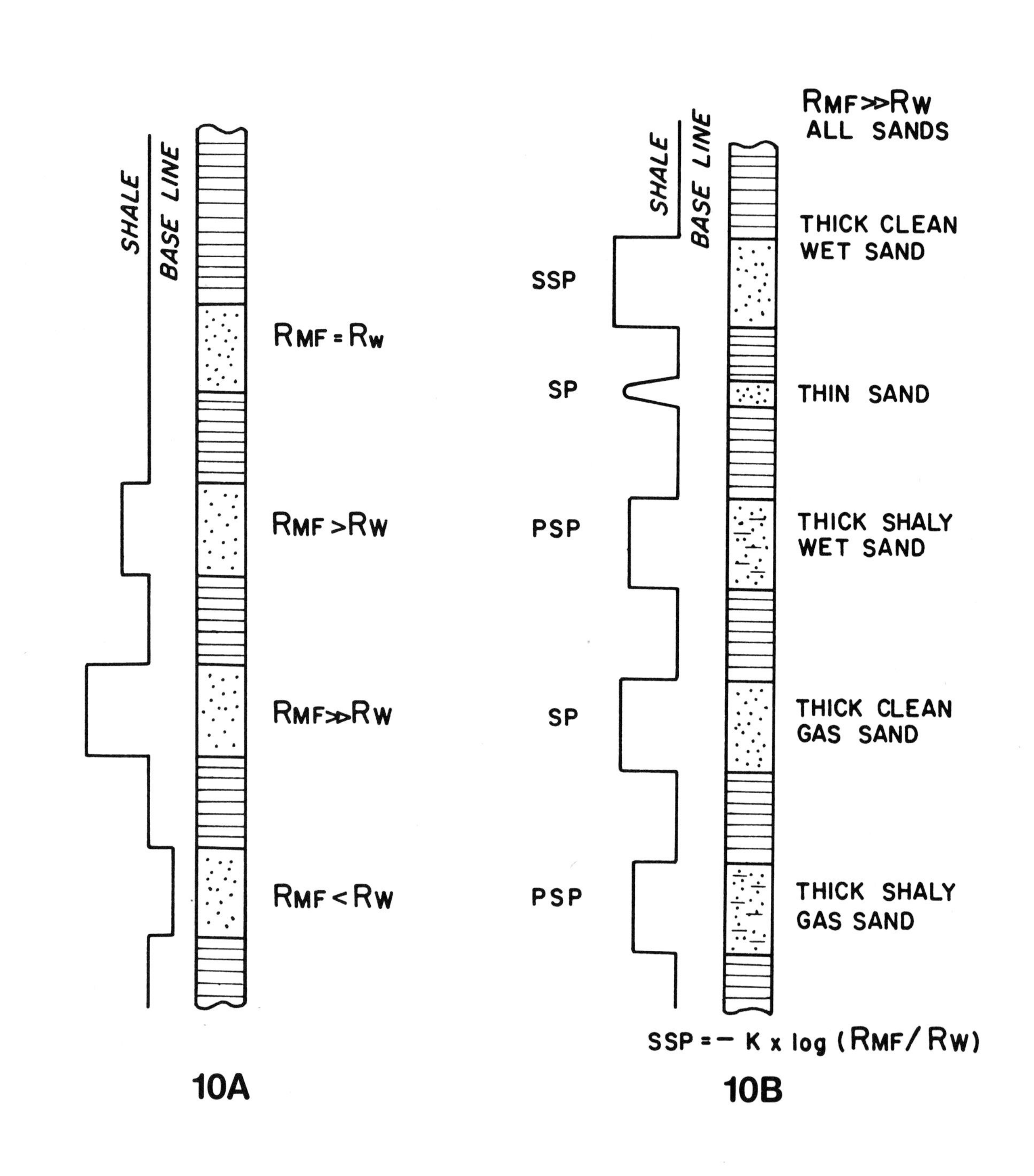
SHALE
BASE LINE
RMF = Rw
RMF > Rw
RMF>>Rw
RMF < Rw
10A
RMF>>Rw
ALL SANDS
SSP
THICK CLEAN
WET SAND
SP
THIN SAND
PSP
THICK SHALY
WET SAND
SP
THICK CLEAN
GAS SAND
PSP
THICK SHALY
GAS SAND
SSP = − K x log (RMF/ Rw)
10B

Figure 10. Examples of SP deflection from the shale baseline.

10A—SP deflection with different resistivities of mud filtrate (R_{mf}) and formation water (R_w). Where resistivity of the mud filtrate (R_{mf}) is equal to the resistivity of the formation water (R_w) there is no deflection, positive or negative, from the shale baseline.

Where R_{mf} is greater than R_w, the SP line kicks to the left of the shale baseline (negative deflection). Where R_{mf} greatly exceeds R_w, the deflection is proportionately greater.

Where R_{mf} is less than R_w, the kick is to the right of the shale baseline. This is called positive deflection.

Remember, the spontaneous potential log (SP) is used only with conductive (saltwater-based) drilling muds.

10B—SP deflection with resistivity of the mud filtrate (R_{mf}) much greater than formation water (R_w). SSP (static spontaneous potential) at the top of the diagram, is the maximum deflection possible in a thick, shale-free, and water-bearing ("wet") sandstone for a given ratio of R_{mf}/R_w. All other deflections are less, and are relative in magnitude.

SP (spontaneous potential) is the SP response due to the presence of thin beds and/or the presence of gas. PSP (pseudo-static spontaneous potential) is the SP response if shale is present.

Note at bottom of diagram: A formula for the theoretical calculated value of SSP is given. $SSP = -K \times \log (R_{mf}/R_w)$, where $K = (.133 \times T_f) + 60$.

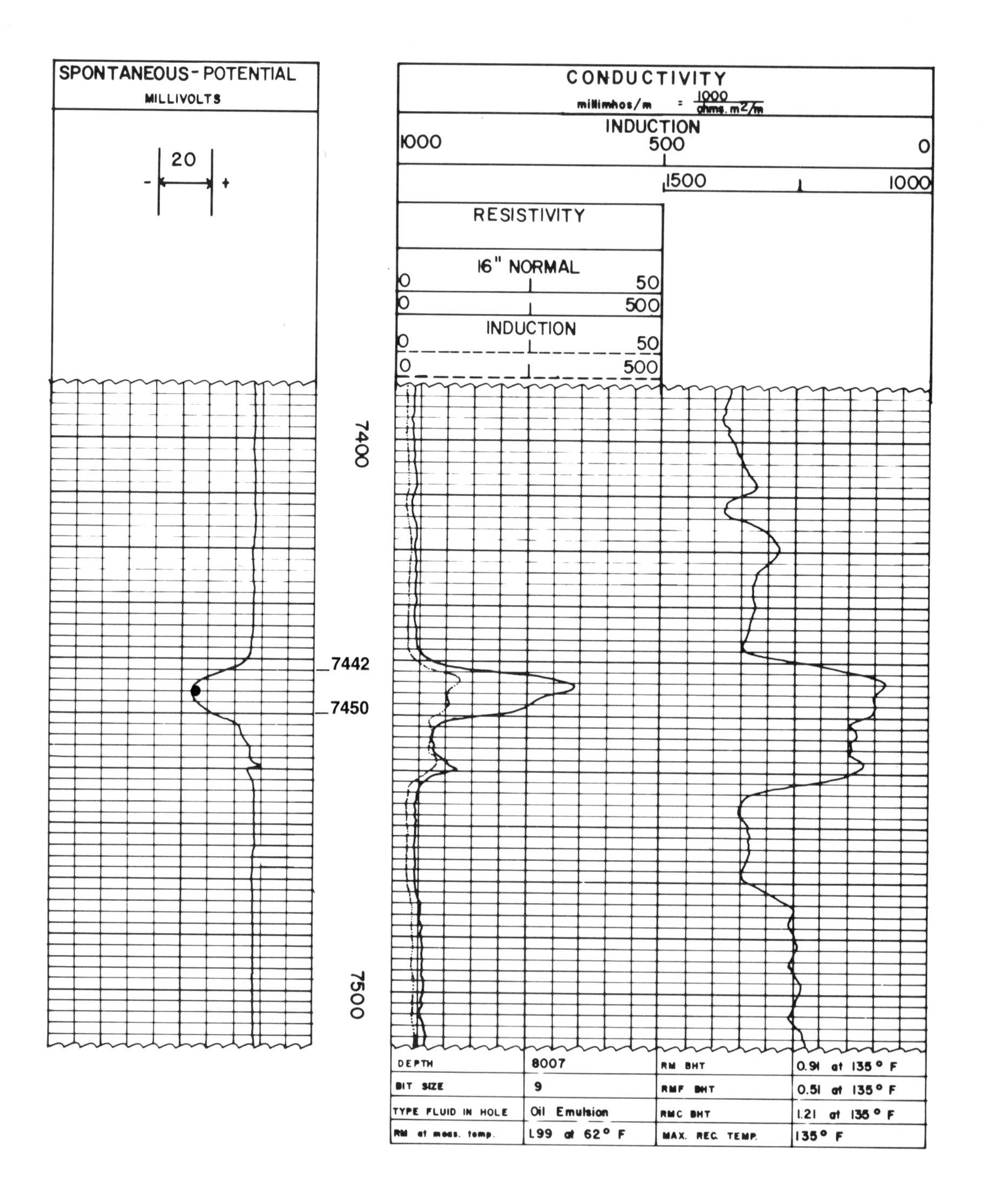
SPONTANEOUS-POTENTIAL
MILLIVOLTS
20
-
+
7400
7442
7450
7500
CONDUCTIVITY
millimhos/m = 1000 / ohms. m2/m
INDUCTION
1000
500
0
1500
1000
RESISTIVITY
16" NORMAL
0
50
0
500
INDUCTION
0
50
0
500
DEPTH
8007
RM BHT
0.91 at 135° F
BIT SIZE
9
RMF BHT
0.51 at 135° F
TYPE FLUID IN HOLE
Oil Emulsion
RMC BHT
1.21 at 135° F
RM at meas. temp.
1.99 at 62° F
MAX. REC. TEMP.
135° F

Figure 11. Determination of formation water resistivity (R_w) from an SP log. This example is an exercise involving the charts on Figures 12 through 14.

Given:

R_{mf} = 0.51 at 135° (BHT)
R_m = 0.91 at 135° (BHT)
Surface temperature = 60°F
Total depth = 8,007 ft
Bottom Hole Temperature (BHT) = 135°F

From the Log track:

1. SP = −40mv (spontaneous potential measured from log at a formation depth of 7,446 ft and uncorrected for bed thickness). It is measured here as two 20mv divisions from the shale baseline. The deflection is negative, so the value (−40mv) is negative.
2. Bed thickness equals 8 ft (7,442 to 7,450 ft).
3. Resistivity short normal (R_i) equals 28 ohm-meters.
4. Formation depth equals 7,446 ft.

Procedure:

1. *Determine T_f*—Use Figure 8 to determine the temperature of the formation (T_f). Use BHT = 135°F, TD = 8,007 ft, surface temperature = 60°F, and formation depth = 7,446 ft. (Your answer should be 130°F).

2. *Correct R_m and R_{mf} to T_f*—Use Figure 9 to correct the values for the resistivity of mud and of mud filtrate, using T_f (130°F) from step 1. Use R_m = 0.91 at 135°F and R_{mf} = 0.51 at 135°F. (Your answers should be: R_m = 0.94 at 130°F and R_{mf} = 0.53 at 130°F).

3. *Determine SP*—Read directly from the SP curve in Fig. 11. It measures two units (at a scale of 20 mv per division) from the shale baseline. The deflection is negative, so your answer is also (−40 mv) negative.

4. *Correct SP to SSP*—Correcting SP for the thin-bed effect will give a value for SSP; use the chart in Figure 12 to find the SP Correction Factor.

   Given R_i/R_m (or R_{sn}/R_m) = 28/0.94 = 30. Bed thickness (read from SP log) Fig. 11 equals 8 ft. Correction factor (from Fig. 12) = 1.3.

   SSP = SP × SP Correction factor (Fig. 12)
   SSP = (−40 mv) × 1.3
   SSP = −52 mv (Your answer)

5. Determine R_{mf}/R_{we} ratio—Use the chart in Figure 13 (Your answer should be 5.0).

6. Determine R_{we}—Divide the corrected value for R_{mf} by the ratio R_{mf}/R_{we} value.

   $R_{we} = R_{mf}/(R_{mf}/R_{we})$
   $R_{we} = 0.53/5.0$
   $R_{we} = 0.106$

7. Correct R_{we} to R_w—Use the chart in Figure 14, and the R_{we} value in step 6 (Your answer should be R_w = 0.11 at T_f).

NOTE: The term short normal describes a log used to measure the shallow formation resistivity, or the resistivity of the invaded zone (R_i). Short normal resistivity (R_{sn}) was used in Procedure Step 4 above, and its use as a logging/resistivity term is common.

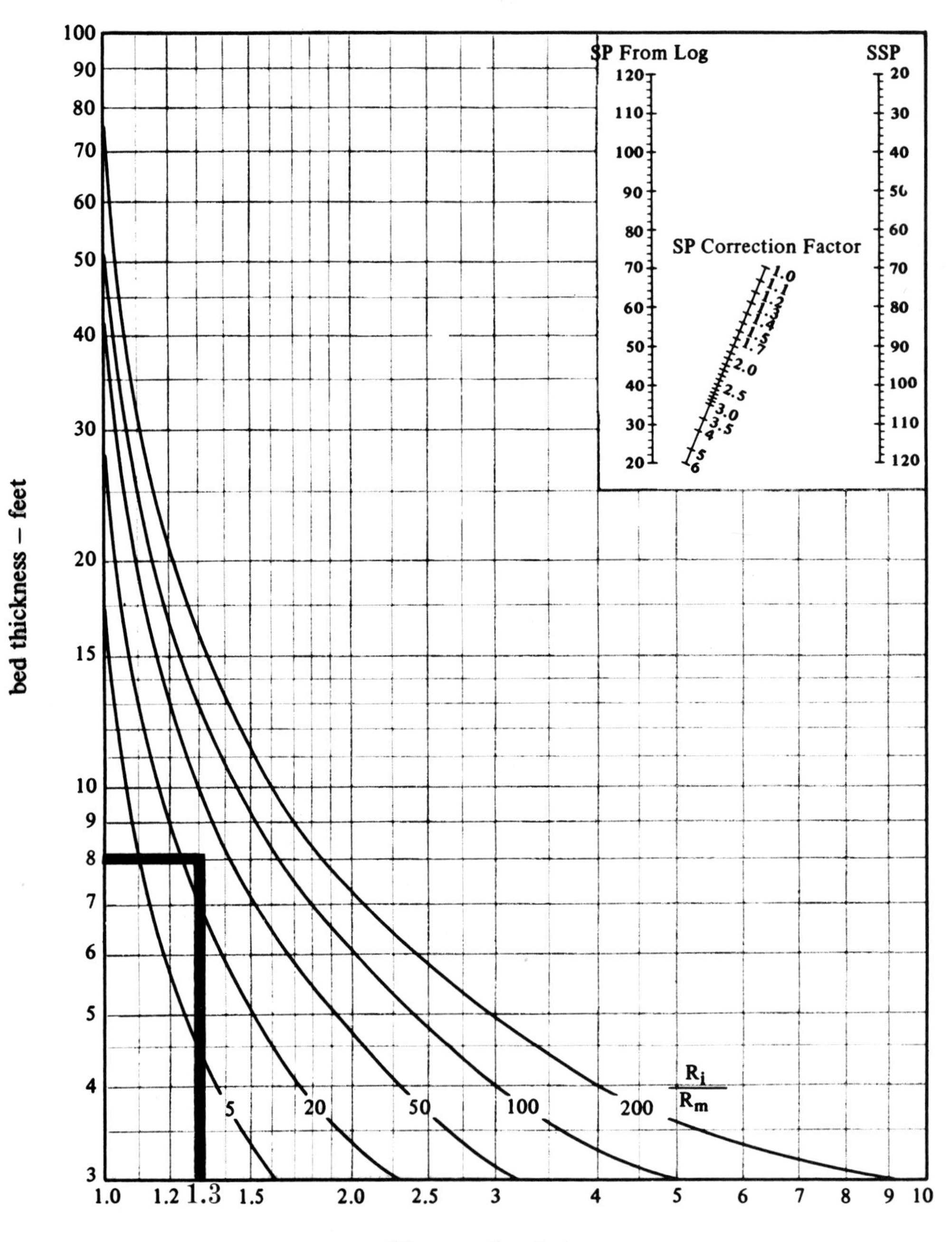
SP CORRECTION
bed thickness – feet
SP correction factor
SP From Log
SSP
SP Correction Factor
R_i / R_m
5
20
50
100
200
1.0
1.2
1.3
1.5
2.0
2.5
3
4
5
6
7
8
9
10
100
90
80
70
60
50
40
30
20
15
10
9
8
7
6
5
4
3

Figure 12. Chart for finding SP Correction Factor used to correct SP to SSP (see exercise, Fig. 11).

Courtesy, Dresser Industries.
Copyright 1975, Dresser Atlas.

Defined:

R_i = shallow resistivity
R_m = resistivity of drilling mud at formation temperature

Example:

$R_i/R_m = 30$
Bed Thickness = 8 ft

Procedure:

1. Locate a bed thickness on the vertical scale (in this case 8 ft).
2. Follow the value horizontally across until it intersects the R_i/R_m curve (in this example $R_i/R_m = 30$, so point will be to the right of the 20 curve).
3. Drop vertically from this intersection and read the SP correction factor on the scale across the bottom (in this example, a value of 1.3).
4. Multiply SP by the SP Correction Factor to find SSP.

For the exercise in Figure 11:
SSP = SP × Correction Factor
SSP = −40mv × 1.3 (−40mv is SP value taken at 7,446 ft, see Fig. 11)
SSP = −52mv

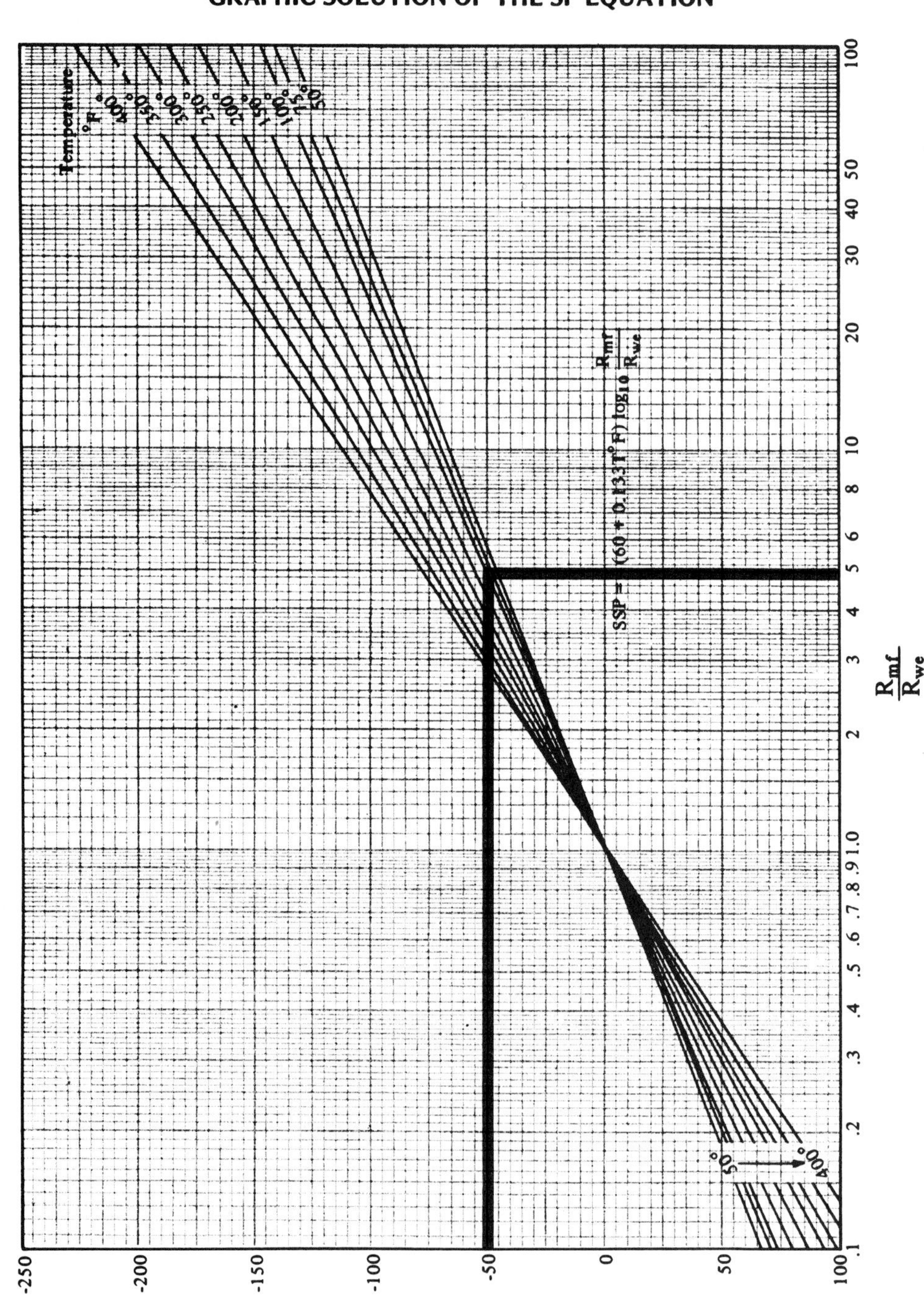
GRAPHIC SOLUTION OF THE SP EQUATION
Temperature °F
400°
350°
300°
250°
200°
150°
100°
75°
50°
SSP = -(60 + 0.133T° F) log10 Rmf/Rwe
Rmf/Rwe
.1
.2
.3
.4
.5
.6
.7
.8
.9
1.0
2
3
4
5
6
8
10
20
30
40
50
100
-250
-200
-150
-100
-50
0
50
100
spontaneous potential (millivolts)

Figure 13. Chart used for determining the R_{mf}/R_{we} ratio from SSP values.

Courtesy, Dresser Industries.
Copyright 1975, Dresser Atlas.

Example:

SSP = −52mv (from SP log and Fig. 12)
T_f = 130°F

Procedure:

1. Locate an SSP value on the vertical scale (in this case −52mv).
2. Follow the value horizontally across until it intersects the sloping formation temperature line (130°F; imagine one between the lines for 100° and 150° temperature lines).
3. Drop vertically from this intersection and read the ratio value on the bottom scale (in this example, the ratio value is 5.0).

R_w FROM R_{we}

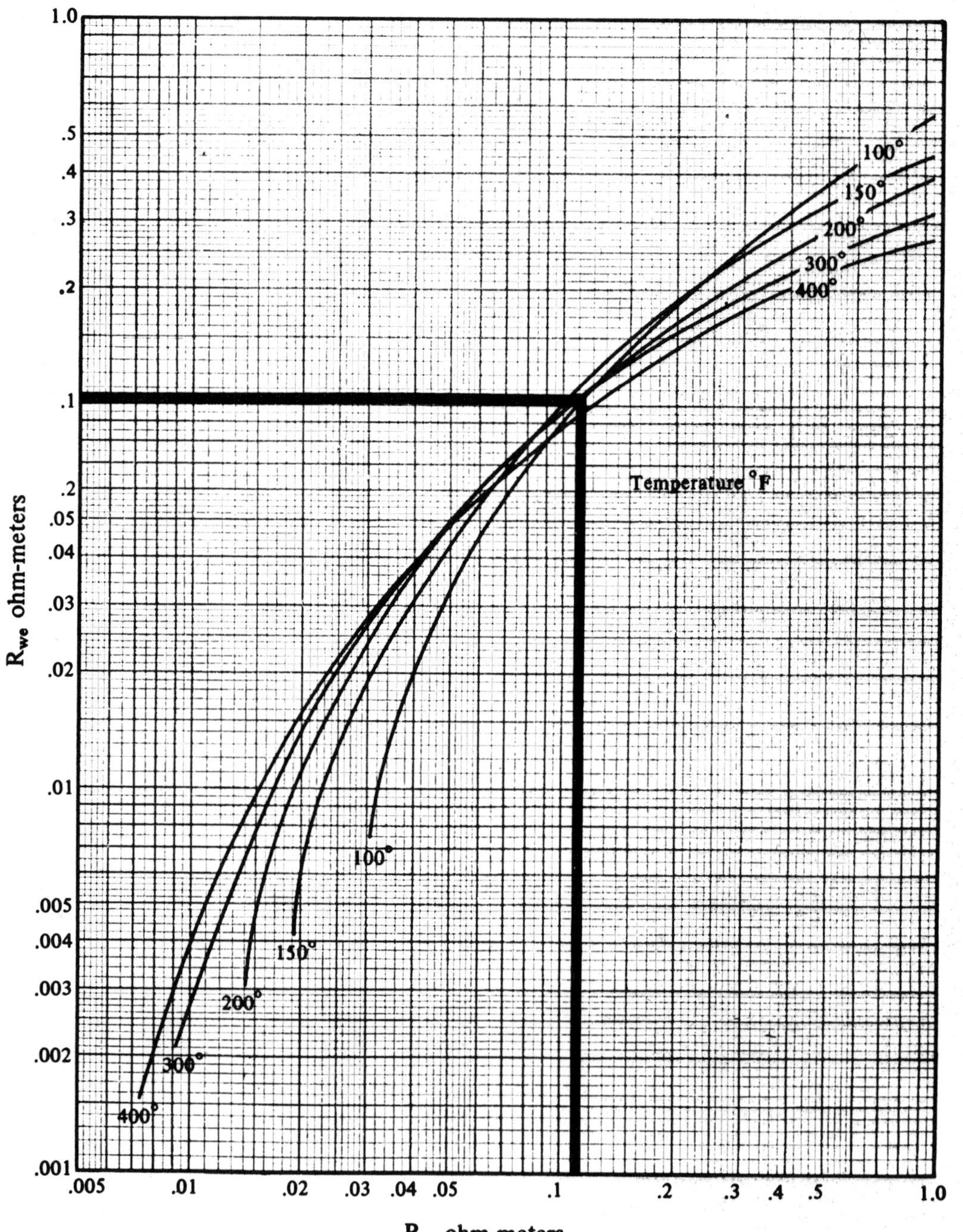

1.0
.5
.4
.3
.2
.1
.2
.05
.04
.03
.02
.01
.005
.004
.003
.002
.001
Rwe ohm-meters
.005
.01
.02
.03
.04
.05
.1
.2
.3
.4
.5
1.0
Rw ohm-meters
Temperature °F
100°
150°
200°
300°
400°

Figure 14. Chart for determining a resistivity value for R_w from R_{we}.

Courtesy, Dresser Industries.
Copyright 1975, Dresser Atlas.

Given:

R_{we} is calculated by dividing R_{mf} corrected to formation temperature (T_f) by the ratio R_{mf}/R_{we}. From the exercise example you can calculate $R_{we} = 0.53/5$ or $R_{we} = 0.106$. R_{mf} at $T_f = 0.53$. $T_f = 130°F$.

Procedure:

1. Locate the value of R_{we} on the vertical scale (in this case 0.106).
2. Follow it horizontally until it intersects the temperature curve desired (in this case 130°F between the 100° and 150° temperature curves).
3. Drop vertically from the intersection and read a value for R_w on the scale at the bottom (in this case 0.11).

value for R_w calculated from SSP. Therefore, to determine R_w from SP it is best, whenever possible, to use the SP curve opposite *known* water-bearing zones.

Volume of Shale Calculation

The SP log can be used to calculate the volume of shale in a permeable zone by the following formula:

$$V_{sh} \text{ (in \%)} = 1.0 - \frac{PSP}{SSP}$$

Where:

V_{sh} = volume of shale

PSP = pseudo static spontaneous potential (SP of shaly formation)

SSP = static spontaneous potential of a thick clean sand or carbonate

$$SSP = -K \times \log (R_{mf}/R_w)$$
$$K = 60 + (0.133 \times T_f)$$

The volume of shale in a sand can be used in the evaluation of shaly sand reservoirs (Chapter VI) and as a mapping parameter for both sandstone and carbonate facies analysis (Chapter VII).

Review - Chapter II

1. The spontaneous potential log (SP) can be used to: (1) detect permeable beds; (2) detect boundaries of permeable beds; (3) determine formation water resistivity (R_w); and (4) determine volume of shale (V_{sh}) in a permeable bed.
2. The variations in the SP are the result of an electric potential that is present between the well bore and the formation as a result of differences in salinities between R_{mf} and R_w.
3. The SP response in shales is relatively constant and its continuity of amplitude is referred to as the shale baseline. In permeable beds the SP will do the following relative to the shale baseline: (1) negative deflection to the left of the shale baseline where $R_{mf} > R_w$; (2) positive deflection to the right of the shale baseline where $R_{mf} < R_w$; (3) no deflection where $R_{mf} = R_w$.
4. The SP curve can be suppressed by thin beds, shaliness, and the presence of gas.

CHAPTER III

RESISTIVITY LOGS

General

Resistivity logs are electric logs which are used to: (1) determine hydrocarbon versus water-bearing zones, (2) indicate permeable zones, and (3) determine resistivity porosity. By far the most important use of resistivity logs is the determination of hydrocarbon versus water-bearing zones. Because the rock's matrix or grains are non-conductive, the ability of the rock to transmit a current is almost entirely a function of water in the pores. Hydrocarbons, like the rock's matrix, are non-conductive; therefore, as the hydrocarbon saturation of the pores increases, the rock's resistivity also increases. A geologist, by knowing a formation's water resistivity (R_w), its porosity (ϕ), and a value for the cementation exponent (m) (Table 1), can determine a formation's water saturation (S_w) from the Archie equation:

$$S_w = \left(\frac{F \times R_w}{R_t}\right)^{1/n}$$

Where:

S_w = water saturation
F = formation factor (a/ϕ^m)
a = tortuosity factor
m = cementation exponent
R_w = resistivity of formation water
R_t = true formation resistivity as measured by a deep reading resistivity log
n = saturation exponent (most commonly 2.0)

The two basic types of logs in use today which measure formation resistivity are induction and electrode logs (Table 4). The most common type of logging device is the induction tool (Dresser Atlas, 1975).

An induction tool consists of one or more transmitting coils that emit a high-frequency alternating current of constant intensity. The alternating magnetic field which is created induces secondary currents in the formation. These secondary currents flow as ground loop currents perpendicular to the axis of the borehole (Fig. 15), and create magnetic fields that induce signals to the receiver coils. The receiver signals are essentially proportional to conductivity†, which is the reciprocal of resistivity (Schlumberger, 1972). The multiple coils are used to focus the resistivity measurement to minimize the effect of materials in the borehole, the invaded zone, and other nearby formations. The two types of induction devices are the Induction Electric Log and the Dual Induction Focused Log.

A second type of resistivity measuring device is the electrode log. Electrodes in the borehole are connected to a power source (generator), and the current flows from the electrodes through the borehole fluid into the formation, and then to a remote reference electrode. Examples of electrode resistivity tools include: (1) normal, (2) Lateral, (3) Laterolog*, (4) Microlaterolog*, (5) Microlog*, (6) Proximity Log*, and (7) spherically focused logs.

Induction logs should be used in non-salt-saturated drilling muds (i.e. $R_{mf} > 3\,R_w$) to obtain a more accurate value of true resistivity (R_t). *Boreholes filled with salt-saturated drilling muds* ($R_{mf} \simeq R_w$) *require electrode logs,* such as the Laterolog* or Dual Laterolog* with or without a Microspherically Focused Log*, to determine accurate R_t values. Figure 16 is a chart which assists in determining when use of an induction log is preferred over an electrode log such as the Laterolog*.

Induction Electric Log

The Induction Electric Log (Fig. 17) is composed of three curves: (1) short normal, (2) induction, and (3) spontaneous potential or SP. These curves are obtained simultaneously during the logging of the well.

Short Normal—The short normal tool measures resistivity at a shallow depth of investigation which is the resistivity of the invaded zone (R_i). When the resistivity of the short normal is compared with the resistivity of the deeper measuring induction tool (R_t), invasion is detected by the separation between the short normal and induction curves (Fig. 17). *The presence of invasion is important because it indicates a formation is permeable.*

The short normal tool has an electrode spacing of 16 inches and can record a reliable value for resistivity from a bed thickness of four feet. The short normal curve is usually recorded in track #2 (Fig. 17). Because the short normal tool works best in conductive, high resistivity muds (where $R_{mf} > 3\,R_w$), salt muds (where $R_{mf} \simeq R_w$) are not a good environment for its use. In addition to providing a value for R_i, the short normal curve can be used to calculate a value for resistivity porosity if a correction is made for unflushed oil in the invaded zone. To obtain a more accurate value of R_i from the short normal curve, an amplified short normal

†conductivity = 1000/resistivity. Conductivity in millimhos/meter; resistivity in ohm-meters.

Table 4. Classification of Resistivity Logs.

INDUCTION LOGS (measure conductivity)

ELECTRODE LOGS (measure resistivity)

A. Normal logs
B. Lateral Log†
C. Laterologs*
D. Spherically Focused Log (SFL)*
E. Microlaterolog (MLL)*
F. Microlog (ML)*
G. Proximity Log (PL)*
H. Microspherically Focused Log (MSFL)*

DEPTH OF RESISTIVITY LOG INVESTIGATION

Flushed Zone (R_{xo})	*Invaded Zone (R_i)*	*Uninvaded Zone (R_t)*
Microlog*	Short Normal ††	Long Normal
Microlaterolog*	Laterolog -8*††	Lateral Log
Proximity* Log	Spherically Focused Log*††	Deep Induction Log
Microspherically Focused Log*	Medium Induction Log	Deep Laterolog*
	Shallow Laterolog*	Laterolog -3*
		Laterolog -7*
		Induction Log 6FF40

†For a review of how to use Lateral logs see: Hilchie (1979).
††When R_{mf} is much greater than R_w, the Laterolog − 8* and Spherically Focused Log* will have a shallower depth of investigation (closer to R_{xo}) than the medium induction, shallow Laterolog*, and the short normal.

curve is sometimes displayed in track #2 along with the short normal curve.

Induction—The induction device (Fig. 17) measures electrical conductivity● using current generated by coils. The transmitting coils produce an electromagnetic signal which induces currents in the formation. These induced currents are recorded as conductivity by receiver coils. Modern induction devices have additional coils which focus the current so that signals are minimized from adjacent formations, the borehole, and the invaded zone. By focusing the current and eliminating unwanted signals, a deeper reading of conductivity is taken, and more accurate values of true formation resistivity (R_t) are determined from the induction log. The induction log has a transmitter/receiver spacing of 40 inches and can measure a reliable value for resistivity down to a bed thickness of five feet.

The induction curve on the Induction Electric Log appears in track #2 (Fig. 17). Because the induction device is a conductivity measuring tool, an induction derived conductivity curve is presented in track #3 (Fig. 17). The track #3 conductivity curve is necessary to more accurately determine the R_t value of low resistivity formations, and to eliminate possible errors when calculating true resistivity from conductivity. Because the induction log does not require the transmission of electricity through drilling fluid, it can be run in air-, oil-, or foam-filled boreholes.

● $C_t = 1000/R_t$ where C_t = conductivity in millimhos/meter, and R_t = true formation resistivity in ohm-meters.

Dual Induction Focused Log

The modern induction log is called the Dual Induction Focused Log (Tixier et al, 1963). This log (Fig. 18) consists of a deep-reading induction device (R_{ILd} which measures R_t), and is similar to an Induction Electric Log. The Dual Induction Focused Log (Fig. 18) also has a medium-reading induction device (R_{ILm} which measures R_i) and a shallow reading (R_{xo}) focused Laterolog* which is similar to the short normal. The shallow reading Laterolog* may be either a Laterolog-8 (LL-8)* or a Spherically Focused Log (SFL)*.

The Dual Induction Focused Log is used in formations that are deeply invaded by mud filtrate. Because of deep invasion, a deep reading induction log (R_{ILd}) may not accurately measure the true resistivity of the formation (R_t). Resistivity values obtained from the three curves on a Dual Induction Focused Log are used to correct deep resistivity (R_{ILd}) to true resistivity (R_t) from a tornado chart (Fig. 19). This tornado chart (Fig. 19) can also help determine the diameter of invasion (d_i) and the ratio of R_{xo}/R_t. An example of the procedure is presented in Figure 19.

The three resistivity curves on the Dual Induction Focused Log are recorded on a four cycle logarithmic scale ranging from 0.2 to 2000 ohm/meters (Fig. 18) and correspond to tracks #2 and #3 on the Induction Electric Log. Normally, a spontaneous potential (SP) curve is placed in track #1 (Fig. 18).

The deep induction log (R_{ILd}) does not always record an accurate value for deep resistivity in thin, resistive (where $R_t > 100$ ohm/meters) zones. Therefore, an alternate method to determine true resistivity (R_t) should be used. The technique is called R_t minimum ($R_{t\,min}$) and is calculated by the following formula:

$$R_{t\,min} = (\text{LL-8* or SFL*}) \times R_w/R_{mf}$$

Where:

$R_{t\,min}$ = true resistivity (also called R_t minimum)
R_{mf} = resistivity of mud filtrate at formation temperature
R_w = resistivity of formation water at formation temperature
LL-8* = shallow resistivity Laterolog-8*
SFL* = shallow resistivity Spherically Focused Log*

The rule for applying $R_{t\,min}$ is to determine R_t from both the Dual Induction Focused Log tornado chart (Fig. 19) and from the $R_{t\,min}$ formula, and use whichever value of R_t is the greater. In addition to the $R_{t\,min}$ method for determining R_t in thin resistive zones, correction curves (Schlumberger, 1979, p. 54-55) are available to correct the deep induction log resistivity (R_{ILd}) to R_t.

Laterolog*

The Laterolog* is designed to measure true formation resistivity (R_t) in boreholes filled with saltwater muds (where $R_{mf} \simeq R_w$). A current from the surveying electrode is forced into the formation by focusing electrodes. The focusing electrodes emit current of the same polarity as the surveying electrode but are located above and below it. The focusing, or guard electrodes, prevent the surveying current from flowing up the borehole filled with saltwater mud (Fig. 20). The effective depth of Laterolog* investigation is controlled by the extent to which the surveying current is focused. Deep reading Laterologs* are therefore more strongly focused than shallow reading Laterologs*.

Invasion can influence the Laterolog*. However, because resistivity of the mud filtrate is approximately equal to the resistivity of formation water ($R_{mf} \simeq R_w$) when a well is drilled with saltwater-based muds, invasion does not strongly affect R_t values derived from a Laterolog*. But, when a well is drilled with freshwater-based muds (where $R_{mf} > 3\,R_w$), the Laterolog* can be strongly affected by invasion. Under these conditions, a Laterolog* should not be used (see Fig. 16). The borehole size and formation thickness affect the Laterolog*, but normally the effect is small enough so that Laterolog* resistivity can be taken as R_t.

The Laterolog* curve (Fig. 21) appears in track #2 of the log and has a linear scale. Because saltwater-based mud where $R_{mf} \simeq R_w$ gives a very poor SP response, a natural gamma ray log is run in track #1 as a lithology and correlation curve (Fig. 21). A Microlaterolog* is sometimes recorded in track #3 (Fig. 21).

Dual Laterolog-Microspherically Focused Log*

The Dual Laterolog* (Fig. 22) consists of a deep reading (R_t) resistivity device (R_{LLd}) and a shallow reading (R_i) resisitivity device (R_{LLS}). Both are displayed in tracks #2 and #3 of the log on a four cycle logarithmic scale. A natural gamma ray log is often displayed in track #1 (Fig. 22).

The Microspherically Focused Log* is a pad type, focused electrode log (a pad type focused electrode log has electrodes mounted in a pad that is forced against the borehole wall) that has a very shallow depth of investigation, and measures resistivity of the flushed zone (R_{xo}). When a Microspherically Focused Log (MSFL*) is run with the Dual Laterolog* (Fig. 22), the resulting three curves (i.e. deep, shallow, and MSFL*) are used to correct (for invasion) the deep resistivity (R_{LLd}) to true formation resistivity (Suau et al, 1972). A tornado chart (Fig. 23) is necessary to correct R_{LLd} to R_t and to determine the diameter of invasion (d_i) and the ratio of R_t/R_{xo}. The procedure is illustrated in Figure 23.

Microlog (ML*)

The Micrólog* (Fig. 24) is a pad type resistivity device that primarily detects mudcake (Hilchie, 1978). The pad is in contact with the borehole and consists of three electrodes spaced one inch apart. From the pad, two resistivity measurements are made; one is called the micro normal and the other is the micro inverse (Fig. 24). The micro normal device investigates three to four inches into the formation (measuring R_{xo}) and the micro inverse investigates approximately one to two inches and measures the resistivity of the mudcake (R_{mc}). The detection of mudcake by the Microlog* indicates that invasion has occurred and the formation is permeable. Permeable zones show up on the Microlog* as *positive separation* when the micro normal curves read higher resistivity than the micro inverse curves (Fig. 24). † Shale zones are indicated by no separation or "negative separation" (i.e. micro normal < micro inverse).

†Positive separation can only occur when $R_{mc} > R_m > R_{mf}$. To verify these values if there is any doubt, check the log heading for resistivity values of the mudcake, drilling mud, and mud filtrate.

Remember that even though the resistivity of the mud filtrate (R_{mf}) is less than the resistivity of the mudcake (R_{mc}), the micro normal curve will read a higher resistivity in a permeable zone than the shallower-reading micro inverse curve. This is because the filtrate has invaded the formation, and part of the resistivity measured by the micro normal curve is read from the rock matrix, whereas the micro inverse curve measures only the mudcake (R_{mc}) which has a lower resistivity than rock.

However, in enlarged boreholes, a shale zone can exhibit minor, positive separation. In order to detect zones of erroneous positive separation, a microcaliper log is run in track #1 (Fig. 24), so that borehole irregularities are detected. Nonporous and impermeable zones have high resistivity values on both the micro normal and micro inverse curves (Fig. 24). Hilchie (1978) states that resistivities of approximately ten times the resistivity of the drilling mud (R_m) at formation temperature indicate an impermeable zone.

The Microlog* does not work well in saltwater-based drilling muds (where $R_{mf} \simeq R_w$) or gypsum-based muds, because the mudcake may not be strong enough to keep the pad away from the formation. Where the pad is in contact with the formation, positive separation cannot occur.

Microlaterolog* and Proximity Log*

The Microlaterolog (MLL)* (Fig. 21) and the Proximity Log (PL)* (Fig. 25), like the Microspherically Focused Log (MSFL)*, are pad type focused electrode logs designed to measure the resistivity in the flushed zone (R_{xo}). Because the Microlaterolog* is strongly influenced by mudcake thicknesses greater than 1/4 inch (Hilchie, 1978), the Microlaterolog* should be run only with saltwater-based drilling muds. The Proximity Log*, which is more strongly focused than the Microlaterolog*, is designed to investigate deeper so it can be used with freshwater-based drilling muds where mudcake is thicker.

Resistivity Derived Porosity

The minerals that make up the grains in the matrix of the rock and the hydrocarbons in the pores are nonconductive. Therefore, the ability of rock to transmit an electrical current is almost entirely the result of the water in the pore space. Thus resistivity measurements can be used to determine porosity. Normally, measurements of a formation's resistivity close to the borehole (flushed zone, R_{xo}, or invaded zone, R_i) are used to determine porosity. Shallow resistivity devices, used to measure R_{xo} and R_i, include the following: (1) Microlaterolog*; (2) Proximity Log*; (3) Laterolog-8*; (4) Microspherically Focused Log*; (5) short normal log; and (6) Spherically Focused Log*.

When a porous and permeable water-bearing formation is invaded by drilling fluid, formation water is displaced by mud filtrate. Porosity in a water-bearing formation can be related to shallow resistivity (R_{xo}) by the following equations:

$$S_{xo} = \sqrt{F \times \frac{R_{mf}}{R_{xo}}}$$

Where $S_{xo} = 1.0$ (100%) in water-bearing zones.

$$1.0 = \sqrt{F \times \frac{R_{mf}}{R_{xo}}}$$

square both sides:

$$1.0 = F \times \frac{R_{mf}}{R_{xo}}$$

solve for F:

$$F = \frac{R_{xo}}{R_{mf}}$$

remember $F = a/\phi^m$

therefore:

$$\frac{a}{\phi^m} = \frac{R_{xo}}{R_{mf}}$$

solve for porosity (ϕ)

$$\phi = \left(\frac{a \times R_{mf}}{R_{xo}}\right)^{1/m}$$

Where:
- ϕ = formation porosity
- R_{mf} = resistivity of mud filtrate at formation temperature
- S_{xo} = water saturation of the flushed zone
- R_{xo} = resistivity of flushed zone from Microlaterolog*, Proximity Log*, Laterolog-8*, or Microspherically Focused Log* values
- a = constant
  - a = 1.0 for carbonates
  - a = 0.62 for unconsolidated sands
  - a = 0.81 for consolidated sands
- m = constant
  - m = 2.0 for consolidated sands and carbonates
  - m = 2.15 for unconsolidated sands
- F = formation factor

In hydrocarbon-bearing zones, the shallow resistivity (R_{xo}) is affected by the unflushed residual hydrocarbons left by the invading mud filtrate. These residual hydrocarbons will result in a value for shallow resistivity (R_{xo}) which is too high because hydrocarbons have a higher resistivity than formation water. Therefore, the calculated resistivity porosity in hydrocarbon-bearing zones will be too low. To correct for residual hydrocarbons in the flushed zone, water saturation of the flushed zone (S_{xo}) must be known or estimated. Then, a formation's shallow resistivity (R_{xo}) can be related to porosity by the following:

$$S_{xo} = \sqrt{F \times \frac{R_{mf}}{R_{xo}}}$$

now square both sides:

remember: $F = a/\phi^m$

$$S_{xo}^2 = F \times \frac{R_{mf}}{R_{xo}}$$

solve for F:

$$F = \frac{S_{xo}^2 \times R_{xo}}{R_{mf}}$$

Therefore:

$$\frac{a}{\phi^m} = \frac{S_{xo}^2 \times R_{xo}}{R_{mf}}$$

solve for porosity (ϕ):

$$\phi = \left[\frac{a(R_{mf}/R_{xo})}{(S_{xo})^2}\right]^{1/m}$$

Where:

ϕ = formation porosity

R_{mf} = resistivity of mud filtrate at formation temperature

R_{xo} = resistivity of the flushed zone

a = constant
- a = 1.0 for carbonates
- a = 0.62 for unconsolidated sands
- a = 0.81 for consolidated sands

m = constant
- m = 2.0 for carbonates and consolidated sands
- m = 2.15 for unconsolidated sands

S_{xo} = water saturation of the flushed zone
- S_{xo} = 1.0 minus residual hydrocarbon saturation (RHS). See Table 5 for examples.

F = formation factor

Table 5. Percentages of Residual Hydrocarbon Saturation as a function of hydrocarbon density and porosity (modified after Hilchie, 1978).

	API°Gravity	RHS%	S_{xo}%
Gas		40 to 5	60 to 95
High gravity oil	40 to 50	10 to 5	90 to 95
Medium gravity oil	20 to 40	20 to 10	80 to 90
Low gravity oil	10 to 20	30 to 20	70 to 80
Porosity %		**RHS%**	**S_{xo}%**
25 to 35		30	70
15 to 20		15	85

Review - Chapter III

1. Resistivity logs are used to: (1) determine hydrocarbon- versus water-bearing zones; (2) indicate permeable zones; and (3) determine resistivity porosity.
2. A formation's resistivity can be measured by either induction or electrode (Laterolog*, normal, Lateral, spherically focused logs, Micrology*, Microlaterolog*, and Proximity*) logs.
3. Induction logs (induction electric log or Dual Induction Focused Log) should be run in non-salt saturated drilling muds (where $R_{mf} > 3 R_w$).
4. Laterologs* or Dual Laterologs* with R_{xo} should be run in salt-saturated drilling muds (where $R_{mf} \simeq R_w$).
5. By use of tornado charts, the deep resistivity log on either the Dual Induction Focused Log or the Dual Laterolog* with R_{xo} can be corrected for the effects of invasion to determine a more accurate value of true formation resistivity (R_t).
6. Most minerals which make up the matrix of the rock and the hydrocarbons in the pores are non-conductive. Therefore, the ability of the rock to transmit an electric current is almost entirely a function of the water in the rock's pores.

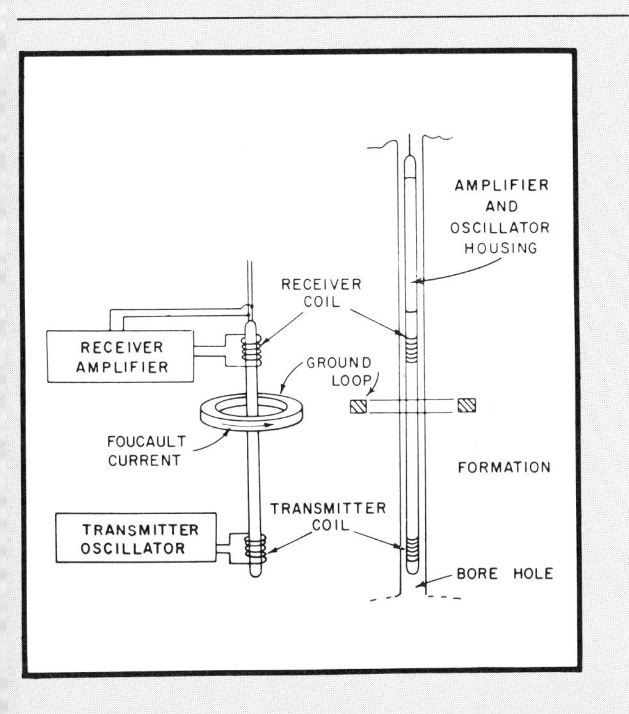

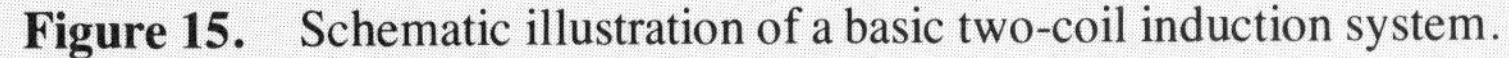

Figure 15. Schematic illustration of a basic two-coil induction system.

Courtesy, Schlumberger Well Services.
Copyright 1972, Schlumberger.

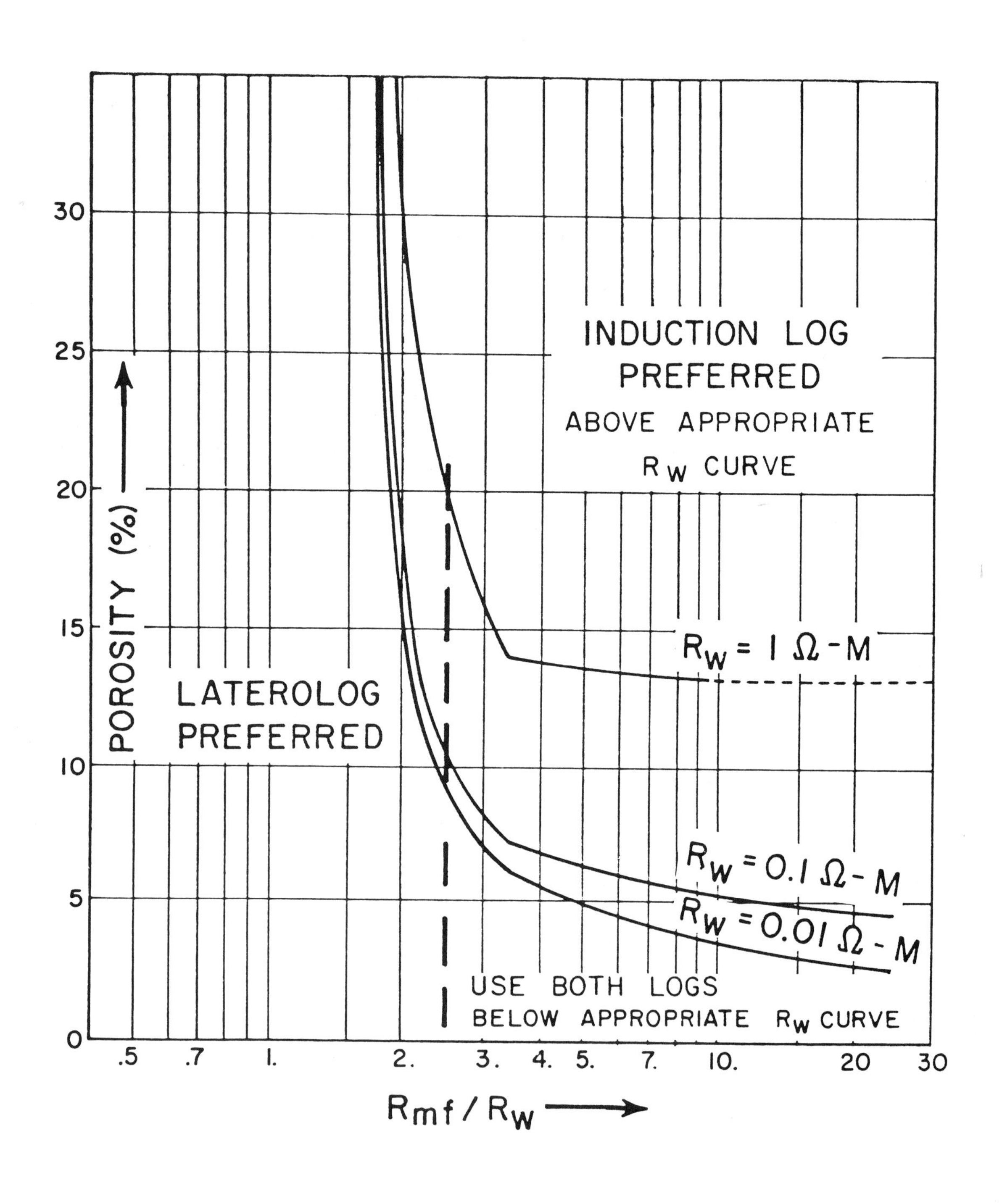
INDUCTION LOG
PREFERRED
ABOVE APPROPRIATE
RW CURVE
LATEROLOG
PREFERRED
RW = 1 Ω-M
RW = 0.1 Ω-M
RW = 0.01 Ω-M
USE BOTH LOGS
BELOW APPROPRIATE RW CURVE
POROSITY (%)
30
25
20
15
10
5
0
.5
.7
1.
2.
3.
4.
5.
7.
10.
20
30
Rmf / RW

Figure 16. Chart for quick determination of preferred conditions for using an induction log versus a Laterolog* (Schlumberger, 1972).

Courtesy, Schlumberger Well Services.
Copyright 1972, Schlumberger.

Selection is a function of the ratio of R_{mf}/R_w and, to some extent, porosity.

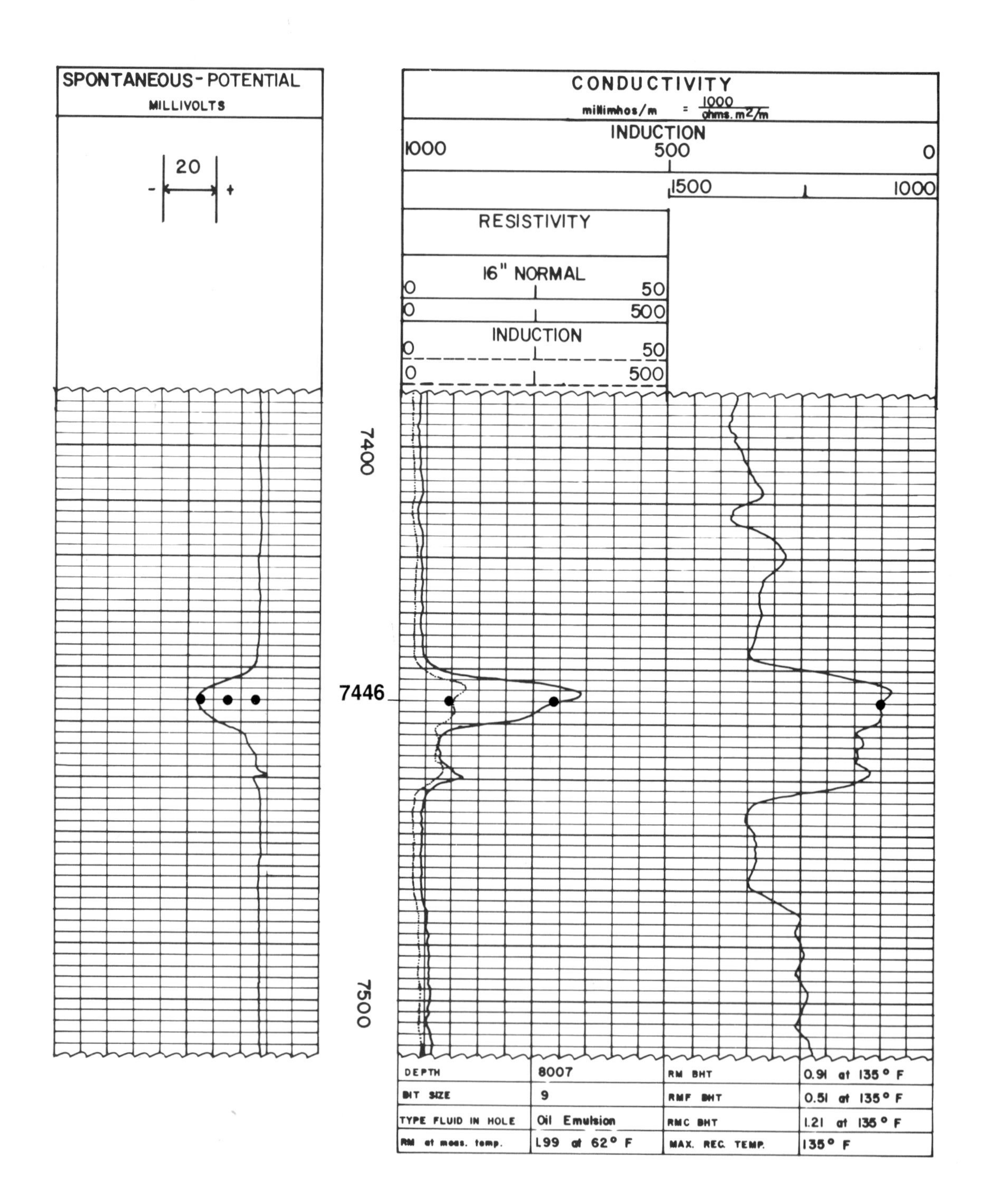
SPONTANEOUS-POTENTIAL
MILLIVOLTS
20
-
+
CONDUCTIVITY
millimhos/m = 1000 / ohms. m2/m
INDUCTION
1000
500
0
1500
1000
RESISTIVITY
16" NORMAL
0
50
0
500
INDUCTION
0
50
0
500
7400
7446
7500
DEPTH
8007
RM BHT
0.91 at 135° F
BIT SIZE
9
RMF BHT
0.51 at 135° F
TYPE FLUID IN HOLE
Oil Emulsion
RMC BHT
1.21 at 135° F
RM at meas. temp.
1.99 at 62° F
MAX. REC. TEMP.
135° F

Figure 17. Example Induction Electric Log.† The purpose for presenting this log is to illustrate the different curves and to give you guidance on picking log values. The Induction Electric Log is normally used when $R_{mf} \gg R_w$.

Note that log scales are shown *horizontally* at top of log.

Track #1—The log track on the far left contains the spontaneous potential (SP) log. Typically, each increment on the scale equals 20 millivolts, so that the value at the sample depth of 7,446 ft is about 40. Because the deflection is to the left (negative deflection) the log value is negative, or approximately −40 mv.

Track #2—The middle log track contains two resistivity curves. One measures shallow resistivity (R_i, 16″ -normal or short normal electrode log represented by solid line) and the other measures deep resistivity (R_t, an induction log represented by dotted line). The scale values increase from left to right, but two scales are present: The first scale measures from 0 to 50 ohm-meters in increment values of 5 ohm-meters. This first scale contains both the R_i and R_t curves. The second-cycle scale measures from 0 to 500 ohm-meters in increment values of 50 ohm-meters. It contains no curves in this example because the second-cycle scale is used only when the resistivity curves in the first-cycle scale exceed the maximum scale values.

At the sample depth of 7,446 ft read a value for the 16″ -normal of 28 ohm-meters. This is counted horizontally as almost 6 increments of 5 ohm-meters per increment (28 is nearly 6 × 5 or 30). The induction reading on track #2 is counted at 10 ohm-meters, or 2 increments of 5 ohm-meters per increment.

Track #3—The log track on the far right contains a conductivity curve measured by the induction log. The induction log actually measures conductivity, not resistivity, but because conductivity is a reciprocal of resistivity, resistivity can be derived. This is done automatically as the log is recorded in track #3. However, the conductivity curve can be used to convert values to resistivity. In this way, track #2 resistivity values can be checked for accuracy.

For example, to convert track #3 values to resistivity the procedure is as follows: The values on the conductivity scale increase from *right-to-left,* and two scales are present: values from 0 to 1,000 are marked in 50 mmhos/meter increments for the first cycle, and values from 1,000 to 1,500 are marked for the second cycle (second cycle values are not necessary on this log). Therefore, at a depth of 7,446 ft, track #3 shows a value of 100 mmhos/meter, or 2 increments (from the right) of 50 mmhos/meter for each division.

Because resistivity equals 1,000 ÷ conductivity, resistivity = 1,000/100 or, in this case, 10. So, conductivity converted to resistivity from the induction log is 10 ohm-meters.

†On this and all subsequent logs in the text, each small division on the depth scale is equal to 2 ft.

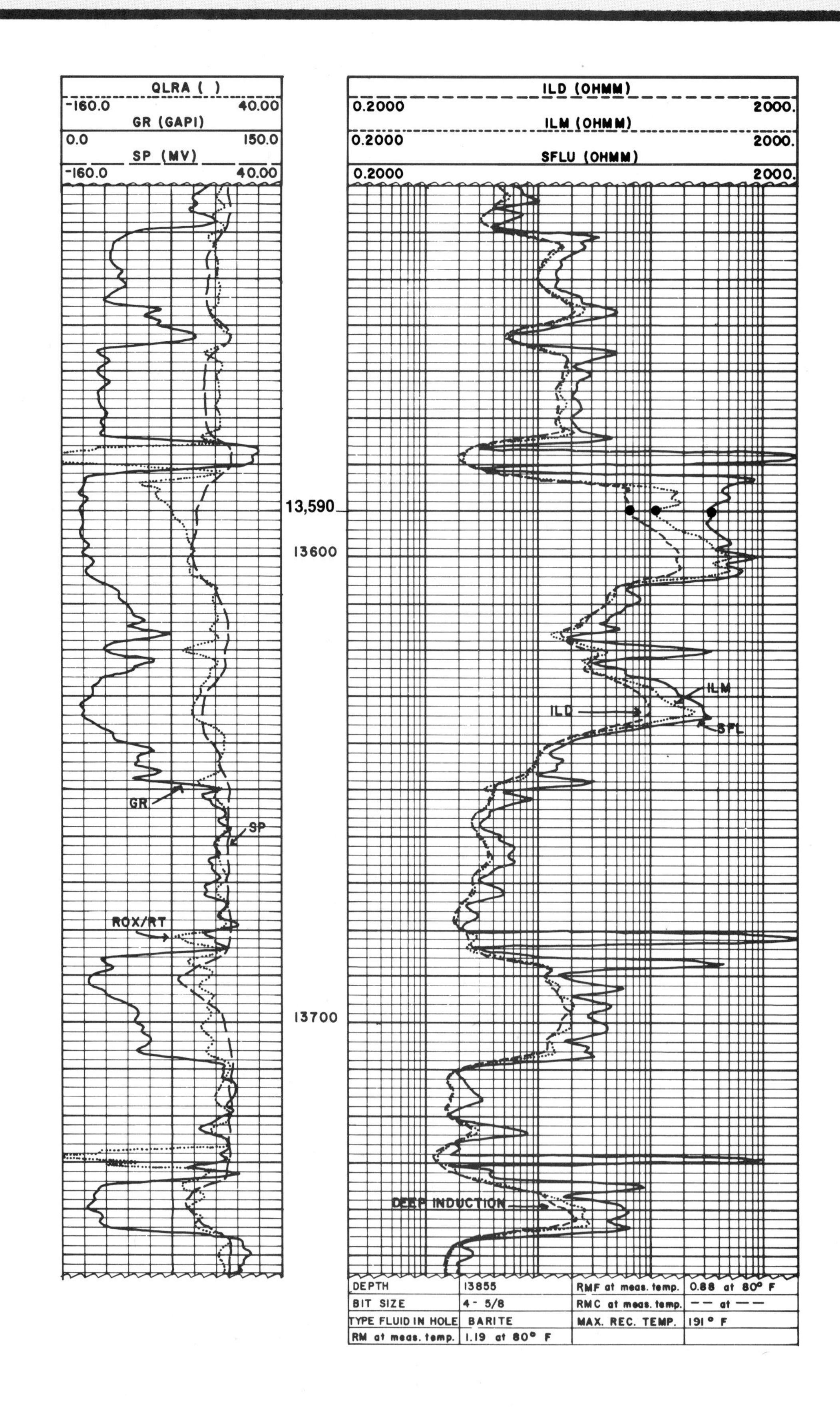
QLRA ( )
-160.0
40.00
GR (GAPI)
0.0
150.0
SP (MV)
-160.0
40.00
ILD (OHMM)
0.2000
2000.
ILM (OHMM)
0.2000
2000.
SFLU (OHMM)
0.2000
2000.
13,590
13600
13700
GR
SP
ROX/RT
ILM
ILD
SFL
DEEP INDUCTION
DEPTH
13855
BIT SIZE
4 - 5/8
TYPE FLUID IN HOLE
BARITE
RM at meas. temp.
1.19 at 80° F
RMF at meas. temp.
0.88 at 80° F
RMC at meas. temp.
-- at --
MAX. REC. TEMP.
191° F

Figure 18. Example of a Dual Induction Focused Log. Use this log to pick values and determine ratios for the tornado chart exercise in Figure 19. The Dual Induction Focused Log is normally used when R_{mf} is much greater than R_w, and also where invasion is deep.

Track #1 in this log suite contains a gamma ray, SP, and R_{xo}/R_t quick look curves. The gamma ray, SP, and R_{xo}/R_t quick look curves will be discussed in subsequent chapters.

The resistivity scale in tracks #2 and #3 is a logarithmic scale from 0 to 2,000 ohm-meters, increasing from left to right. Note the following logs.

Deep induction log resistivity—The dashed line ILD represents R_{ILd} and measures the deep resistivity of the formation, or close to true resistivity (R_t). At the sample depth in this exercise (13,590 ft), deep resistivity (ILD) reads a value of 70.

Medium induction log resistivity—The dotted-and-dashed line ILM represents R_{ILm} and measures the medium resistivity of the formation, or resistivity of the invaded zone (R_i). At the sample depth in this exercise (13,590 ft), medium resistivity (R_i) reads a value of 105.

Spherically Focused Log* resistivity—the solid line SFL* represents R_{SFL}* and measures the shallow resistivity of the formation, or resistivity of the flushed zone (R_{xo}). At the sample depth in this exercise (13,590 ft), resistivity of the flushed zone (R_{xo}) reads a value of 320.

The following ratios are needed for work on the tornado chart (Fig. 19), and the values are picked from the example log:

$R_{SFL}/R_{ILd} = 320/70 = 4.6$
$R_{ILm}/R_{ILd} = 105/70 = 1.5$

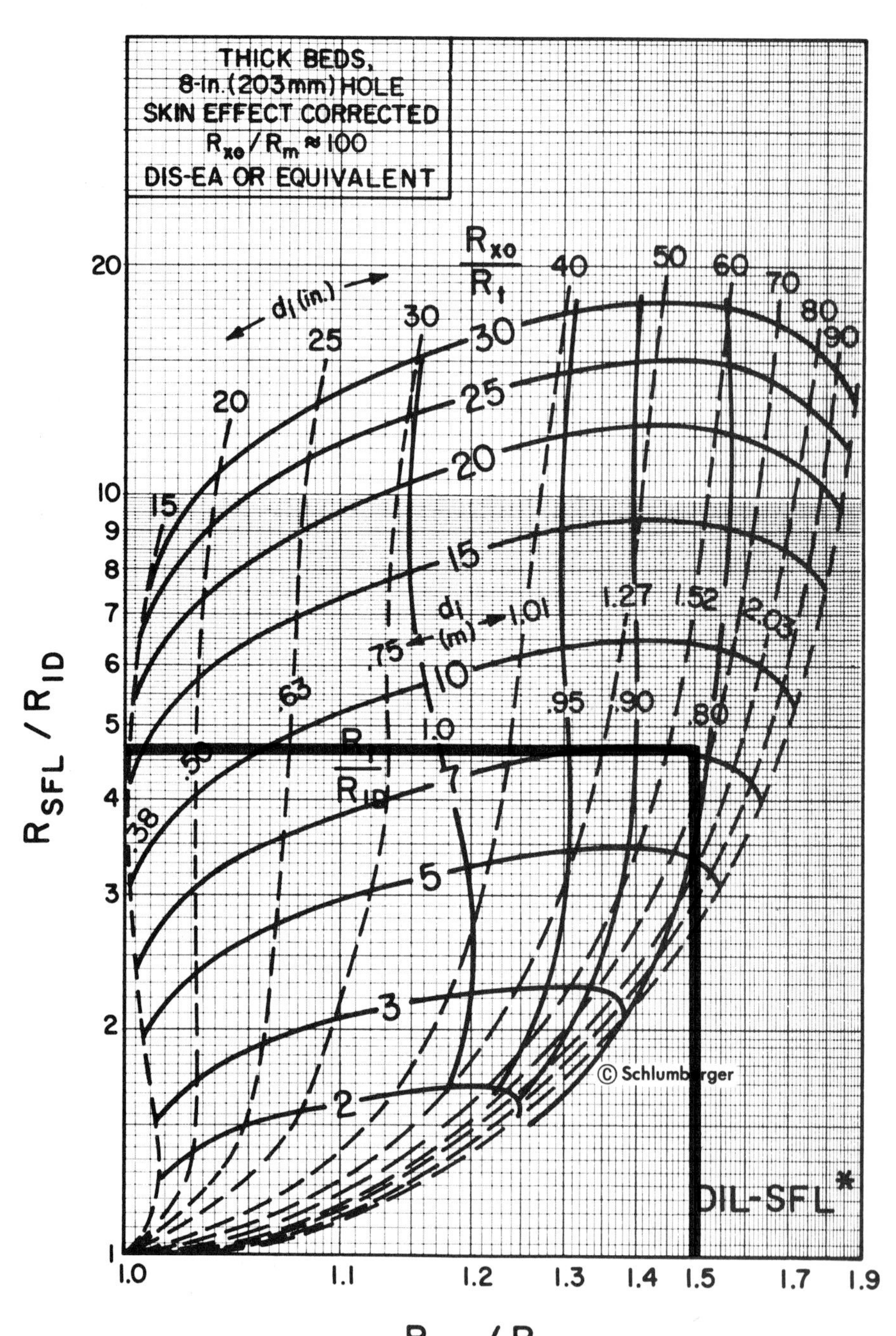
DUAL INDUCTION – SPHERICALLY FOCUSED LOG
ILd — ILm — SFL*
THICK BEDS,
8-in.(203mm) HOLE
SKIN EFFECT CORRECTED
$R_{xo}/R_m \approx 100$
DIS-EA OR EQUIVALENT
R_{xo}/R_t
d_i (in.)
d_i (m)
R_t/R_{ID}
R_{SFL}/R_{ID}
R_{IM}/R_{ID}
© Schlumberger
DIL-SFL*

Figure 19. Dual Induction-(SFL*) tornado chart used for correcting R_{ILd} values to R_t, as an indicator of true resistivity. Log values used in this exercise are picked from the example Dual Induction Log in Figure 18.

Courtesy, Schlumberger Well Services.
Copyright 1979, Schlumberger

Given:

$R_{ILd} = 70$
$R_{ILm} = 105$
$R_{SFL} = 320$
$R_{SFL}/R_{ILd} = 4.6$
$R_{ILm}/R_{ILd} = 1.5$

Procedure:

By using the tornado chart, pick the following values:

R_t/R_{ILd}—Plot the ratio values for R_{SFL}/R_{ILd} and R_{ILm}/R_{ILd} by using the scales on the vertical axis (R_{SFL}/R_{ILd}) and horizontal axis (R_{ILm}/R_{ILd}). Where the values cross, read the R_t/R_{ILd} value from the tornado chart scale depicted by solid, vertically oriented lines. Note that the scale decreases in intensity from left to right (from 1.0 to 0.80) and that the R_t/R_{ILd} value falls just to the left of the 0.80 line, giving us a value of 0.82.

d_i—Find the diameter of invasion surrounding the borehole by locating the same point used above, on the d_i scale of the chart. The scale is indicated by the dashed, vertically oriented lines on the tornado chart (note that the d_i scale is given in inches across the top of the tornado, and is given in meters through the midpart of the tornado chart), and the scale reads horizontally. In this example, the sample value is plotted midway between the 60- and 70-inch value line, so we determine that the diameter of invasion (d_i) is 65 inches.

R_{xo}/R_t—Ratio of resistivity of the flushed zone (R_{xo}) over the true resistivity of the formation (uncorrected, R_t). This ratio, derived from the chart, is used in later calculations. The scale is represented by the solid, horizontally oriented lines, and the scale values are shown as whole numbers midway across the lines. In this example, the plotted sample falls on the scale with a value of 7.0.

Finally, with values taken from the chart as outlined above, calculate corrected values for R_t and R_{xo} .

$(R_t/R_{ILd}) \times R_{ILd} = R_t$ (corrected)
(Ratio value from chart) $\times$ log value = R_t (corrected)
$0.82 \times 70 = 57.4$ (R_t corrected, or true formation resistivity).

and

$(R_{xo}/R_t) \times R_t = R_{xo}$ (corrected)
(ratio value from chart) $\times$ (corrected R_t value) = R_{xo} (corrected)
$7 \times 57.4 = 401.8$ (corrected resistivity of the flushed zone)

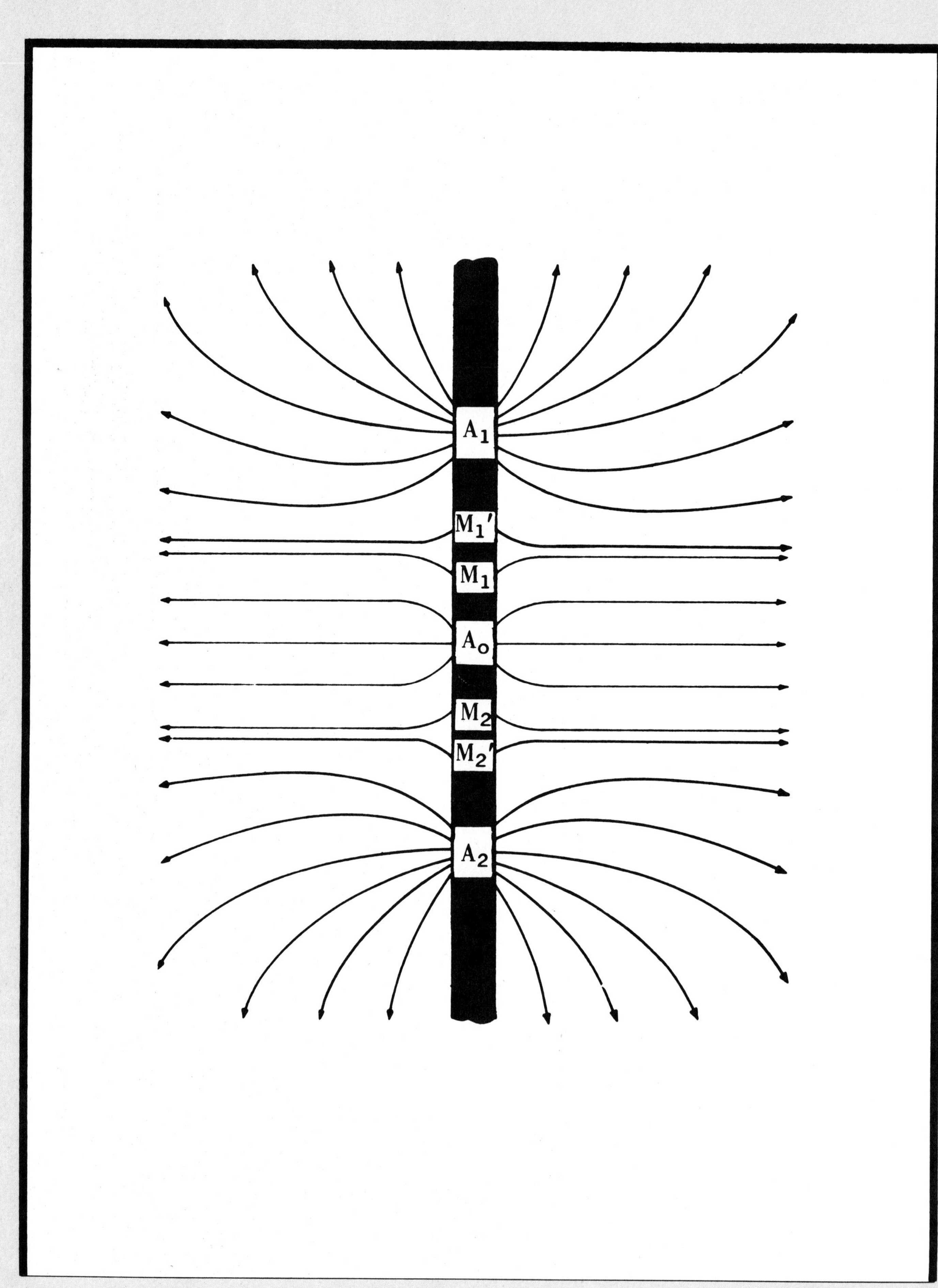
A1
M1'
M1
Ao
M2
M2'
A2

Figure 20. Schematic illustration of a focused Laterolog* illustrating current flow.

Courtesy, Dresser Industries.
Copyright 1974, Dresser Atlas.

As cited in the example, A_1 (above) and A_2 (below) are the focusing (or guard) electrodes which direct and force the current from the A_o electrode into the formation. The monitoring electrodes ($M_1 - M_1'$ and $M_2 - M_2'$) are brought to the same potential by adjusting the current that emits from the focusing electrodes A_1 and A_2.

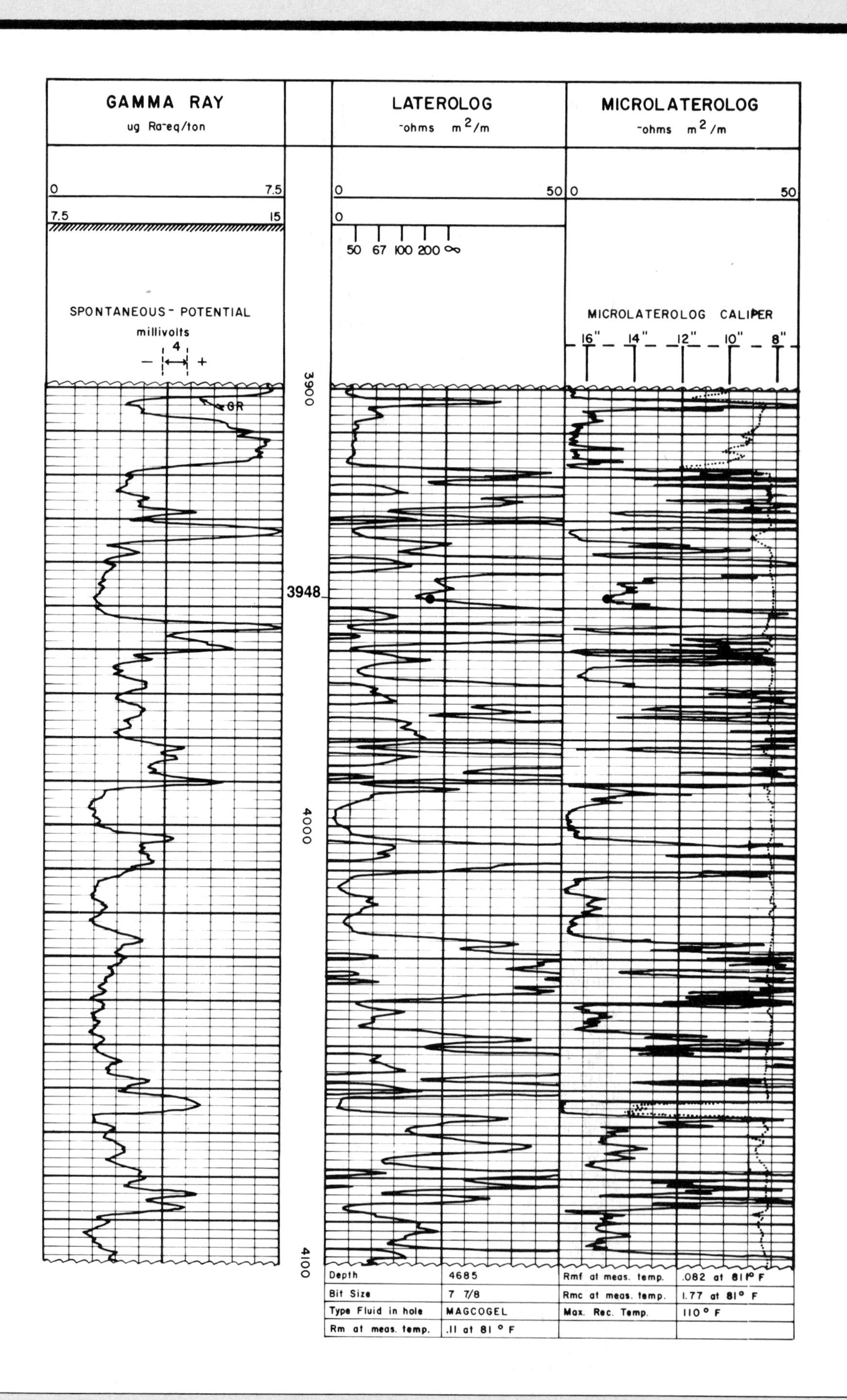
GAMMA RAY
ug Ra-eq/ton
0
7.5
7.5
15
SPONTANEOUS - POTENTIAL
millivolts
4
— +
GR
LATEROLOG
-ohms m2/m
0
50
0
50 67 100 200 ∞
MICROLATEROLOG
-ohms m2/m
0
50
MICROLATEROLOG CALIPER
16" 14" 12" 10" 8"
3900
3948
4000
4100
Depth
4685
Bit Size
7 7/8
Type Fluid in hole
MAGCOGEL
Rm at meas. temp.
.11 at 81 °F
Rmf at meas. temp.
.082 at 81°F
Rmc at meas. temp.
1.77 at 81° F
Max. Rec. Temp.
110° F

Figure 21. Example Laterolog* and Microlaterolog*. The purpose for presenting this log is to illustrate the log curves, and to give you guidance on picking log values. These logs are used when $R_{mf} \simeq R_w$.

Track #1—The log track on the far left in this example is a gamma ray log. Gamma ray logs are discussed in a later chapter, but they commonly accompany Laterologs*.

Track #2—The middle log track here is the Laterolog* which measures the deep resistivity (R_t) or true resistivity of the formation. Note that the scale increases from left to right, in increments of 5 ohm-meters from 0 to 50 on the first cycle, and in hybrid increments from 0 to ∞ on the second cycle.

At the sample depth of 3,948 ft the Laterolog* value reads 21 ohm-meters.

Track #3—The right-hand log in this suite is the Microlaterolog* which measures the resistivity of the flushed zone (R_{xo}). Note that the scale starts with zero between tracks #2 and #3—that is, zero for the Microlaterolog* is not the same point as zero for the Laterolog* farther to the left. The scale ranges from 0 to 50 ohm-meters in increments of 5 ohm-meters. There is no second cycle recorded.

At the sample depth of 3,948 ft the Microlaterolog* reads 10 ohm-meters, or the depth line intersects the log curve at two increments from zero.

Note: In order to correct (for invasion) the Laterolog* to true resistivity (R_t), do the following (use the example at 3,948 ft):

$R_t = 1.67\,(R_{LL}) - 0.67\,(R_{xo})$ (Hilchie, 1979)

$R_t = 1.67\,(21) - 0.67\,(10)$

$R_t = 28.4$ ohm-meters

Where:

R_t = resistivity of the uninvaded zone

R_{LL} = Laterolog* resistivity (21 ohm-meters at 3,948 ft)

R_{xo} = Microlaterolog* resistivity (10 ohm-meters at 3,948 ft)

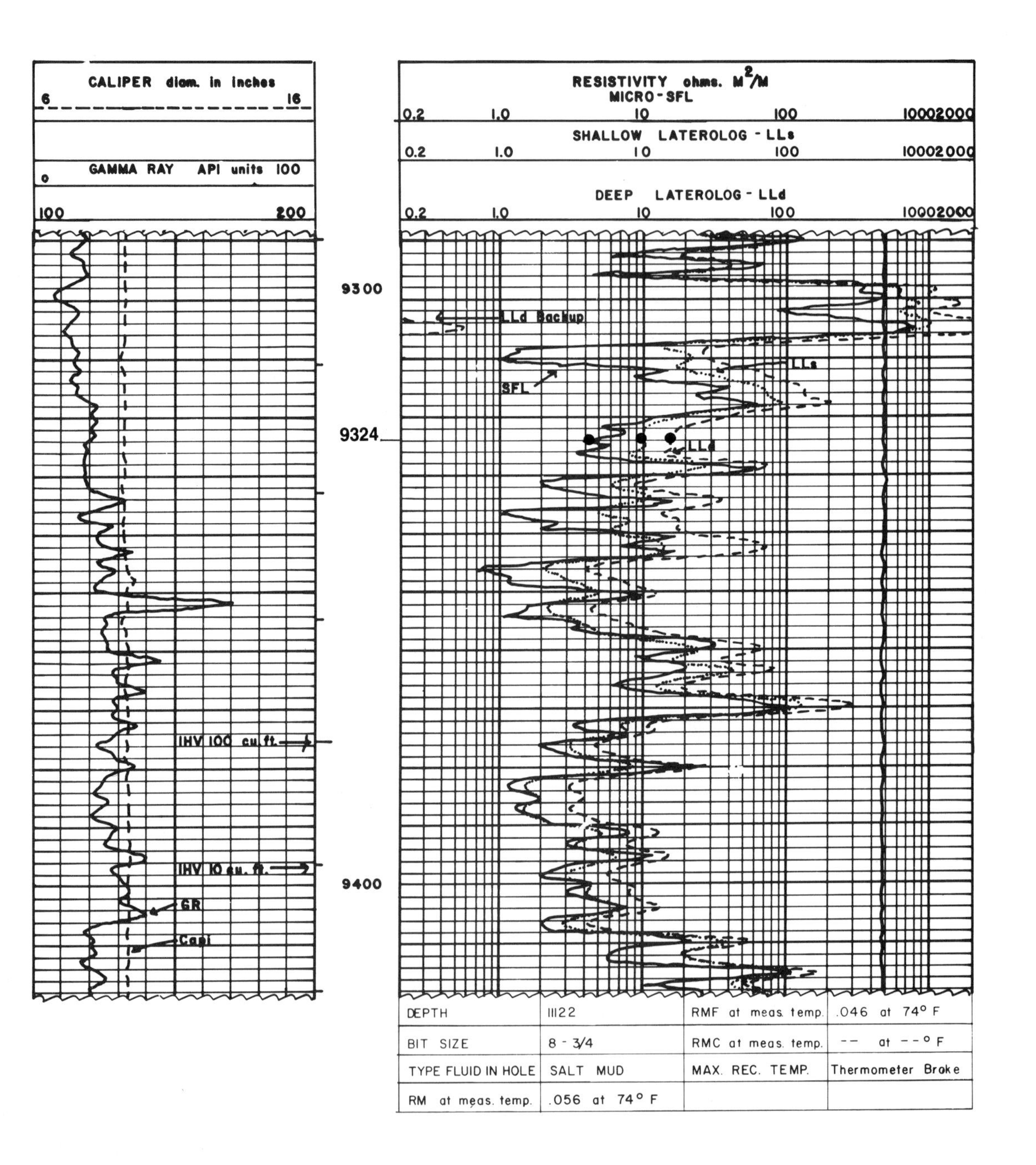

DEPTH	11122	RMF at meas. temp.	.046 at 74° F
BIT SIZE	8 - 3/4	RMC at meas. temp.	-- at --° F
TYPE FLUID IN HOLE	SALT MUD	MAX. REC. TEMP.	Thermometer Broke
RM at meas. temp.	.056 at 74° F		

Figure 22. Example of Dual Laterolog* with Microspherically Focused Log (MSFL)*. Use this log to pick values and determine ratios for the tornado chart in Figure 23. These logs are used when $R_{mf} \simeq R_w$ and invasion is deep.

The resistivity scale in tracks #2 and #3 is a four-cycle logarithmic scale ranging from 0 to 2,000; the values increase from left-to-right.

Deep Laterolog* resistivity—The dashed line LLd represents R_{LLd} and measures the deep resistivity of the formation, or true resistivity (R_t). At the sample depth of this exercise (9,324 ft), true resistivity (R_t) reads a value of 16.0.

Shallow Laterolog* resistivity—The dotted-and-dashed line LLs represents R_{LLs} and measures the shallow resistivity of the formation or the resistivity of the invaded zone (R_i). At the sample depth of this exercise (9,324 ft), resistivity (R_i) reads a value of 10.0.

Microspherically Focused Log (MSFL)* resistivity—The solid line SFL* represents R_{MSFL}* and measures the resistivity of the flushed zone (R_{xo}). At the sample depth in this exercise (9,324 ft), resistivity of the flushed zone (R_{xo}) reads a value of 4.5.

The following ratios are needed for work on the tornado chart (Fig. 23), and the values represented are picked from the log as shown above:

$R_{LLd}/R_{MSFL}* = 16/4.5 = 3.6$
$R_{LLd}/R_{LLs} = 16/10 = 1.6$

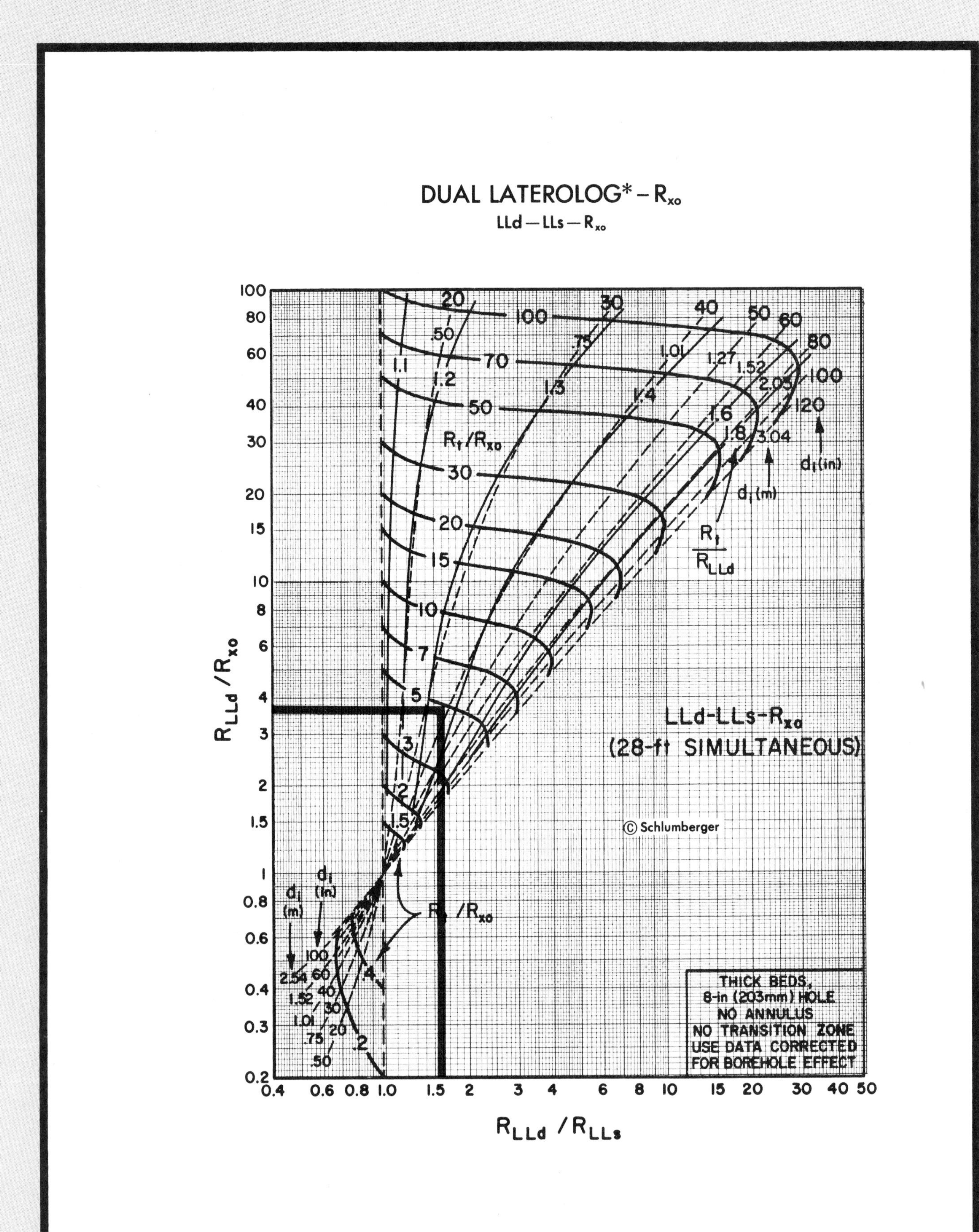
DUAL LATEROLOG* – R_{xo}
LLd — LLs — R_{xo}
R_{LLd}/R_{xo}
R_{LLd}/R_{LLs}
R_t/R_{xo}
R_t/R_{LLd}
d_i (in)
d_i (m)
LLd-LLs-R_{xo}
(28-ft SIMULTANEOUS)
© Schlumberger
THICK BEDS,
8-in (203mm) HOLE
NO ANNULUS
NO TRANSITION ZONE
USE DATA CORRECTED
FOR BOREHOLE EFFECT

Figure 23. Dual Laterolog*—Microspherically Focused Log* tornado chart for correcting R_{LLd} to R_t. Log values in this exercise are picked from the example Dual Laterolog*-MSFL* in Figure 22.

Courtesy, Schlumberger Well Services.
Copyright 1979, Schlumberger.

Given:

$R_{LLd} = 16.0$
$R_{LLs} = 10.0$
$R_{MSFL}* = 4.5$
$R_{LLd}/R_{MSFL}* = 3.6$
$R_{LLd}/R_{LLs} = 1.6$

Procedure:

Plot the values for $R_{LLd}/R_{MSFL}*$ (3.6) and R_{LLd}/R_{LLs} (1.6) using the vertical and horizontal scales at the side and bottom of the chart. Determine subsequent ratio values from the tornado chart.

R_t/R_{LLd}—The scale for this value is represented by the solid, vertically oriented lines. The scale values read across the top part of the tornado chart, and range from 1.1 to 1.8. Our value falls between the scale values 1.3 and 1.4, so we assign a value of 1.35.

d_i—The diameter of invasion around the borehole is picked from the chart; the scale is represented by the dashed, vertically oriented lines, and the scale values read across the top of the tornado chart ranging from 20 to 120 (inches) or 0.5 to 3.04 (meters). Our value falls between the scale values of 30 and 40 (inches), so we assign a value of 36 inches.

R_t/R_{xo}—The scale for this ratio value is represented by the solid, horizontally oriented lines. The scale values read from bottom to top on the left part of the chart, and range from 1.5 to 100. Our value falls between the scale values 3 and 5 (much closer to 5), so we assign a value of 4.5.

Finally, corrected values for true resistivity of the formation (R_t) and resistivity of the flushed zone (R_{xo}) are determined using these ratios.

$(R_t/R_{LLd}) \times R_{LLd} = R_t$ (corrected R_t)
(ratio) $\times$ log value = corrected R_t
$1.35 \times 16.0 = 21.6$ (R_t corrected, or true formation resistivity).

And:

(corrected R_t)/$(R_t/R_{xo}) = R_{xo}$ (corrected R_{xo})
(corrected R_t)/(ratio value from chart) = corrected R_{xo}
$21.6/4.5 = 4.8$ (corrected resistivity of flushed zone)

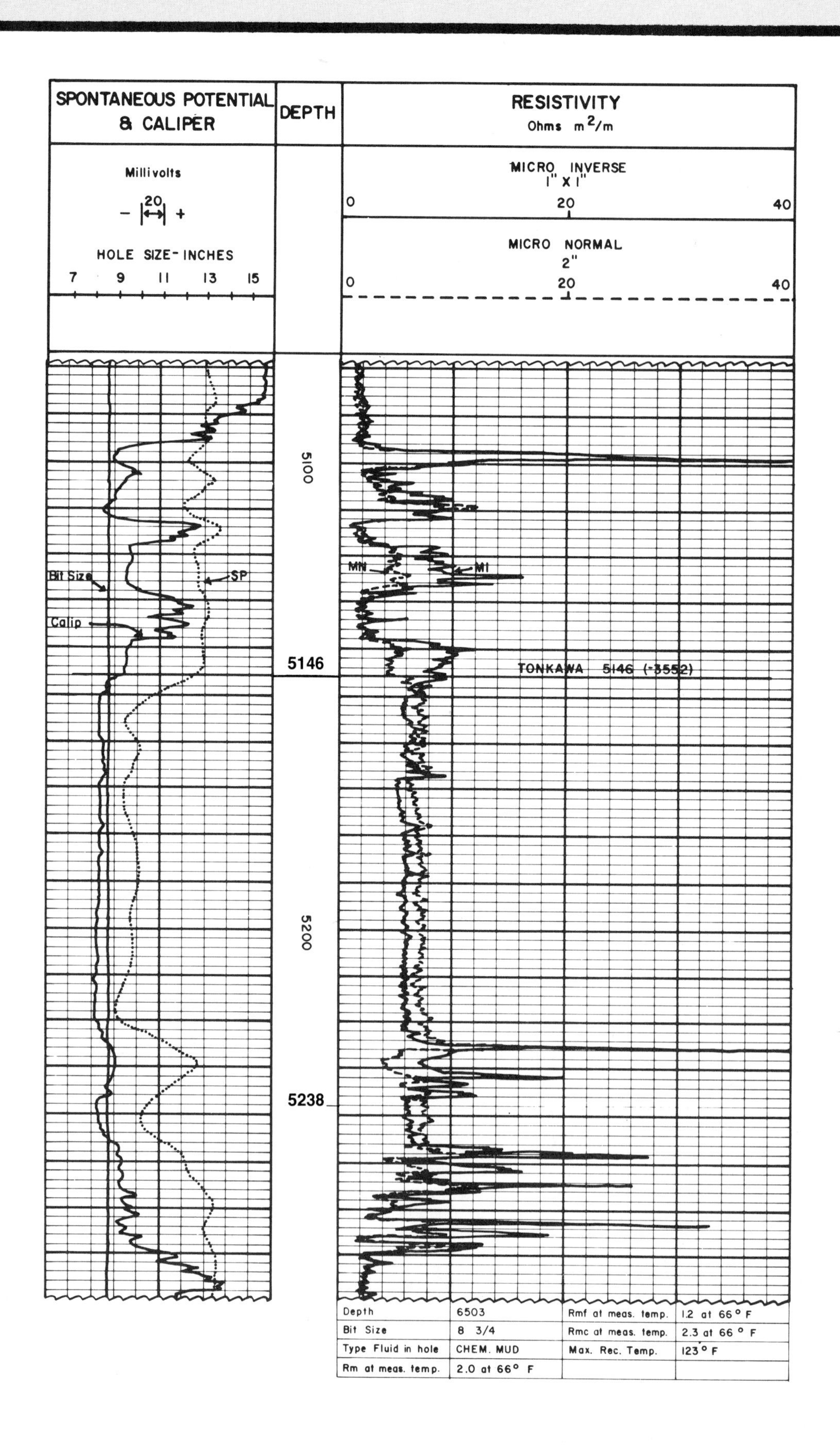

SPONTANEOUS POTENTIAL & CALIPER
DEPTH
RESISTIVITY
Ohms m²/m
Millivolts
20
− +
HOLE SIZE- INCHES
7 9 11 13 15
MICRO INVERSE
1" X 1"
0 20 40
MICRO NORMAL
2"
0 20 40
5100
Bit Size
SP
Calip
MN
MI
5146
TONKAWA 5146 (-3552)
5200
5238
Depth
6503
Rmf at meas. temp.
1.2 at 66° F
Bit Size
8 3/4
Rmc at meas. temp.
2.3 at 66° F
Type Fluid in hole
CHEM. MUD
Max. Rec. Temp.
123° F
Rm at meas. temp.
2.0 at 66° F

Figure 24. Example Microlog* with spontaneous potential log and caliper. This log demonstrates permeability two ways: positive separation between the micro normal and micro inverse logs in tracks #2 and #3 and decreased borehole size due to mudcake, detected by the caliper log in track #1.

Examine the log from a sample depth 5,146 ft to 5,238 ft.

Track #1—Note that the caliper shows a borehole diameter of approximately 11 inches just above the sample depth, but the hole size decreases to about 8.5 inches within the sample interval (the caliper measurement is shown by the solid line in track #1), thus indicating the presence of mudcake and a permeable zone.

Track #2—Note the positive separation between the micro normal log and the micro inverse log; the separation is about 2 ohm-meters. Positive separation is indicated where the resistivity value of the micro normal log (shown by the dashed line) is greater than the resistivity value for the micro inverse log (shown by the solid line).

This higher micro normal resistivity value is because the micro normal curve reads deeper into the flushed zone. The combination of mud filtrate, formation water and/or residual hydrocarbons and rock in the flushed zone gives a higher resistivity reading than the mudcake (measured by the micro inverse curve).

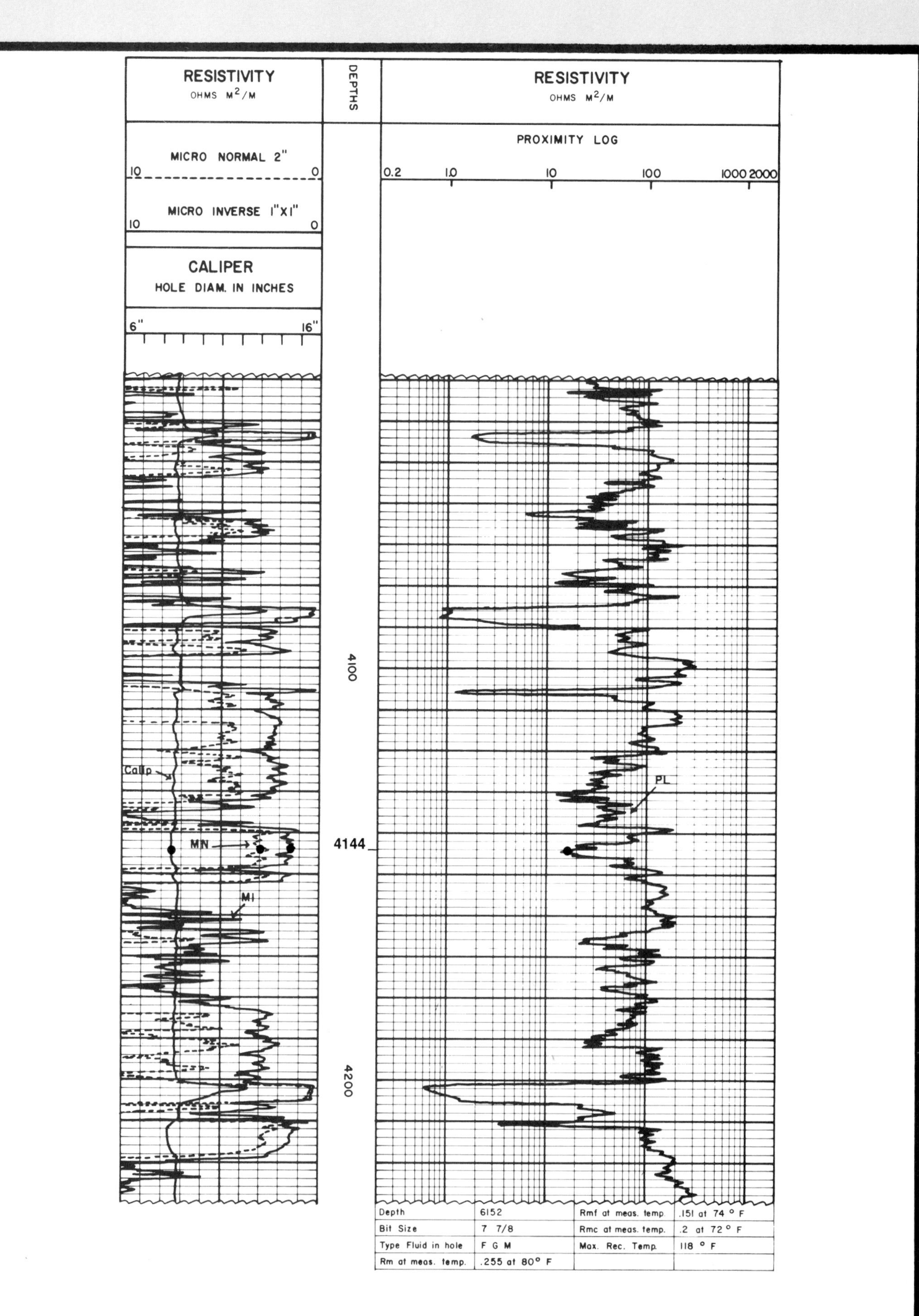
RESISTIVITY
OHMS M2/M
MICRO NORMAL 2"
10
0
MICRO INVERSE 1"X1"
10
0
CALIPER
HOLE DIAM. IN INCHES
6"
16"
DEPTHS
RESISTIVITY
OHMS M2/M
PROXIMITY LOG
0.2
1.0
10
100
1000 2000
4100
4144
4200
Calip
MN
MI
PL
Depth
6152
Rmf at meas. temp.
.151 at 74 °F
Bit Size
7 7/8
Rmc at meas. temp.
.2 at 72 °F
Type Fluid in hole
F G M
Max. Rec. Temp.
118 °F
Rm at meas. temp.
.255 at 80° F

Figure 25. Example of a Proximity Log* with a Microlog* and caliper. The Proximity Log* is designed to read the resistivity of the flushed zone (R_{xo}). This particular log package includes: a Proximity Log* to read R_{xo}, a Microlog* to determine permeable zones, and a caliper to determine the size of the borehole.

Examine the log curves at the sample depth of 4,144 ft.

Track #1—Track #1 depicts both a Microlog* and a caliper log. At the sample depth of 4,144 ft note that micro normal (shown by the dashed line) shows higher resistivity than micro inverse (shown by the solid line). Note: on this example, the resistivity values for micro normal and micro inverse increase from *right-to-left*. Micro inverse has a value of about 1.5, and micro normal has a value of about 3.0; the Microlog* indicates a permeable zone. The caliper log indicates a borehole slightly less than 9 inches.

Tracks #2 and #3—The Proximity Log* measures resistivity of the flushed zone (R_{xo}). In this example the scale is logarithmic, reading from left-to-right. At the sample depth of 4,144 ft we read a proximity curve value (R_{xo}) of 18 ohm-meters.

POROSITY LOGS

Sonic Log

The sonic log is a porosity log that measures interval transit time (Δt) of a compressional sound wave traveling through one foot of formation. The sonic log device consists of one or more sound transmitters, and two or more receivers. Modern sonic logs are borehole compensated devices (BHC*). These devices greatly reduce the spurious effects of borehole size variations (Kobesh and Blizard, 1959), as well as errors due to tilt of the sonic tool (Schlumberger, 1972).

Interval transit time (Δt) in microseconds per foot is the reciprocal of the velocity of a compressional sound wave in feet per second. Interval transit time (Δt) is recorded in tracks #2 and #3 (example Fig. 26). A sonic derived porosity curve is sometimes recorded in tracks #2 and #3, along with the Δt curve (Fig. 26). Track #1 normally contains a caliper log and a gamma ray log or an SP log (Fig. 26).

The interval transit time (Δt) is dependent upon both lithology and porosity. Therefore, a formation's matrix velocity (Table 6) must be known to derive sonic porosity either by chart (Fig. 27) or by the following formula (Wyllie et al, 1958):

Table 6. Sonic Velocities and Interval Transit Times for Different Matricies. These constants are used in the Sonic Porosity Formula (after Schlumberger, 1972).

	V_{ma} (ft/sec)	Δt_{ma} (μsec/ft)	Δt_{ma} (μsec/ft) commonly used
Sandstone	18,000 to 19,500	55.5 to 51.0	55.5 to 51.0
Limestone	21,000 to 23,000	47.6 to 43.5	47.6
Dolomite	23,000 to 26,000	43.5 to 38.5	43.5
Anhydrite	20,000	50.0	50.0
Salt	15,000	66.7	67.0
Casing (Iron)	17,500	57.0	57.0

$$\phi_{sonic} = \frac{\Delta t_{log} - \Delta t_{ma}}{\Delta t_f - \Delta t_{ma}}$$

Where:

ϕ_{sonic} = sonic derived porosity
Δt_{ma} = interval transit time of the matrix (Table 6)
Δt_{log} = interval transit time of formation
Δt_f = interval transit time of the fluid in the well bore (fresh mud = 189; salt mud = 185)

The Wyllie et al (1958) formula for calculating sonic porosity can be used to determine porosity in consolidated sandstones and carbonates with intergranular porosity (grainstones) or intercrystalline porosity (sucrosic dolomites). However, when sonic porosities of carbonates with vuggy or fracture porosity are calculated by the Wyllie formula, porosity values will be too low. This will happen because the sonic log only records matrix porosity rather than vuggy or fracture secondary porosity. The percentage of vuggy or fracture secondary porosity can be calculated by subtracting *sonic porosity* from *total porosity.* Total porosity values are obtained from one of the nuclear logs (i.e. density or neutron). The percentage of secondary porosity, called SPI or secondary porosity index, can be a useful mapping parameter in carbonate exploration.

Where a sonic log is used to determine porosity in unconsolidated sands, an empirical compaction factor or Cp should be added to the Wyllie et al (1958) equation:

$$\phi_{sonic} = \left(\frac{\Delta t_{log} - \Delta t_{ma}}{\Delta t_f - \Delta t_{ma}}\right) \times 1/Cp$$

Where:

ϕ_{sonic} = sonic derived porosity
Δt_{ma} = interval transit time of the matrix (Table 6)
Δt_{log} = interval transit time of formation
Δt_f = interval transit time of the fluid in the well bore (fresh mud = 189; salt mud = 185)
Cp = compaction factor

The compaction factor is obtained from the following formula:

$$Cp = \frac{\Delta t_{sh} \times C}{100}$$

Where:

Cp = compaction factor
Δt_{sh} = interval transit time for adjacent shale
C = a constant which is normally 1.0 (Hilchie, 1978).

The interval transit time (Δt) of a formation is increased due to the presence of hydrocarbons (i.e. *hydrocarbon effect*). If the effect of hydrocarbons is not corrected, the sonic derived porosity will be too high. Hilchie (1978) suggests the following empirical corrections for hydrocarbon effect:

$$\phi = \phi_{sonic} \times 0.7 \text{ (gas)}$$
$$\phi = \phi_{sonic} \times 0.9 \text{ (oil)}$$

Density Log

The formation density log is a porosity log that measures *electron density* of a formation. It can assist the geologist to: (1) identify evaporite minerals, (2) detect gas-bearing zones, (3) determine hydrocarbon density, and (4) evaluate shaly sand reservoirs and complex lithologies (Schlumberger, 1972).

The density logging device is a contact tool which consists of a medium-energy gamma ray source that emits gamma rays into a formation. The gamma ray source is either Cobalt-60 or Cesium-137.

Gamma rays collide with electrons in the formation; the collisions result in a loss of energy from the gamma ray particle. Tittman and Wahl (1965) called the interaction between incoming gamma ray particles and electrons in the formation, *Compton Scattering*. Scattered gamma rays which reach the detector, located a fixed distance from the gamma ray source, are counted as an indicator of formation density. The number of Compton Scattering collisions is a direct function of the number of electrons in a formation (electron density). Consequently, electron density can be related to bulk density (ρ_b) of a formation in gm/cc.

The bulk density curve is recorded in tracks #2 and #3 (Fig. 28), along with a correction curve (Δ_ρ). Because the modern density log is a compensated log (dual detectors), the correction curve (Δ_ρ; Fig. 28) records how much correction has been applied to the bulk density curve (ρ_b), due to borehole irregularities. *Whenever the correction curve (Δ_ρ) exceeds 0.20 gm/cc, the value of the bulk density obtained from the bulk density curve (ρ_b) should be considered invalid.* A density derived porosity curve is sometimes present in tracks #2 and #3 along with the bulk density (ρ_b) and correction (Δ_ρ) curves. Track #1 contains a gamma ray log and a caliper (example, Fig. 28).

Formation bulk density (ρ_b) is a function of matrix density, porosity, and density of the fluid in the pores (salt mud, fresh mud, or hydrocarbons). To determine density porosity, either by chart (Fig. 29) or by calculation, the matrix density (Table 7) and type of fluid in the borehole must be known. The formula for calculating density porosity is:

$$\phi_{den} = \frac{\rho_{ma} - \rho_b}{\rho_{ma} - \rho_f}$$

Where:

ϕ_{den} = density derived porosity

ρ_{ma} = matrix density (see Table 7)

ρ_b = formation bulk density

ρ_f = fluid density (1.1 salt mud, 1.0 fresh mud, and 0.7 gas)

Table 7. Matrix Densities of Common Lithologies. Constants presented here are used in the Density Porosity Formula (after Schlumberger, 1972).

	ρ_{ma}(gm/cc)
Sandstone	2.648
Limestone	2.710
Dolomite	2.876
Anhydrite	2.977
Salt	2.032

Where invasion of a formation is shallow, low density of the formation's hydrocarbons will increase density porosity. Oil does not significantly affect density porosity, but gas does (gas effect). Hilchie (1978) suggests using a gas density of 0.7 gm/cc for fluid density (ρ_f) in the density porosity formula if gas density is unknown.

Neutron Logs

Neutron logs are porosity logs that measure the hydrogen ion concentration in a formation. In clean formations (i.e. shale-free) where the porosity is filled with water or oil, the neutron log measures liquid-filled porosity.

Neutrons are created from a chemical source in the neutron logging tool. The chemical source may be a mixture of americium and beryllium which will continuously emit neutrons. These neutrons collide with the nuclei of the formation material, and result in a neutron losing some of its energy. Because the hydrogen atom is almost equal in mass to the neutron, maximum energy loss occurs when the neutron collides with a hydrogen atom. Therefore, the maximum amount of energy loss is a function of a formation's hydrogen concentration. Because hydrogen in a porous formation is concentrated in the fluid-filled pores, energy loss can be related to the formation's porosity.

Whenever pores are filled with gas rather than oil or water, neutron porosity will be lowered. This occurs because there is less concentration of hydrogen in gas compared to oil or water. A lowering of neutron porosity by gas is called *gas effect*.

Neutron log responses vary, depending on: (1) differences in detector types, (2) spacing between source and detector, and (3) lithology—i.e. sandstone, limestone, and dolomite. These variations in response can be compensated for by using the appropriate charts (Figs. 30 and 31). A geologist should remember that neutron logs (unlike all other logs) *must* be interpreted from the specific chart designed for a specific log (i.e. Schlumberger charts for Schlumberger logs and Dresser Atlas charts for Dresser Atlas logs). The reason for this is that while other logs are calibrated in basic physical units, neutron logs are not (Dresser Atlas, 1975).

The first modern neutron log was the Sidewall Neutron Log. The Sidewall Neutron Log has both the source and detector in a pad which is pushed against the side of the borehole. The most modern of the neutron logs is a Compensated Neutron Log which has a neutron source and two detectors. The advantage of Compensated Neutron logs over Sidewall Neutron logs is that they are less affected by borehole irregularities. Both the Sidewall and Compensated Neutron logs can be recorded in apparent limestone, sandstone, or dolomite porosity units. If a formation is limestone, and the neutron log is recorded in apparent limestone porosity units, apparent porosity is equal to true porosity. However, when the lithology of a formation is sandstone or dolomite, apparent limestone porosity *must* be corrected to true porosity by using the appropriate chart (Fig. 30 for Sidewall Neutron Log; or Fig. 31 for Compensated Neutron Log). The procedure is identical for each of the charts and is shown in Figures 30 and 31.

Combination Neutron-Density Log

The Combination Neutron-Density Log is a combination porosity log. Besides its use as a porosity device, it is also used to determine lithology and to detect gas-bearing zones.

The Neutron-Density Log consists of neutron and density curves recorded in tracks #2 and #3 (example, Fig. 32), and a caliper and gamma ray log in track #1. Both the neutron and density curves are normally recorded in limestone porosity units with each division equal to either two percent or three percent porosity; however, sandstone and dolomite porosity units can also be recorded.

True porosity can be obtained by, first, reading apparent limestone porosities from the neutron and density curves (example: Fig. 32 at 9,324 ft, $\phi_N = 8\%$ and $\phi_D = 3.5\%$). Then, these values are crossplotted on a neutron-density porosity chart (Figs. 33 or 34) to find true porosity. In the example from Figures 32 and 34, the position of the crossplotted neutron-density porosities at 9,324 ft (Fig. 34) indicates that the lithology is a limey dolomite and the porosity is 6%.

Examination of the neutron-density porosity chart (Fig. 34) reveals that the porosity values are only slightly affected by changes in lithology. Therefore, porosity from a Neutron-Density Log can be calculated mathematically. The alternate method of determining neutron-density porosity is to use the root mean square formula.

$$\phi_{N\text{-}D} \approx \sqrt{\frac{\phi_N^2 + \phi_D^2}{2}}\ ^{\dagger}$$

Where:
$\phi_{N\text{-}D}$ = neutron-density porosity
ϕ_N = neutron porosity (limestone units)
ϕ_D = density porosity (limestone units)

If the neutron and density porosities from Figure 32 at a depth of 9,324 ft are entered into the root mean square formula, we calculate a porosity of 6.2%. This calculated porosity value compares favorably with the value obtained from the crossplot method.

Whenever a Neutron-Density Log records a density porosity of less than 0.0—a common value in anhydritic dolomite reservoirs (Fig. 32; depth 9,328 ft)—the following formula should be used to determine neutron-density porosity:

$$\phi_{N\text{-}D} \approx \frac{\phi_N + \phi_D}{2}$$

Where:
$\phi_{N\text{-}D}$ = neutron-density porosity
ϕ_N = neutron porosity (limestone units)
ϕ_D = density porosity (limestone units)

Figure 35 is a schematic illustration of how lithology affects the Combination Gamma Ray Neutron-Density log. The relationship between log responses on the Gamma Ray Neutron-Density Log and rock type provides a powerful tool for the subsurface geologist. By identifying rock type from logs, a geologist can construct facies maps.

Figure 35 also illustrates the change in neutron-density response between an oil- or water-bearing sand and a gas-bearing sand. The oil- or water-bearing sand has a density log reading of four porosity units more than the neutron log. In contrast, the gas-bearing sand has a density reading of up to 10 porosity units more than the neutron log.

Where an *increase* in density porosity occurs along with a *decrease* in neutron porosity in a gas-bearing zone, it is called *gas effect*. Gas effect is created by gas in the pores. Gas in the pores causes the density log to record too high a porosity (i.e. gas is lighter than oil or water), and causes the neutron log to record too low a porosity (i.e. gas has a lower concentration of hydrogen atoms than oil or water). The effect of gas on the Neutron-Density Log is a very important log response because it helps a geologist to detect gas-bearing zones.

Figure 36 is a schematic illustration of a Gamma Ray Neutron-Density Log through several gas sands. It illustrates how changes in porosity, invasion, hydrocarbon density, and shale content alter the degree of gas effect observed on the Neutron-Density Log.

† Slight variations of this formula may be used in some areas. Also, some log analysts restrict the use of this formula to gas-bearing formations, and use $\phi_{N\text{-}D} = (\phi_N + \phi_D)/2$ in oil- or water-bearing formations.

Review - Chapter IV

1. The three types of porosity logs are: (1) sonic, (2) density, and (3) neutron.

2. The sonic log is a porosity log that measures the interval transit time (Δt) of a compressional sound wave through one foot of formation. The unit of measure is microseconds per foot (μsec/ft). Interval transit time is related to formation porosity.

3. The density log is a porosity log that measures the electron density of a formation. The formation's electron density is related to a formation's bulk density (ρ_b) in gm/cc. Bulk density, in turn, can be related to formation porosity.

4. The neutron log is a porosity log that measures the hydrogen ion concentration in a formation. In shale-free formations where porosity is filled with water, the neutron log can be related to water-filled porosity.

5. In gas reservoirs, the neutron log will record a lower porosity than the formation's true porosity because gas has a lower hydrogen ion concentration than oil or water (gas effect).

6. The Neutron-Density Log is a combination porosity log. Porosity can be determined from a Neutron-Density Log either by a crossplot chart or by formula.

7. Additional uses of the combination Neutron-Density Log are: (1) detection of gas bearing zones; and (2) determination of lithology.

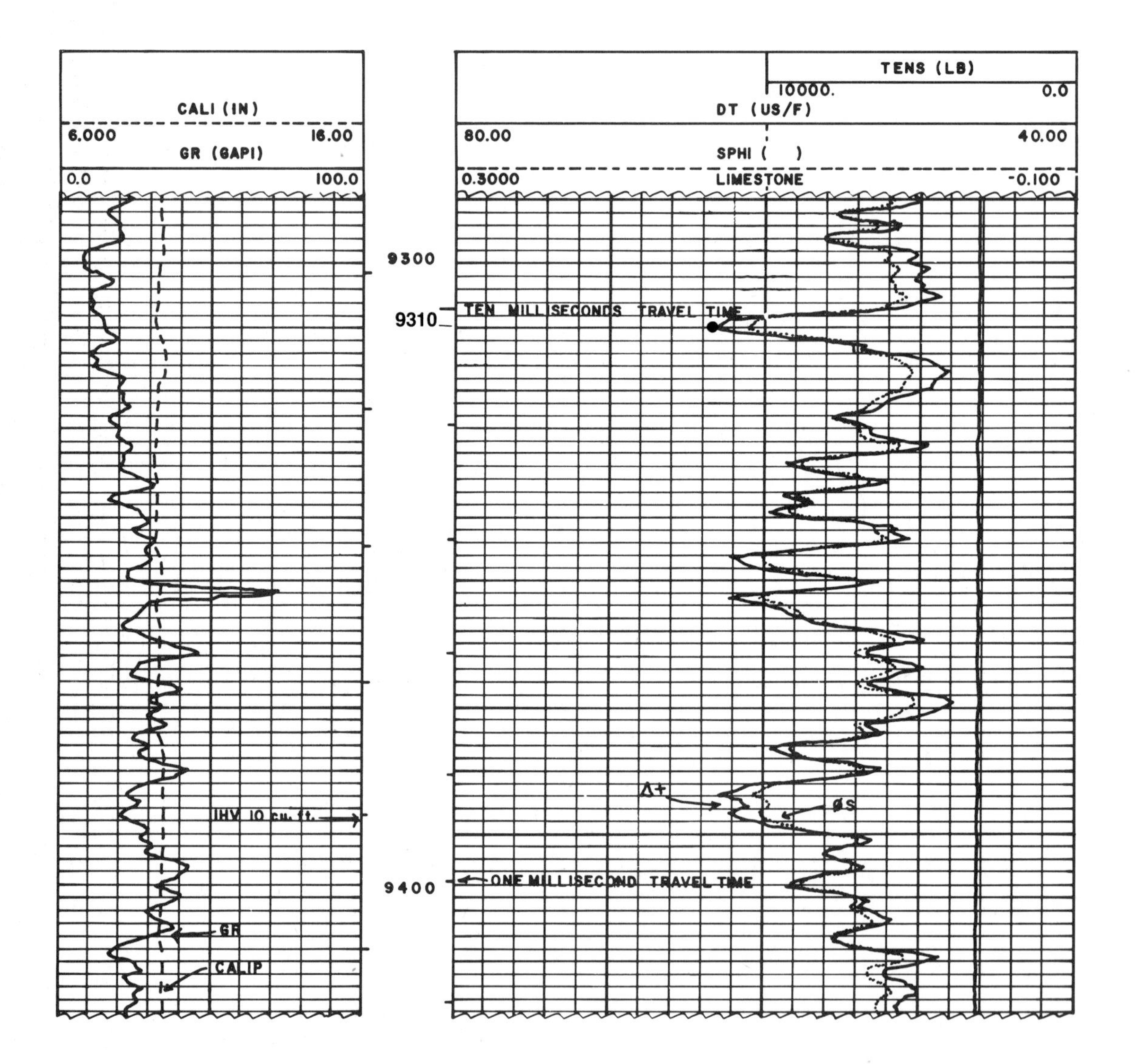
CALI (IN)
6.000 16.00
GR (GAPI)
0.0 100.0
TENS (LB)
10000. 0.0
DT (US/F)
80.00 40.00
SPHI ( )
0.3000 LIMESTONE -0.100
9300
9310
TEN MILLISECONDS TRAVEL TIME
IHV 10 cu. ft.
Δt
ØS
9400
ONE MILLISECOND TRAVEL TIME
GR
CALIP

Figure 26. Example sonic log with gamma ray log and caliper. This example is shown to display the scales of a sonic log, and to be used in picking an interval transit time (Δt) value for Figure 27.

Track #1—This track includes both the gamma ray and caliper curves. Note that the gamma ray scale reads from 0 to 100 API gamma ray units, increasing from left-to-right in increments of 10 units. The gamma ray scale is represented by a solid line.

The caliper scale ranges from 6 to 16 inches, from left-to-right in one-inch increments, and is represented by a dashed line.

Tracks #2 and #3—Both the interval transit time (Δt) scale and the porosity scale are shown in this track. Sonic log interval transit time (Δt) is represented by a solid line, on a scale ranging from 40 to 80 μsec/ft increasing from *right-to-left*.

The sonic porosity measurement (limestone matrix) is shown by a dashed line, on a scale ranging from −10% to +30% porosity increasing from *right-to-left*.

At the sample depth used in Figure 27 (9,310 ft), read a sonic log interval transit time (Δt) value of 63 μsec/ft.

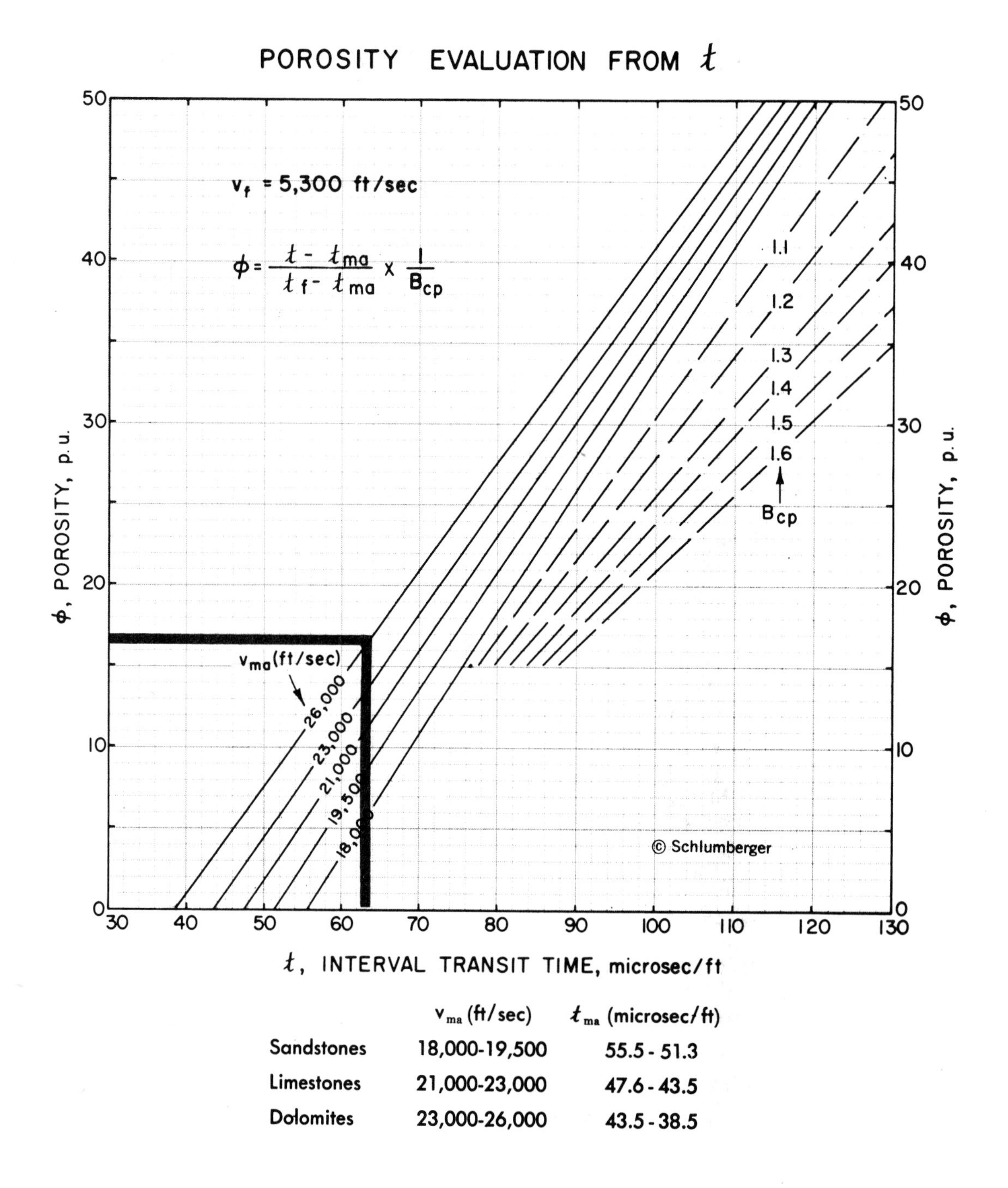

	v_{ma} (ft/sec)	t_{ma} (microsec/ft)
Sandstones	18,000-19,500	55.5 - 51.3
Limestones	21,000-23,000	47.6 - 43.5
Dolomites	23,000-26,000	43.5 - 38.5

igure 27. Chart used for converting interval transit time (Δt) values to sonic porosity, using values picked from a sonic log.

Courtesy, Schlumberger Well Services.
Copyright 1977, Schlumberger.

Given:

V_{ma} = 26,000 ft/sec where V_{ma} is the sonic velocity of the matrix (in this case dolomite; see Table 6). Δt (from log) = 63 μsec/ft at a depth of 9,310 ft (see Fig. 26).

Procedure:

1. Find an interval transit time value (Δt) taken from the sonic log in Figure 26 (in this example 63 μsec/ft) on the scale at the bottom of the chart.
2. Follow the value (63) vertically until it intersects the diagonal line representing 26,000 ft/sec (dolomite, in this case).
3. From that point, follow the value horizontally to the left, and read the porosity value from the porosity scale. In this case, the value is 16.5% (ϕ = 16.5%).

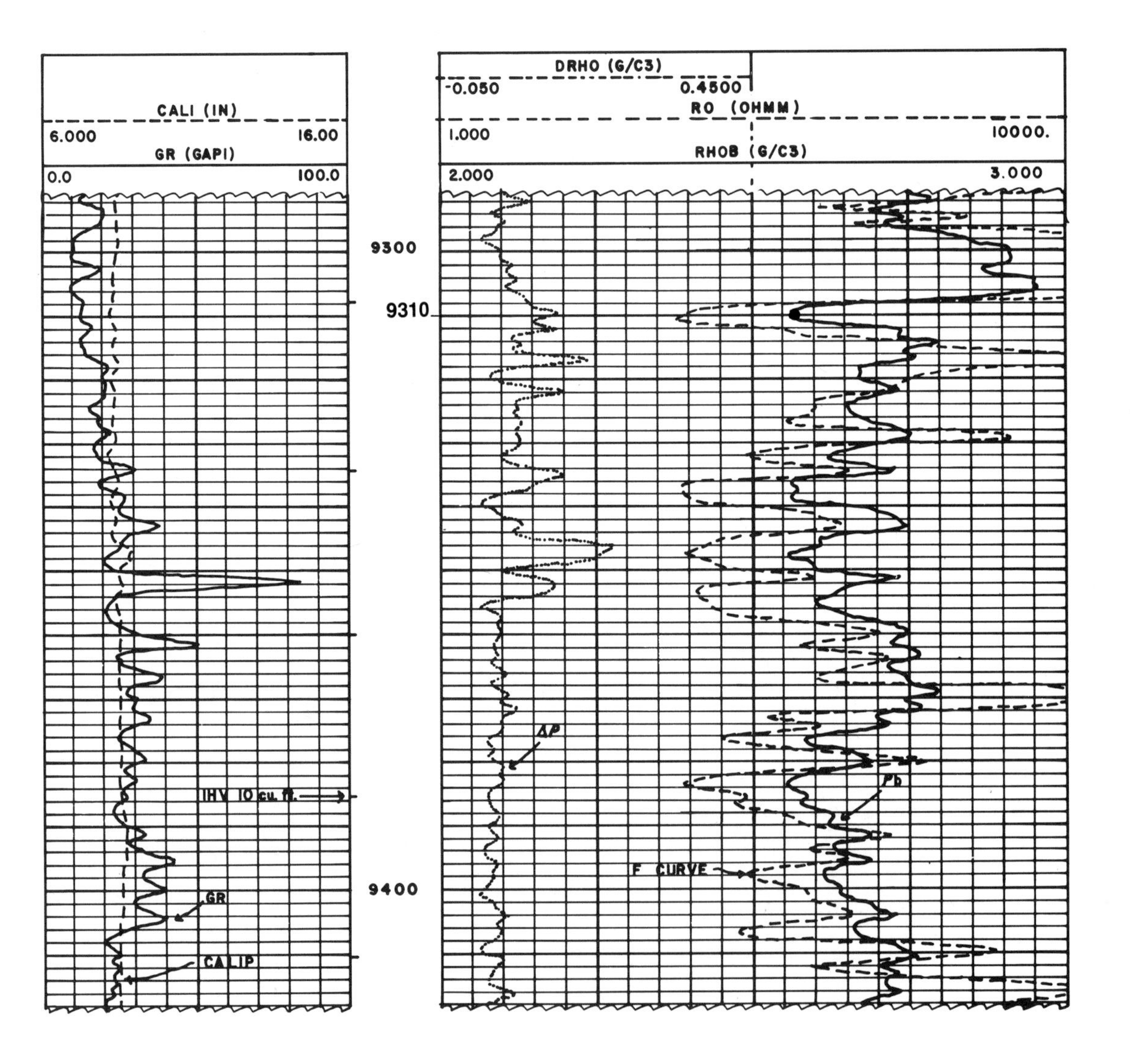
CALI (IN)
6.000
16.00
GR (GAPI)
0.0
100.0
DRHO (G/C3)
-0.050
0.4500
RO (OHMM)
1.000
10000.
RHOB (G/C3)
2.000
3.000
9300
9310
9400
IHV 10 cu. ft.
GR
CALIP
ΔP
Pb
F CURVE

Figure 28. Example of a bulk density log with a gamma ray log and caliper, and formation factor curve (F). This log is presented to show you the scales of a density log, and is used in picking values for Figure 29.

Track #1—This track includes both the gamma ray and caliper logs. Note that both scales read left-to-right; the gamma ray values range from 0 to 100 API gamma ray units, and the caliper measures the borehole size from 6 to 16 inches.

Tracks #2 and #3—The bulk density curve (ρ_b), correction curve (Δ_ρ), and formation factor curve (F) are recorded in this track. The correction (Δ_ρ), formation factor, and the bulk density scales increase in value from left to right.

The bulk density (ρ_b) scale ranges in value from 2.0 gm/cc to 3.0 gm/cc and is represented by a solid line. The density log correction curve (Δ_ρ) ranges in value from -0.05 gm/cc to $+0.45$ gm/cc in increments of 0.05 gm/cc, but only uses the left half of the log track. The formation factor curve (F) ranges in value from 1 to 10,000 (discussed later) and is represented by a dashed line.

At the sample depth used in Figure 29 (9,310 ft) read a bulk density value (ρ_b) of 2.56 gm/cc.

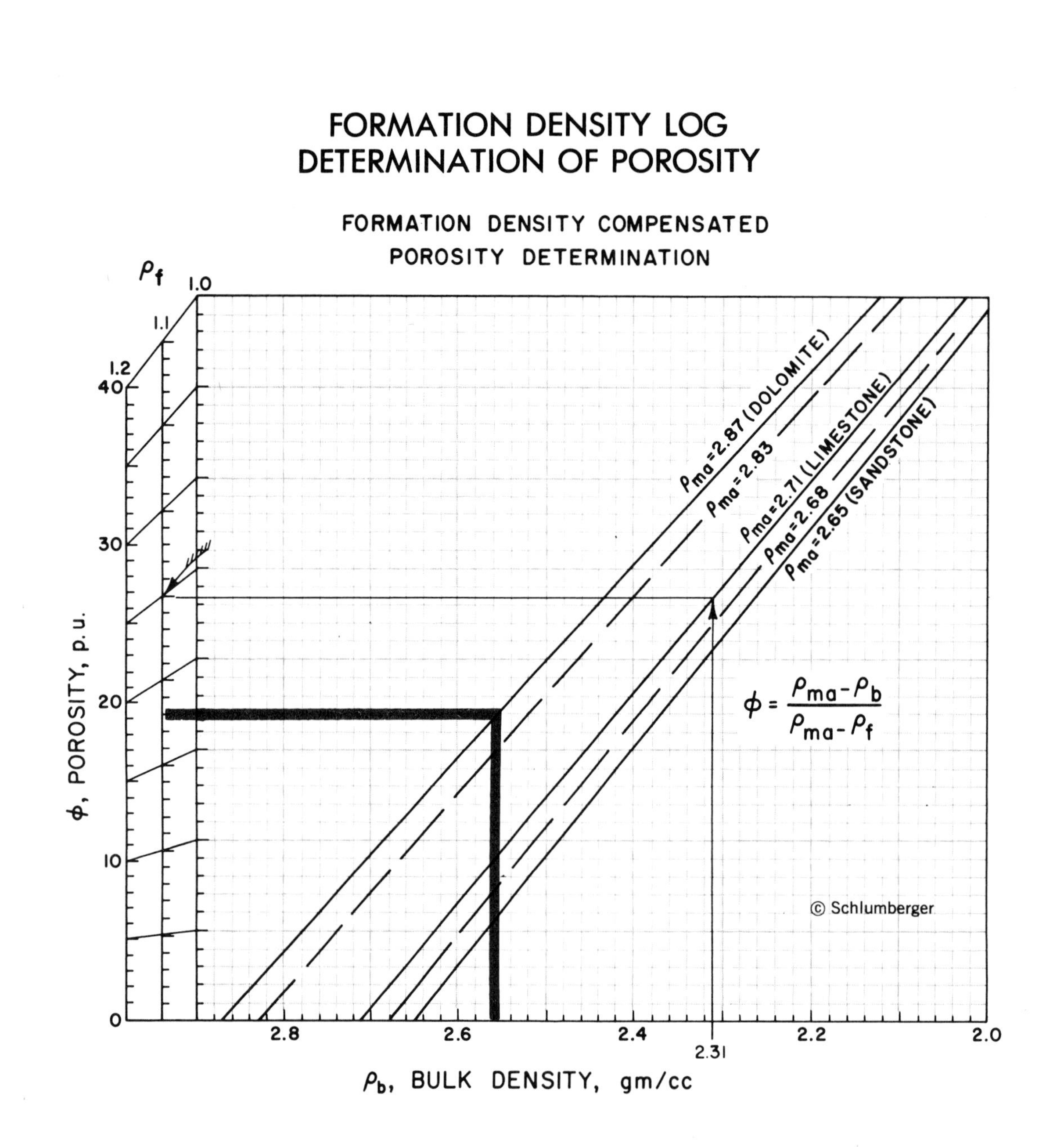

EXAMPLE: ρ_b = 2.31 gm/cc in limestone lithology
ρ_{ma} = 2.71 (limestone)
ρ_f = 1.1 (salt mud)

SOLUTION: ϕ_D = 25 p.u.

Figure 29. Chart for converting bulk density (ρ_b) to porosity (ϕ) using values picked from a density log.

Courtesy, Schlumberger Well Services.
Copyright 1977, Schlumberger.

Given:

ρ_{ma} = 2.87 gm/cc (dolomite; Table 7)
ρ_f = 1.1 gm/cc (suggested constant fluid density for salt mud; see text)
ρ_b = 2.56 gm/cc at a depth of 9,310 ft (from log; Fig. 28)

Procedure:

1. Find a value for bulk density (ρ_b) on the horizontal scale at the bottom of Figure 29 (in this example 2.56 gm/cc).
2. Follow the value vertically until it intersects the diagonal line representing the matrix density (ρ_{ma}) used (in this case 2.87 for dolomite).
3. From that point, follow the horizontal line to the left where the porosity (ϕ) value is represented on the porosity scale at a fluid density (ρ_f) of 1.1. In this case, the porosity (ϕ) is 18%.

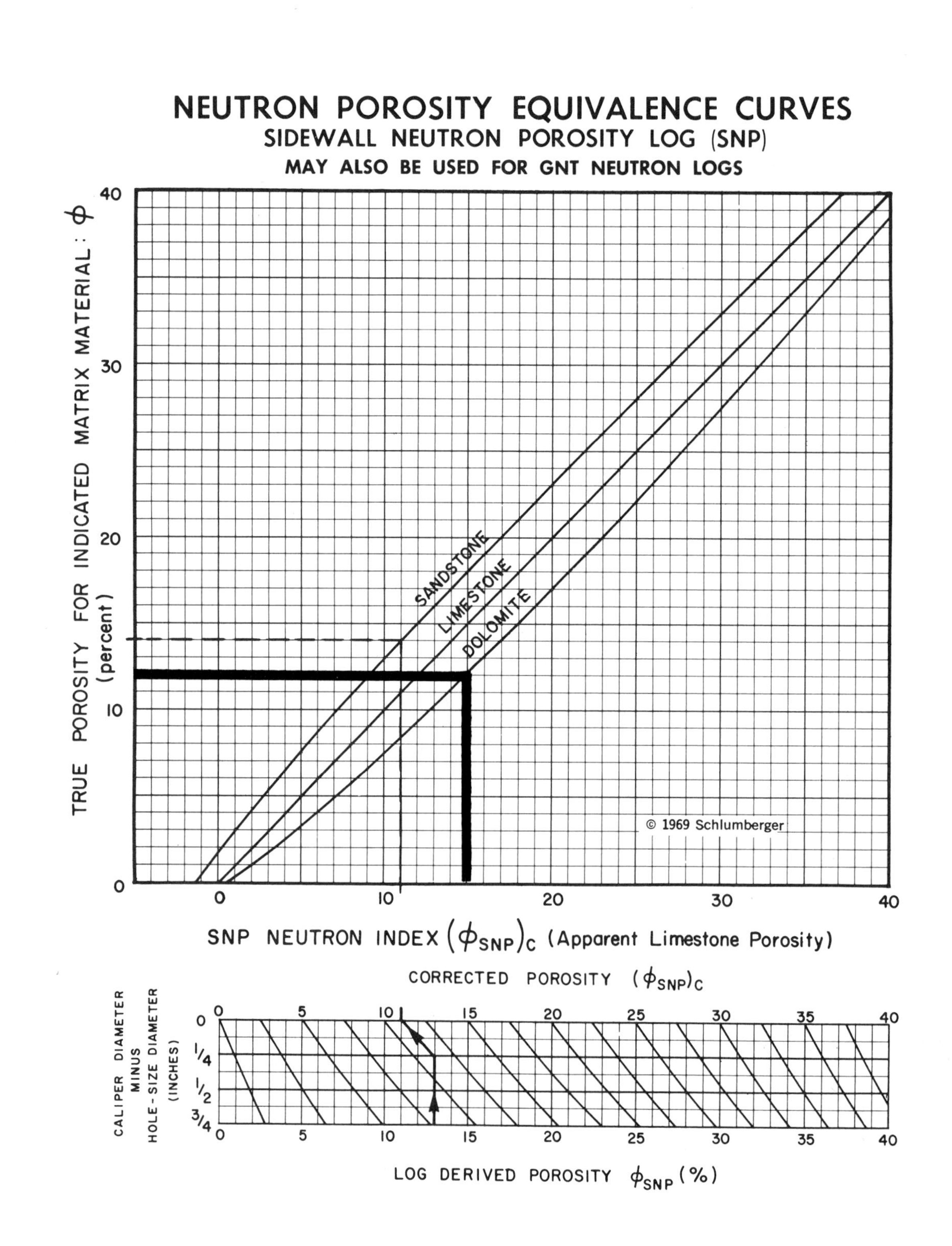
NEUTRON POROSITY EQUIVALENCE CURVES
SIDEWALL NEUTRON POROSITY LOG (SNP)
MAY ALSO BE USED FOR GNT NEUTRON LOGS
TRUE POROSITY FOR INDICATED MATRIX MATERIAL : φ (percent)
SANDSTONE
LIMESTONE
DOLOMITE
© 1969 Schlumberger
SNP NEUTRON INDEX (φSNP)C (Apparent Limestone Porosity)
CORRECTED POROSITY (φSNP)C
CALIPER DIAMETER MINUS HOLE-SIZE DIAMETER (INCHES)
LOG DERIVED POROSITY φSNP (%)

Figure 30. Chart for correcting Sidewall Neutron Porosity Log (SNP*) for lithology. Note: this is a Schlumberger example; do not use this chart with another type neutron log (see text).

Courtesy, Schlumberger Well Services.
Copyright 1969, Schlumberger.

Given: The lithology is dolomite. Also, the apparent limestone porosity is 15%. The value for apparent limestone porosity is read directly from a Sidewall Neutron Porosity Log (SNP*). A Sidewall Neutron Log (SNP*) is not shown here; instead the value is given to you.

Procedure:

1. Find the value for apparent limestone porosity (read from an SNP* log) along the scale at the bottom of the correction chart. In this example, the value is 15%.
2. Follow the value vertically until it intersects the diagonal curve representing dolomite.
3. From that point, follow the value horizontally to the left, and read the true porosity (ϕ) on the left-hand scale: 12%.

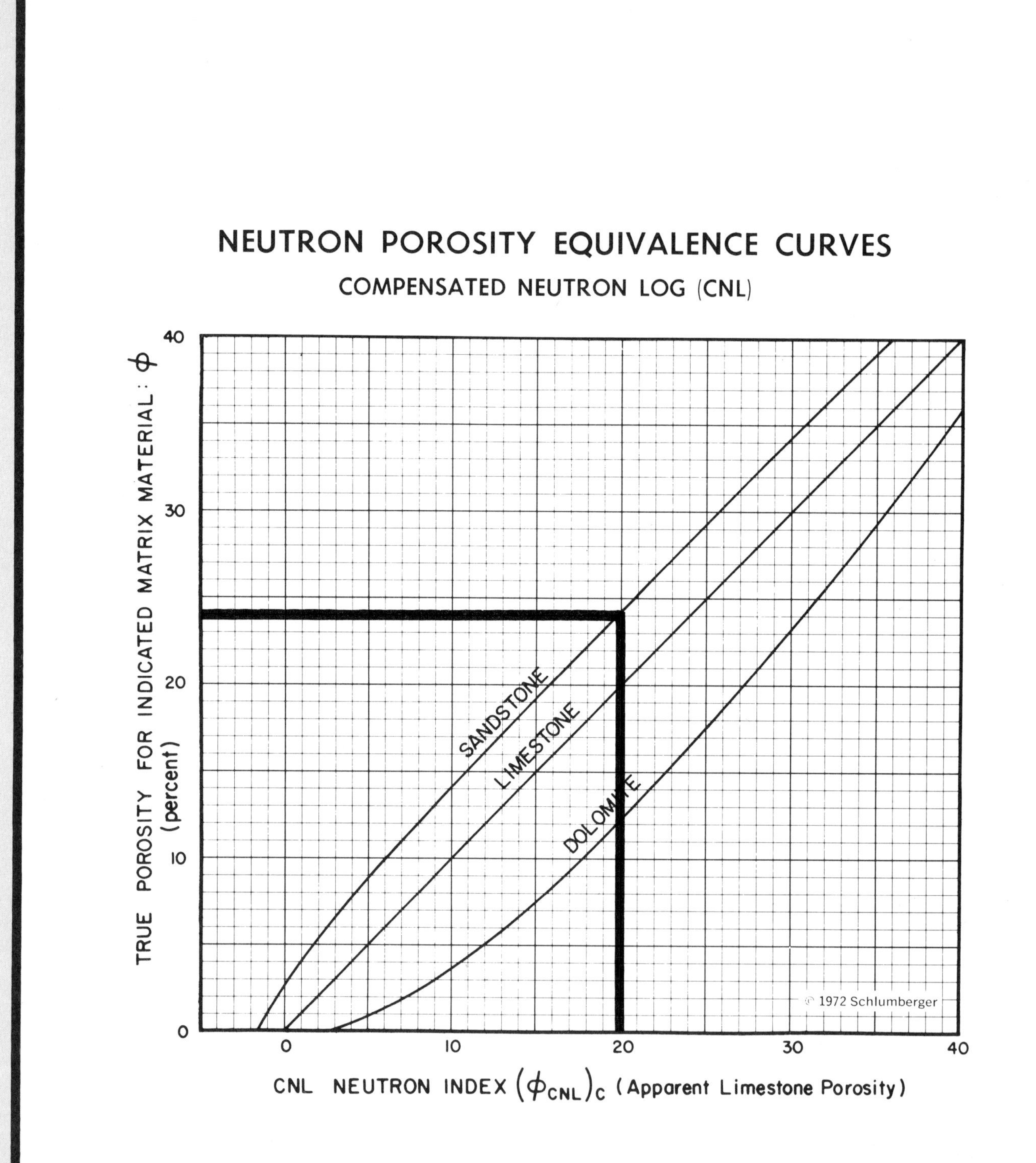
NEUTRON POROSITY EQUIVALENCE CURVES
COMPENSATED NEUTRON LOG (CNL)
TRUE POROSITY FOR INDICATED MATRIX MATERIAL: ϕ (percent)
40
30
20
10
0
SANDSTONE
LIMESTONE
DOLOMITE
© 1972 Schlumberger
0
10
20
30
40
CNL NEUTRON INDEX $(\phi_{CNL})_C$ (Apparent Limestone Porosity)

Figure 31. Chart for correcting Compensated Neutron Log (CNL*) for lithology. Note: this is a Schlumberger chart; do not use this chart with another type of neutron log (see text).

Courtesy, Schlumberger Well Services.
Copyright 1972, Schlumberger.

Given: The lithology is sandstone. Also, the apparent limestone porosity is 20%. The value for apparent limestone porosity is read directly from a Compensated Neutron Log (CNL*). A Compensated Neutron Log (CNL*) is not shown here; instead the value is given to you.

Procedure:

1. Find the value for apparent limestone porosity (read from a CNL* log) along the scale at the bottom of the chart. In this example, the value is 20%.
2. Follow the value vertically until it intersects the diagonal curve representing lithology (in this case, sandstone).
3. From that point, follow the value horizontally to the left, and read the true porosity (ϕ) on the left-hand scale. Porosity = 24%.

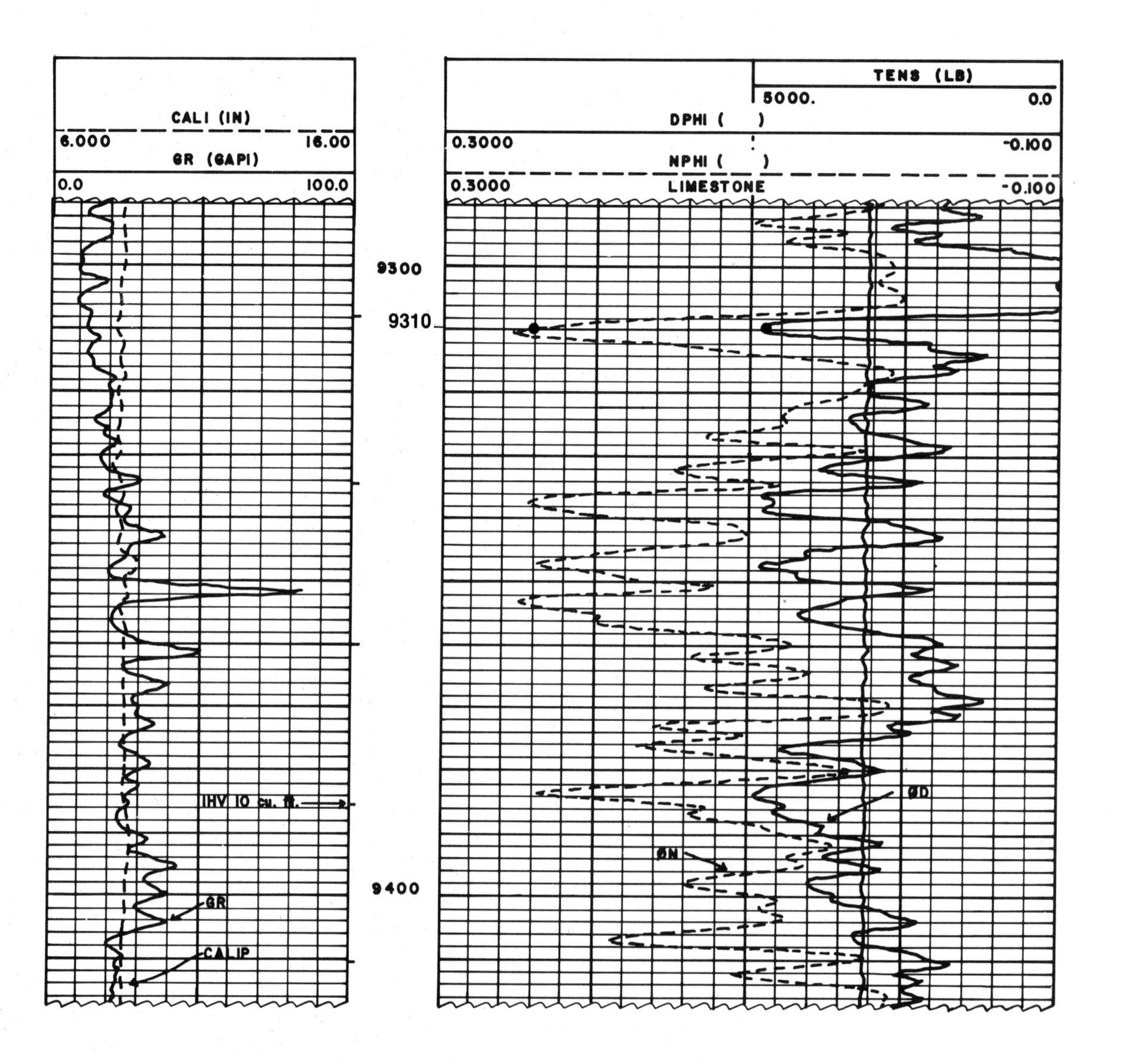
CALI (IN)
6.000
16.00
GR (GAPI)
0.0
100.0
TENS (LB)
5000.
0.0
DPHI ( )
0.3000
-0.100
NPHI ( )
0.3000
LIMESTONE
-0.100
9300
9310
9400
IHV 10 cu. ft.
GR
CALIP
ØD
ØN

Figure 32. Example of a Combination Neutron-Density Log with gamma ray log and caliper. This log illustrates the log curves and scales of a combination log, but is also used here for picking values for exercises in Figure 33 and Figure 34.

Track #1—This track contains both gamma ray and caliper curves. Note that the gamma ray scale reads from 0 to 100 API gamma ray units and the caliper measures a borehole size from 6 to 16 inches.

Tracks #2 and #3—Both neutron porosity (ϕ_N) and density porosity (ϕ_D) curves are in tracks #2 and #3. The scale for both is the same, ranging from -10% to $+30\%$ in increments of 2%, and is measured in limestone porosity units. On this log the density porosity (ϕ_D) is represented by a solid line, and the neutron porosity (ϕ_N) is represented by a dashed line.

Figures 33 and 34 are charts and examples for correcting Neutron-Density Log porosities for lithology. Because salt versus freshwater drilling muds can affect the porosity values, two different charts are used. Figure 33 is used to correct porosity for lithology where there is freshwater-based drilling mud (where $R_{mf} > 3\ R_w$); and the other (Fig. 34) is used where there is saltwater-based drilling mud (where $R_{mf} \simeq R_w$).

At the sample depth of 9,310 ft, the neutron porosity value (ϕ_N) is 24%, and the density porosity value (ϕ_D) is 9%.

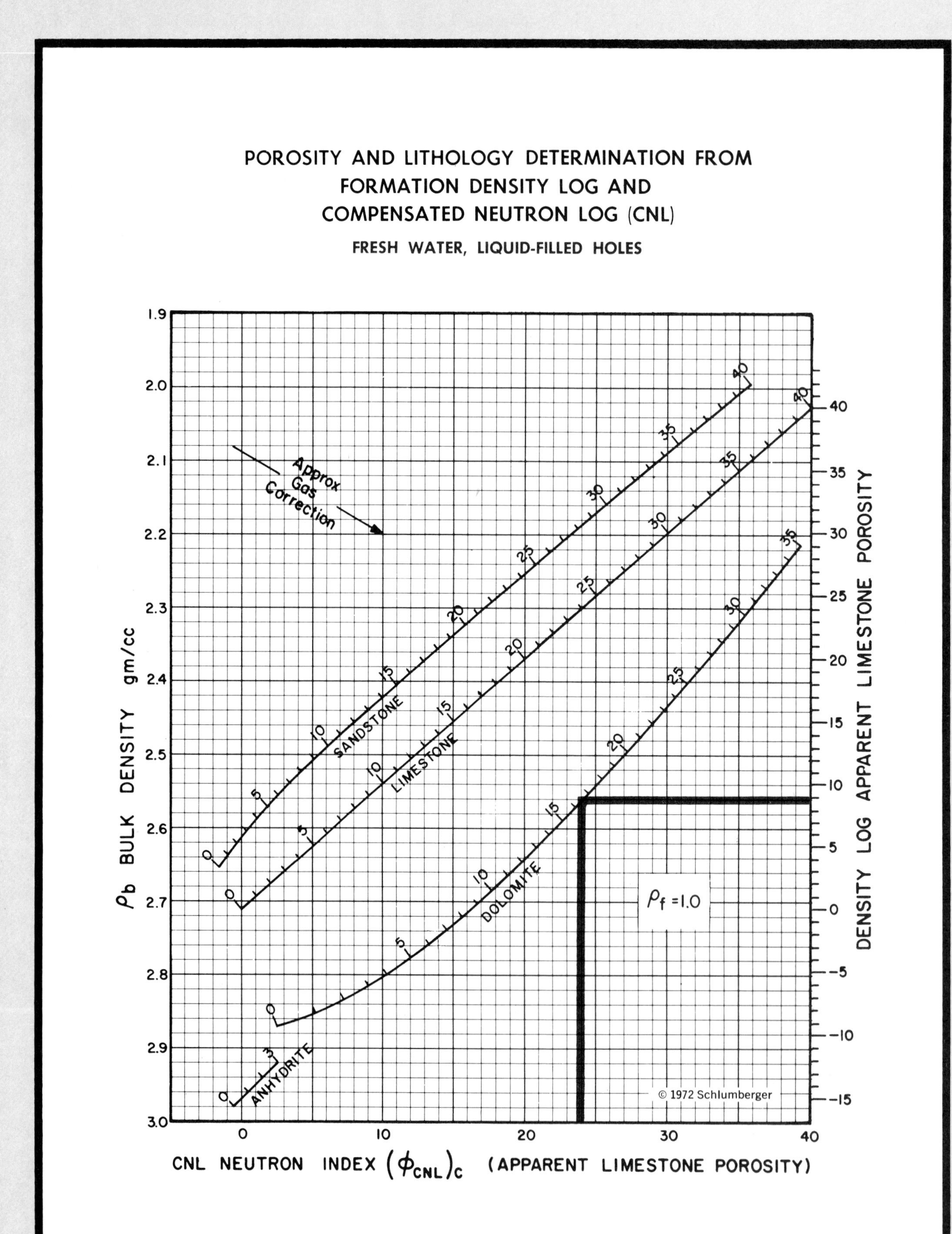
POROSITY AND LITHOLOGY DETERMINATION FROM
FORMATION DENSITY LOG AND
COMPENSATED NEUTRON LOG (CNL)
FRESH WATER, LIQUID-FILLED HOLES
ρb BULK DENSITY gm/cc
DENSITY LOG APPARENT LIMESTONE POROSITY
CNL NEUTRON INDEX (φCNL)c (APPARENT LIMESTONE POROSITY)
Approx Gas Correction
SANDSTONE
LIMESTONE
DOLOMITE
ANHYDRITE
ρf =1.0
© 1972 Schlumberger

Figure 33. Chart for correcting Neutron-Density Log porosities for lithology where freshwater-based drilling mud is used (where $R_{mf} > 3\ R_w$).

Courtesy, Schlumberger Well Services.
Copyright 1972, Schlumberger.

Given: $\rho_f = 1.0$ gm/cc (suggested fluid density of fresh muds; see text under the heading: Density Logs). $\phi_N = 24\ \%$, and $\phi_D = 9\%$ at a depth of 9,310 ft (from log; see Fig. 32).

Procedure:

1. Locate the neutron porosity value (ϕ_N) on the bottom scale (24%) and find the density porosity value (ϕ_D) on the right-hand scale (9%).
2. Follow the values until they intersect on the chart. In this example, the values meet on the lithology curve for dolomite, and the intersection shows a true porosity value of 16.5%.

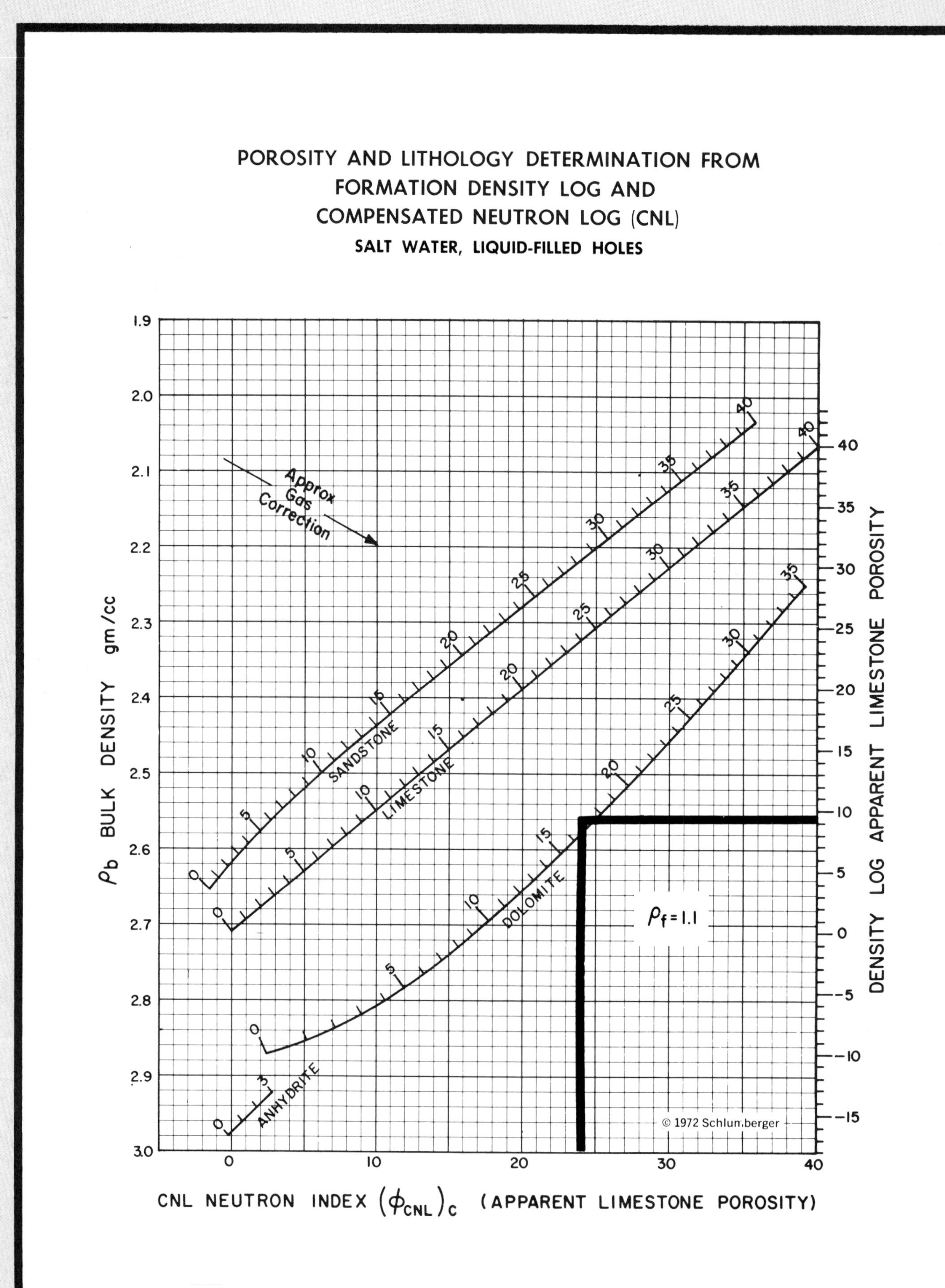
POROSITY AND LITHOLOGY DETERMINATION FROM
FORMATION DENSITY LOG AND
COMPENSATED NEUTRON LOG (CNL)
SALT WATER, LIQUID-FILLED HOLES
ρb BULK DENSITY gm/cc
1.9
2.0
2.1
2.2
2.3
2.4
2.5
2.6
2.7
2.8
2.9
3.0
Approx Gas Correction
SANDSTONE
LIMESTONE
DOLOMITE
ANHYDRITE
ρf = 1.1
DENSITY LOG APPARENT LIMESTONE POROSITY
40
35
30
25
20
15
10
5
0
-5
-10
-15
0
10
20
30
40
CNL NEUTRON INDEX (φCNL)c (APPARENT LIMESTONE POROSITY)
© 1972 Schlumberger

Figure 34. Chart for correcting Neutron-Density Log porosities for lithology where saltwater-based drilling mud is used (where $R_{mf} \simeq R_w$).

Courtesy, Schlumberger Well Services.

Copyright 1972, Schlumberger.

Given: $\rho_f = 1.1$ gm/cc (suggested fluid density of salt muds; see text under the heading: Density Logs). $\phi_N = 24\%$, and $\phi_D = 9\%$ at a depth of 9,310 ft (from log; Fig. 32).

Procedure:

1. Locate the neutron porosity value (ϕ_N) on the bottom scale (24%), and find the density porosity value (ϕ_D) on the right-hand scale (9%).
2. Follow the values until they intersect on the chart. In this example, the values meet at a point just off the lithology curve for dolomite, and the intersection shows a true porosity value of 17%.

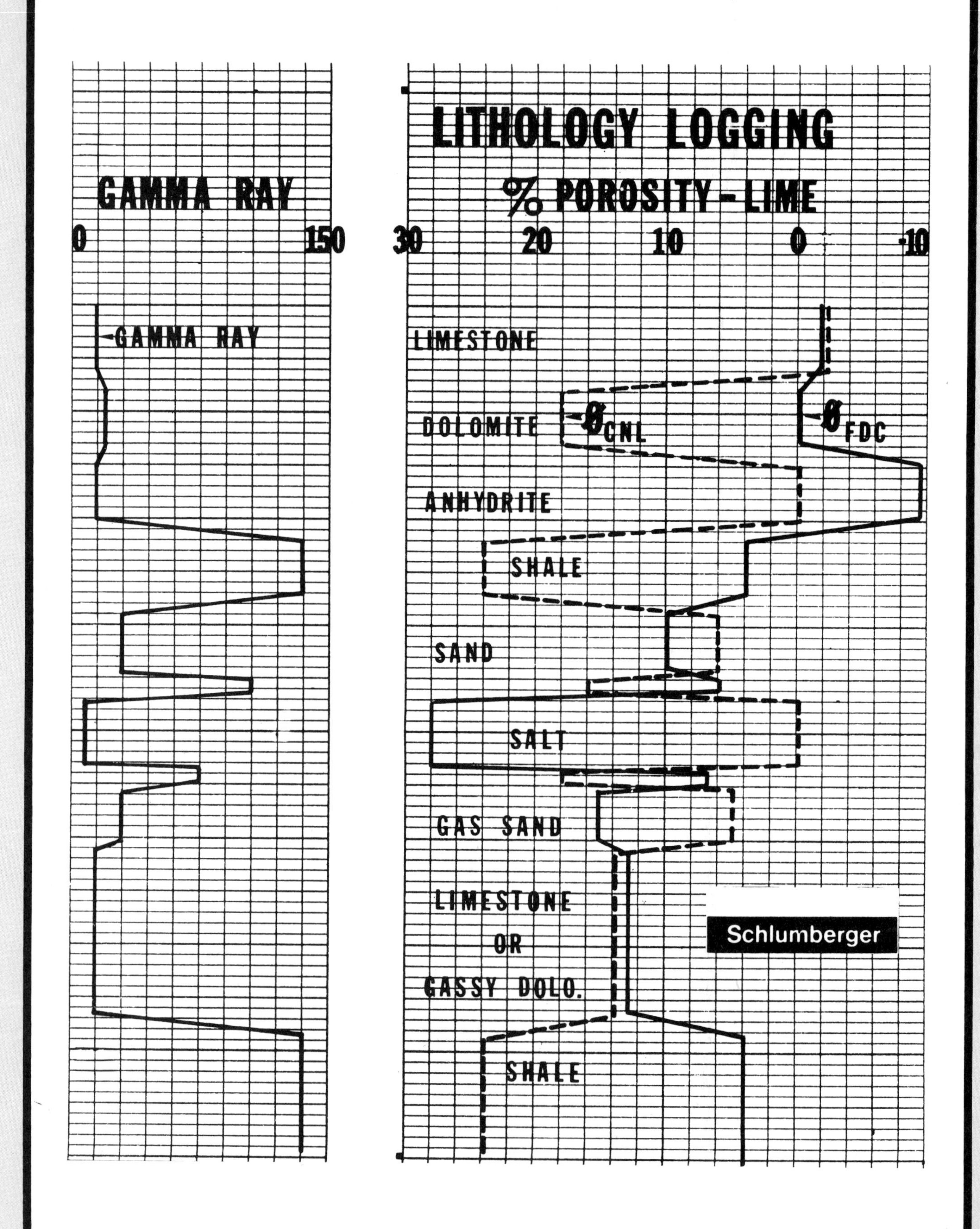
LITHOLOGY LOGGING
GAMMA RAY
% POROSITY - LIME
0
150
30
20
10
0
-10
GAMMA RAY
LIMESTONE
DOLOMITE
ØCNL
ØFDC
ANHYDRITE
SHALE
SAND
SALT
GAS SAND
LIMESTONE
OR
GASSY DOLO.
SHALE
Schlumberger

Figure 35. Example of generalized lithology logging with Combination Gamma Ray Neutron-Density Log. This figure shows sample relationships between log responses and the rock type, and also shows changes in the log response from oil- or water-bearing rock units compared to gas-bearing units.

This briefly shows how the Combination Gamma Ray Neutron-Density Log is used as a tool for determining lithology.

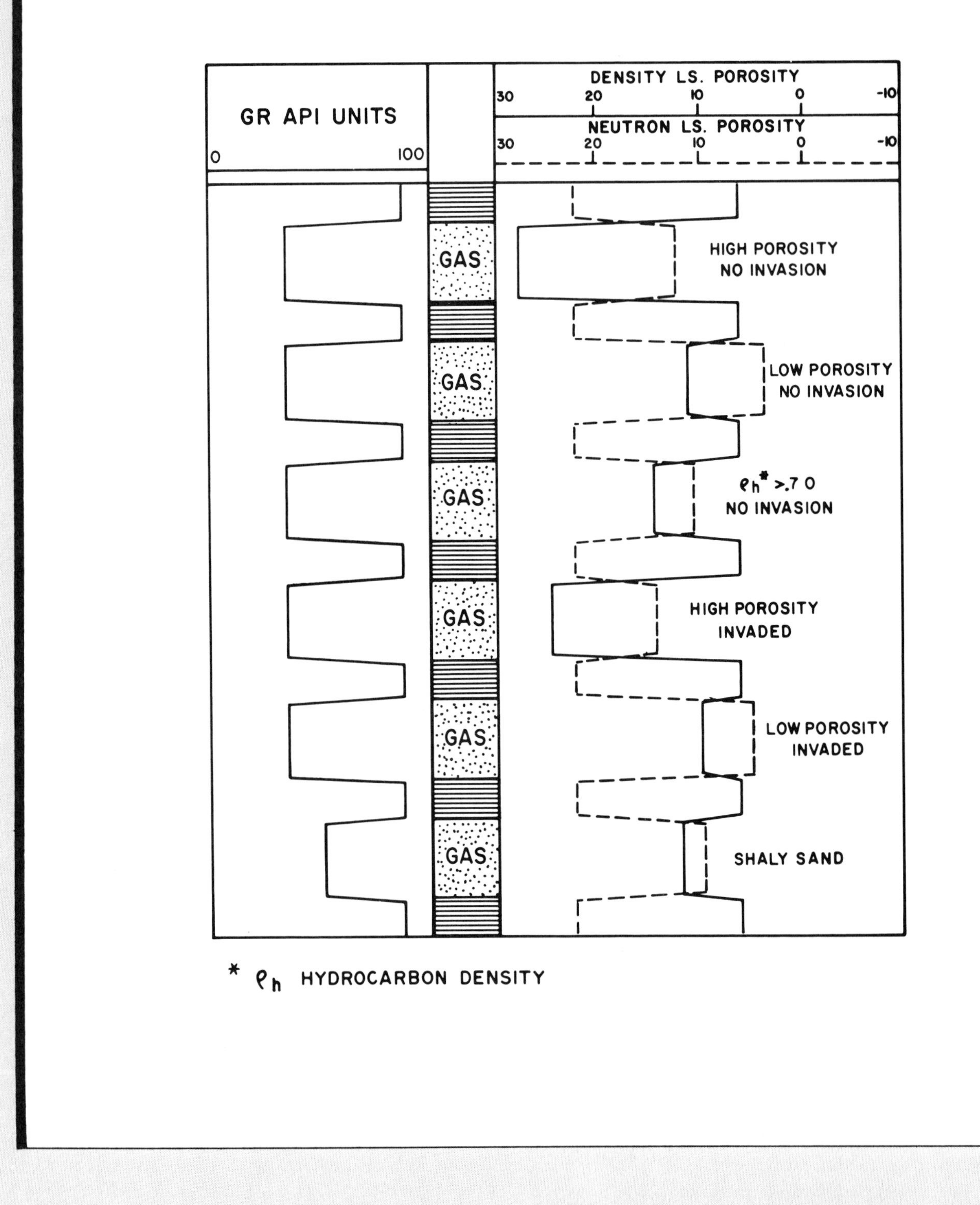

Figure 36. Schematic illustration of neutron-density responses in gas-bearing sandstones (modified after Truman et al, 1972). Generalized neutron-density log responses show how gas effect varies with depth of invasion, porosity, hydrocarbon density, and shale content.

CHAPTER V

GAMMA RAY LOGS

General

Gamma ray logs measure natural radioactivity in formations and because of this measurement, they can be used for identifying lithologies and for correlating zones. Shale-free sandstones and carbonates have low concentrations of radioactive material, and give low gamma ray readings. As shale content increases, the gamma ray log response increases because of the concentration of radioactive material in shale. However, clean sandstone (i.e. low shale content) may also produce a high gamma ray response if the sandstone contains potassium feldspars, micas, glauconite, or uranium-rich waters.

In zones where the geologist is aware of the presence of potassium feldspars, micas, or glauconite, a Spectralog** can be run in addition to the gamma ray log. The Spectralog** breaks the natural radioactivity of a formation into the different types of radioactive material: (1) thorium, (2) potassium, and (3) uranium.

If a zone has a high potassium content coupled with a high gamma ray log response, the zone may not be shale. Instead, it could be a feldspathic, glauconitic, or micaceous sandstone.

Besides their use with identifying lithologies and correlating zones, gamma ray logs provide information for calculating the volume of shale in a sandstone or carbonate. The gamma ray log is recorded in track #1 (example, Fig. 37), usually with a caliper. Tracks #2 and #3 often contain either a porosity log or a resistivity log.

Volume of Shale Calculation

Because shale is more radioactive than sand or carbonate, gamma ray logs can be used to calculate volume of shale in porous reservoirs. The volume of shale can then be applied for analysis of shaly sands (see Chapter VI).

Calculation of the gamma ray index is the first step needed to determine the volume of shale from a gamma ray log (the following formula from Schlumberger, 1974).

$$I_{GR} = \frac{GR_{log} - GR_{min}}{GR_{max} - GR_{min}}$$

Where:

I_{GR} = gamma ray index
GR_{log} = gamma ray reading of formation
GR_{min} = minimum gamma ray (clean sand or carbonate)
GR_{max} = maximum gamma ray (shale)

As an example of this calculation, pick these values from the gamma ray log in Figure 37 (they will be used in Figure 38):

GR_{log} = 28 at 13,570 ft (formation reading)
GR_{min} = 15 at 13,590 ft
GR_{max} = 128 at 13,720 ft

Then,

$$I_{GR} = \frac{28 - 15}{128 - 15} = \frac{13}{113}$$

$$I_{GR} = 0.115$$

Finally, the calculated value of the gamma ray index (I_{GR}) is located on the chart in Figure 38, and a corresponding value for volume of shale (V_{sh}) in either consolidated or unconsolidated sands is determined.

From Figure 38, and using a value for I_{GR} of 0.115, find:

V_{sh} = 0.057 older rocks (consolidated)
V_{sh} = 0.028 Tertiary rocks (unconsolidated)

The volume of shale is also calculated mathematically from the gamma ray index (I_{GR}) by the following Dresser Atlas (1979) formulas:

Older rocks, consolidated:

$$V_{sh} = 0.33\,[2^{(2 \times I_{GR})} - 1.0]$$

or, Tertiary rocks, unconsolidated:

$$V_{sh} = 0.083\,[2^{(3.7 \times I_{GR})} - 1.0]$$

Where:

V_{sh} = volume of shale
I_{GR} = gamma ray index

Review - Chapter V

1. Gamma ray logs are lithology logs that measure the natural radioactivity of a formation.
2. Because radioactive material is concentrated in shale, shale has high gamma ray readings. Shale-free sandstones and carbonates, therefore, have low gamma ray readings.
3. Gamma ray logs are used to: (1) identify lithologies; (2) correlate between formations; and (3) calculate volume of shale.

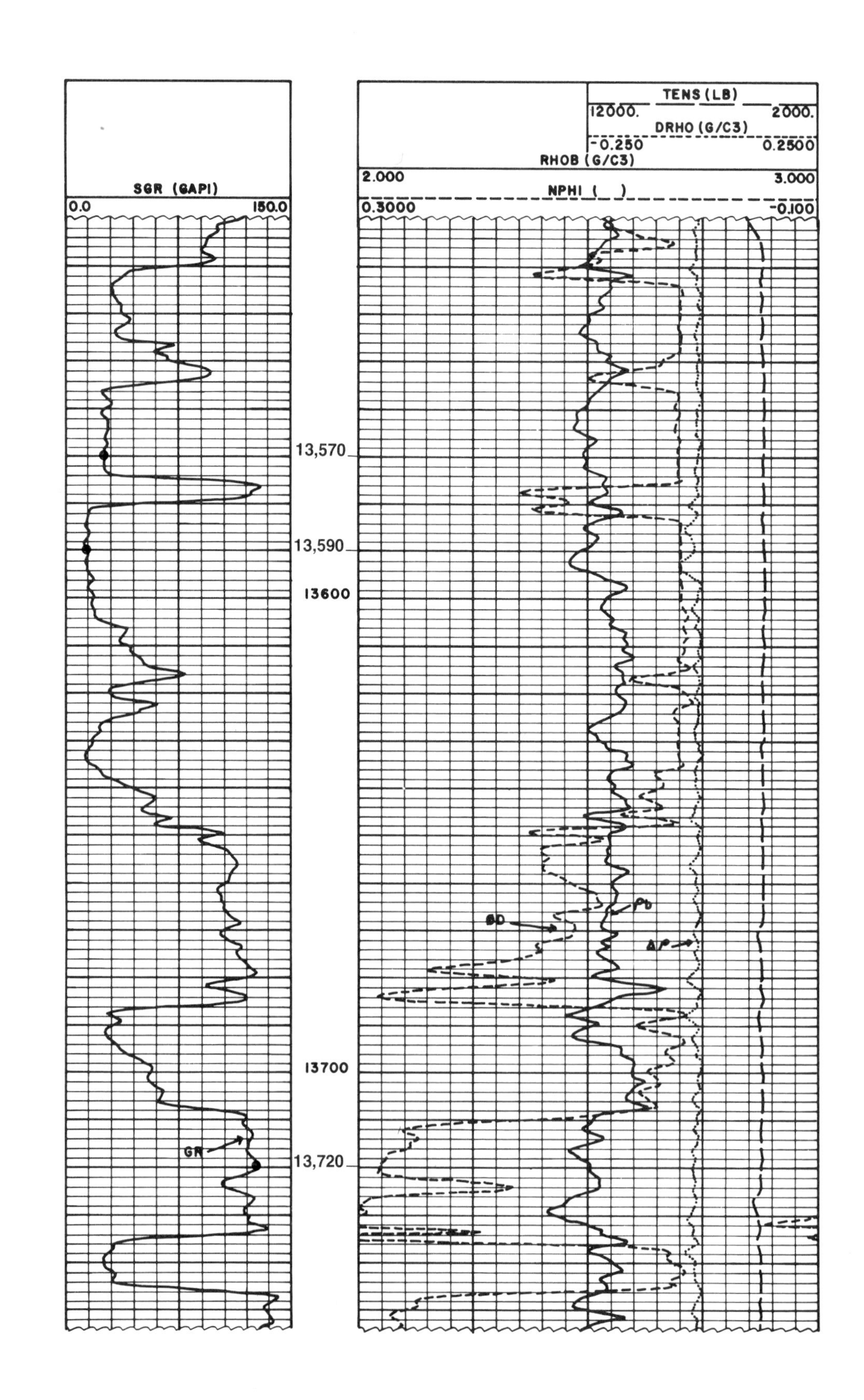
SGR (GAPI)
0.0
150.0
TENS (LB)
12000.
2000.
DRHO (G/C3)
-0.250
0.2500
RHOB (G/C3)
2.000
3.000
NPHI ( )
0.3000
-0.100
13,570
13,590
13600
13700
13,720
GR
BD
ρb
Δρ

Figure 37. Example density log with gamma ray log. This example illustrates the curves and scales of a gamma ray log, and is also used to pick values for Figure 38.

Track #1—The gamma ray log is the only one represented on this track. Note that the scale increases from left-to-right, and ranges from 0 to 150 API gamma ray units.

Tracks #2 and #3—These tracks include logs representing bulk density (ρ_b), density porosity (ϕ_D), density correction curve (Δ_ρ), and a tension curve.

Bulk density (ρ_b) is represented by a solid line and ranges from 2.0 to 3.0 gm/cc increasing from *left-to-right*. Density porosity (ϕ_D) is represented by a dashed line and ranges from -10% to $+30\%$ increasing from *right-to-left*. The correction curve (Δ_ρ) is represented by a dotted-and-dashed line and ranges from -0.25 to $+0.25$ gm/cc increasing from *left-to-right,* but only uses the right half of the track. The tension curve is a log that measures how much weight is being pulled on the wireline during logging. It is represented by a broken line and ranges from 2,000 to 12,000 lbs increasing from *right-to-left,* but it only uses the right half of the track.

At the example depth of Figure 38 (13,570 ft), pick the gamma ray reading of the formation. It is 28 gamma ray units (the scale measures in increments of 15 units; slightly less than two units from 0).

Next pick the minimum gamma ray reading from the log (13,590 ft; GR_{min} = 15 gamma ray units), and the maximum gamma ray reading from the log (13,720 ft; GR_{max} = 128 gamma ray units).

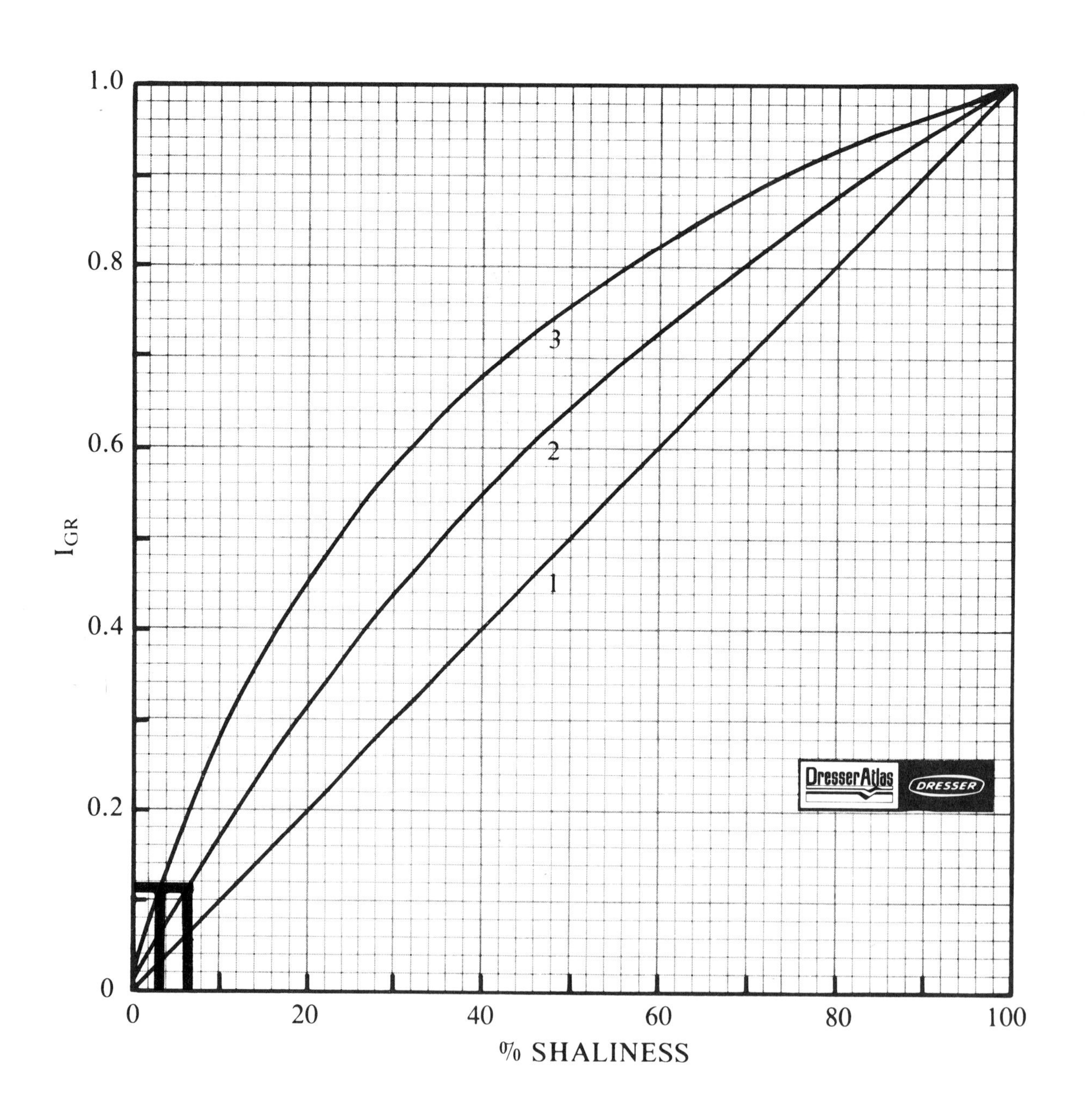
1.0
0.8
0.6
0.4
0.2
0
I_{GR}
0
20
40
60
80
100
% SHALINESS
1
2
3
DresserAtlas
DRESSER

Figure 38. Chart for correcting the gamma ray index (I_{GR}) to the volume of shale (V_{sh}).

Courtesy, Dresser Industries.
Copyright 1979, Dresser Atlas.

Given:

From Figure 37: GR_{log} = 28, GR_{min} = 15, and GR_{max} = 128. Using the formula (see text), calculate the gamma ray index (I_{GR}) to be 0.115.

Procedure:

1. Find the gamma ray index value (I_{GR}) on the vertical scale on the left (in this case I_{GR} = 0.115).
2. Follow the value horizontally to where it intersects curve 3 (representing unconsolidated, Tertiary rocks) and curve 2 (representing consolidated, older rocks). Curve 1 represents where $V_{sh} = I_{GR}$.
3. Drop to the scale at the bottom and read the values for the two intersections (measured in percent of shaliness).

Where I_{GR} = 0.115, volume of shale (V_{sh}) equals 5.7% for older, consolidated rocks (curve 2) and 2.8% for Tertiary, unconsolidated rocks (curve 3).

LOG INTERPRETATION

General

Once porosity and true resistivity of a potential zone are determined, a geologist is ready to calculate and use log parameters. Log parameters can help evaluate a zone and determine whether a well completion attempt is warranted. This section will cover some of the different methods which help establish these important parameters. Methods discussed are: S_w, S_{xo}, bulk volume water, Pickett and Hingle crossplots, and "quick look" analysis. Also, determination of log derived permeability (K_e) shaly sand analysis will be discussed.

As important as log parameters are, however, *they should not be applied to the exclusion of other data.* This statement is, perhaps obvious to the reader, but nevertheless, it can't be over-emphasized. A geologist should always consider every item of relevant data, such as drill stem tests, sample shows, mud log analysis, nearby production, etc., before making a decision to "set pipe."

Archie Equation S_w and S_{xo}

Water saturation (S_w) of a reservoir's uninvaded zone is calculated by the Archie (1942) formula.

$$S_w = \left(\frac{a}{\phi^m} \times \frac{R_w}{R_t}\right)^{1/n}$$

Where:

S_w = water saturation of the uninvaded zone (Archie method)
R_w = resistivity of formation water at formation temperature
R_t = true resistivity of formation (i.e. R_{ILd} or R_{LLd} corrected for invasion)
ϕ = porosity
a = tortuosity factor (Table 1; Chapter I)
m = cementation exponent (Table 1; Chapter I)
n = saturation exponent which varies from 1.8 to 2.5, but is normally equal to 2.0

The uninvaded zone's water saturation (S_w), determined by the Archie equation, is the most fundamental parameter used in log evaluation. But, merely knowing a zone's water saturation (S_w) will not provide enough information to completely evaluate a zone's potential productivity. A geologist must also know whether: (1) water saturation is low enough for a water-free completion, (2) hydrocarbons are moveable, (3) the zone is permeable, and (4) whether (volumetrically) there are economic, recoverable hydrocarbon reserves.

Water saturation of a formation's flushed zone (S_{xo}) is also based on the Archie equation, but two variables are changed:

$$S_{xo} = \left(\frac{a}{\phi^m} \times \frac{R_{mf}}{R_{xo}}\right)^{1/n}$$

Where:

S_{xo} = water saturation of the flushed zone
R_{mf} = resistivity of the mud filtrate at formation temperature
R_{xo} = shallow resistivity from Laterolog-8*, Micropherically Focused Log*, or Microlaterolog*
ϕ = porosity
a = tortuosity factor (Table 1; Chapter I)
m = cementation exponent (Table 1; Chapter I)
n = saturation exponent which varies from 1.8 to 2.5, but is normally equal to 2.0

Water saturation of the flushed zone (S_{xo}) can be used as an *indicator of hydrocarbon moveability.* For example, if the value of S_{xo} is much larger than S_w, then hydrocarbons in the flushed zone have probably been moved or flushed out of the zone nearest the borehole by the invading drilling fluids (R_{mf}).

Ratio Method

The Ratio Method identifies hydrocarbons from the difference between water saturations in the flushed zone (S_{xo}) and the uninvaded zone (S_w). When water saturation of the uninvaded zone (S_w) is divided by water saturation of the flushed zone (S_{xo}), the following results:

$$\left(\frac{S_w}{S_{xo}}\right)^2 = \frac{R_{xo}/R_t}{R_{mf}/R_w}$$

Where:

S_w = water saturation uninvaded zone
S_{xo} = water saturation flushed zone
R_{xo} = formation's shallow resistivity from Laterolog-8*, Microspherically Focused Log*, or Microlaterolog*
R_t = formation's true resistivity (R_{ILd} or R_{LLd} corrected for invasion)

R_{mf} = resistivity of the mud filtrate at formation temperature
R_w = resistivity of the formation water at formation temperature

When S_w is divided by S_{xo}, the formation factor ($F = a/\phi^m$) is cancelled out of the equation because formation factor is used to calculate both S_w and S_{xo} (Table 2). This can be very helpful in log analysis because, from the ratio of $(R_{xo}/R_t)/(R_{mf}/R_w)$, the geologist can determine a value for both the moveable hydrocarbon index (S_w/S_{xo}) and water saturation by the Ratio Method without knowing porosity. Therefore, a geologist can still derive useful formation evaluation log parameters even though porosity logs are unavailable.

Formulas for calculating the moveable hydrocarbon index and water saturation by the Ratio Method are:

$$\frac{S_w}{S_{xo}} = \left(\frac{R_{xo}/R_t}{R_{mf}/R_w}\right)^{1/2}$$

Where:
S_w/S_{xo} = moveable hydrocarbon index
R_{xo} = shallow resistivity from Laterolog-8*, Microspherically Focused Log*, or Microlaterolog*
R_t = true resistivity (R_{ILd} or R_{LLd} corrected for invasion)
R_{mf} = resistivity of mud filtrate at formation temperature
R_w = resistivity of formation water at formation temperature

If the ratio S_w/S_{xo} is equal to 1.0 or greater, then hydrocarbons were not moved during invasion. This is true regardless of whether or not a formation contains hydrocarbons. Whenever the ratio of S_w/S_{xo} is less than 0.7 for sandstones or less than 0.6 for carbonates, moveable hydrocarbons are indicated (Schlumberger, 1972).

To determine water saturation (S_w) by the Ratio Method, you must know the flushed zone's water saturation. In the flushed zone of formations with moderate invasion and "average" residual hydrocarbon saturation, the following relationship is normally true:

$$S_{xo} = (S_w)^{1/5}$$

Where:
S_{xo} = water saturation of the flushed zone
S_w = water saturation of the uninvaded zone

However, by substituting the above equation in the relationship:

$$\left(\frac{S_w}{S_{xo}}\right)^2 = \frac{R_{xo}/R_t}{R_{mf}/R_w}$$

The following results:

$$\frac{(S_w)^2}{(S_w)^{2/5}} = S_w^{8/5} = \frac{R_{xo}/R_t}{R_{mf}/R_w}$$

Therefore:

$$S_{wr} = \left(\frac{R_{xo}/R_t}{R_{mf}/R_w}\right)^{5/8}$$

or:

$$S_{wr} = \left(\frac{R_{xo}/R_t}{R_{mf}/R_w}\right)^{0.625}$$

Where:
S_{wr} = water saturation uninvaded zone, Ratio Method
R_{xo} = shallow resistivity from Laterolog-8*, Microspherically Focused Log*, or Microlaterolog*
R_t = true resistivity (R_{ILd} or R_{LLd} corrected for invasion)
R_{mf} = resistivity of mud filtrate at formation temperature
R_w = resistivity of formation water at formation temperature

After the geologist has calculated water saturation of the uninvaded zone by both the Archie and Ratio methods, he should compare the two values using the following observations:

1. If S_w (Archie) $\simeq S_w$ (Ratio), the assumption of a step-contact invasion profile is indicated to be correct, and all values determined (S_w, R_t, R_{xo}, and d_i) are correct.
2. If S_w (Archie) $> S_w$ (Ratio) then the value for R_{xo}/R_t is too low. R_{xo} is too low because invasion is very shallow, or R_t is too high because invasion is very deep. Also, a transition type invasion profile may be indicated and S_w (Archie) is considered a good value for S_w.
3. If S_w (Archie) $< S_w$ (Ratio) then the value for R_{xo}/R_t is too high. R_{xo} is too high because of the effect of adjacent, high resistivity beds, or R_{ILd} (R_t) is too low because R_{xo} is less than R_t. Also, an annulus type invasion profile may be indicated and/or $S_{xo} < S_w^{1/5}$. In this case a more accurate value for water saturation can be estimated using the following equation (from Schlumberger, 1977):

$$(S_w)_{COR} = S_{wa} \times \left(\frac{S_{wa}}{S_{wr}}\right)^{0.25}$$

Where:
$(S_w)_{COR}$ = corrected water saturation of the uninvaded zone
S_{wa} = water saturation of the uninvaded zone (Archie Method)
S_{wr} = water saturation of the uninvaded zone (Ratio Method)

4. If S_w (Archie) $< S_w$ (Ratio), the reservoir may be a

carbonate with moldic (i.e. oomoldic, fossil-moldic, etc.) porosity and low permeability.

Bulk Volume Water

The product of a formation's water saturation (S_w) and its porosity (ϕ) is the bulk volume of water (BVW).

$$BVW = S_w \times \phi$$

Where:

BVW = bulk volume water

S_w = water saturation of uninvaded zone (Archie equation)

ϕ = porosity

If values for bulk volume water, calculated at several depths in a formation, are constant or very close to constant, they indicate that the zone is homogeneous and at irreducible water saturation ($S_{w\ irr}$). When a zone is at irreducible water saturation, water calculated in the uninvaded zone (S_w) will not move because it is held on grains by capillary pressure. Therefore, hydrocarbon production from a zone at irreducible water saturation should be water-free (Morris and Biggs, 1967).

A formation *not* at irreducible water saturation ($S_{w\ irr}$) will exhibit wide variations in bulk volume water values. Figure 39 illustrates three crossplots of porosity (ϕ) versus $S_{w\ irr}$ for three wells from the Ordovician Red River B-zone, Beaver Creek Field, North Dakota. Note, that with increasing percentages of produced water, scattering of data points from a constant value of BVW (hyperbolic lines) occurs.

Because the amount of water a formation can hold by capillary pressure increases with decreasing grain size, the bulk volume water also increases with decreasing grain size. Table 8 illustrates the relationship of bulk volume water values to decreasing grain size and lithology.

Table 8. Bulk Volume Water as a Function of Grain Size and Lithology. A comparative chart.

Grain Size (millimeters)		Bulk Volume Water
Coarse	1.0 to 0.5 mm	0.02 to 0.025
Medium	0.5 to 0.25 mm	0.025 to 0.035
Fine	0.25 to 0.125 mm	0.035 to 0.05
Very Fine	0.125 to 0.062 mm	0.05 to 0.07
Silt	(<0.0625 mm)	0.07 to 0.09

(Modified after: Fertl and Vercellino, 1978)

CARBONATES*		
Vuggy		0.005 to 0.015
Vuggy and intercrystalline (intergranular)		0.015 to 0.025
Intercrystalline (intergranular)		0.025 to 0.04
Chalky		0.05

*Carbonate values (for BVW) are to be used as a general guide to different types of porosity.

Quick Look Methods

General—Quick look methods are helpful to the geologist because they provide "flags" which point to possible hydrocarbon zones requiring further investigation. The four quick look methods which will be discussed are: (1) R_{xo}/R_t curve, (2) R_{wa} curve, (3) conductivity derived porosity curve, and (4) R_o curve.

R_{xo}/R_t curve—The R_{xo}/R_t curve is presented in track #1 as an overlay to the spontaneous potential curve (SP). From Chapter II, remember that the SP equation is:

$$SP = -K \times \log (R_{mf}/R_w)$$

Where:

SP = spontaneous potential

K = 60 + (0.133 × formation temperature)

R_{mf} = resistivity of mud filtrate at formation temperature

R_w = resistivity of formation water at formation temperature

In water zones ($S_w = 1.0$):

$$R_{xo} = F \times R_{mf} \text{ and } R_o = F \times R_w$$

Where:

R_{mf} = resistivity of mud filtrate at formation temperature

R_{xo} = shallow resistivity

F = formation factor (i.e. a/ϕ^m)

R_w = resistivity of formation water at formation temperature

R_o = wet resistivity (i.e. resistivity of a zone 100% water saturated with water of a certain R_w. From Chapter I, $R_o = R_t$ in wet zones).

From the above equations, the SP equation can be rewritten as:

$$SP = -K \times \log (R_{xo}/R_o)$$

Where:

R_{xo} = shallow resistivity from Laterolog-8*, Microspherically Focused Log* or Microlaterolog*

R_o = wet resistivity ($R_o = R_t$ when $S_w = 100\%$)

In water-bearing zones, the measured values for R_{xo} and R_o (R_t for $S_w = 100\%$; R_{ILd} or R_{LLd}) can be used to calculate a value for SP. This calculated value for SP should duplicate the measured value of SP from the spontaneous potential log in a wet zone. The presence of hydrocarbons results in R_t values which are greater than R_o. This means that when SP is calculated from the R_{xo} and R_t values, it will be lower than the measured value of SP.

On the log, the R_{xo}/R_t curve is plotted as a dashed line. The dashed line tracks the SP curve in wet zones (see Fig. 6A, Chapter I), but deflects to the right, away from the SP curve, in hydrocarbon bearing zones (see Fig. 99, Chapter VIII).

R_{wa} Curve—The R_{wa} curve is presented in track #1 as an overlay to the SP curve, similar to the R_{xo}/R_t curve. In water-bearing zones, the Archie equation for the uninvaded zone can be rewritten as follows:

$$S_w = \sqrt{F \times (R_w/R_t)}$$

Or:

$$1.0 = \sqrt{F \times (R_w/R_t)}$$
(where S_w = 100% or 1.0)

Next, square both sides:

$$1.0 = F \times (R_w/R_t)$$

Now, solve for R_w:

$$R_{wa} = R_o/F$$
(remember: $R_t = R_o$ when S_w = 100%)

Where:
- S_w = water saturation of the uninvaded zone
- R_w = resistivity of formation water at formation temperature
- R_t = true formation resistivity ($R_t = R_o$ when S_w = 100%)
- F = formation factor (a/ϕ^m)
  - $F = 1/\phi^2$ carbonates
  - $F = 0.81/\phi^2$ consolidated sands
  - $F = 0.62/\phi^{2.15}$ unconsolidated sands
- R_{wa} = apparent water resistivity ($R_{wa} = R_w$ in water-bearing zones)

In water-bearing zones (S_w = 100%), the calculated R_{wa} value is equal to R_w. However, if hydrocarbons are present, R_t will be greater than R_o, and R_{wa} will be greater than R_w (Fertl, 1978). The R_{wa} curve is plotted as a dashed line along with the SP curve. Low R_{wa} values are recorded on the left-hand side of the log. The R_{wa} curve will deflect to the left in wet zones and to the right in hydrocarbon-bearing zones (see Fig. 79, Chapter VIII). This deflection is similar to the behavior of the R_{xo}/R_t curve for hydrocarbon or wet zones. *An advantage* of an R_{wa}curve, rather than an R_{xo}/R_t curve, is that R_{wa} values can be converted to a quantitative value for water saturation (S_w). The procedure is as follows:

$$S_w = \sqrt{R_w/R_{wa}}$$

Where:
- S_w = water saturation of the uninvaded zone
- R_w = resistivity of formation water at formation temperature
- R_{wa} = apparent formation water resistivity from R_{wa} curve
  - Note: When $R_w = R_{wa}$ then S_w = 100%

Besides the use of R_{wa} as a quick look method for hydrocarbon detection, R_{wa} can also be applied as a calculated value for formation water resistivity (R_w) in water-bearing zones. R_{wa} is a value for R_w whenever S_w equals 100%. In hydrocarbon-bearing zones, the value of R_{wa} from a water-bearing zone can be used as R_w to calculate water saturation (S_w) if both zones have a constant formation water resistivity.

Conductivity Derived Porosity Curve—The conductivity derived porosity curve is a Dresser Atlas (1975) quick look curve, plotted in track #1 along with the SP curve. Chapter III discussed how resistivity (remember that resistivity = 1,000/conductivity) assists in determining porosity. Here, resistivities of the uninvaded zone (R_t), rather than flushed zone (R_{xo}) resistivities, are applied to find resistivity porosity. The formulas are as follows:

Water-bearing zones:

$$\phi = \left(\frac{a \times R_w}{R_t}\right)^{1/m}$$
$$S_w = 100\% \text{ and } R_t = R_o$$

Hydrocarbon-bearing zones:

$$\phi = \left[\frac{a \times (R_w/R_t)}{S_w^2}\right]^{1/m}$$
$$S_w < 1.0 \text{ and } R_t > R_o$$

Where:
- ϕ = resistivity (conductivity) derived porosity
- R_w = resistivity of formation water at formation temperature
- R_o = formation resistivity when S_w = 100%
- R_t = true formation resistivity (R_{ILd} or R_{LLd})
  - Remember: $R_t = R_o$ when S_w = 100%
- S_w = water saturation of the uninvaded zone
- a = constant (Dresser Atlas uses 1.0 for carbonates and 0.62 for sandstones)
- m = constant (Dresser Atlas uses 2.0 for carbonates and 2.15 for sandstones)

The calculated resistivity porosity of water-bearing zones (S_w = 1.0) is close to true porosity. However, if hydrocarbons are present, the calculated resistivity porosity will be less than true porosity. This apparent porosity loss results because hydrocarbons have greater resistivity than formation water. The resistivity porosity formula for hydrocarbon-bearing zones corrects for the hydrocarbons in the pores when water saturation (S_w) is known.

The Dresser Atlas conductivity derived porosity curves are calculated by assuming all zones are water-bearing (i.e. $R_t = R_o$). Therefore, hydrocarbon-bearing zones show up as a loss of conductivity derived porosity because R_t is greater than R_o. A scale is constructed with higher porosity values on the left. Water-bearing zones then show up as a deflection to the left, and hydrocarbon zones appear as a deflection to the right, similar to deflections on R_{xo}/R_t and R_{wa} curves.

Like the R_{wa} curve, the conductivity derived porosity curve can be converted to a quantitative value for water saturation (S_w). The Dresser Atlas (1975) formula is:

$$S_w = \frac{\phi_w}{\phi} \times 100$$

Where:

S_w = water saturation of the uninvaded zone
ϕ_w = conductivity derived or water-filled porosity
ϕ = true porosity from a porosity log
100 = constant to convert calculated S_w to percent

R_o Curve—the R_o curve is probably one of the oldest quick look curves. But, unlike all other quick look curves, the R_o curve is plotted as an overlay on the resistivity log in tracks #2 and #3, and not on the SP curve. R_o (wet resistivity) is derived by the following formula.

$$R_o = F \times R_w$$

Where:

R_o = wet resistivity or, resistivity the formation should have when S_w = 100%
R_w = resistivity of formation water at formation temperature
F = formation factor (a/ϕ^m; Table 1)

Because an R_o curve represents wet resistivity (S_w = 100%), the difference between the R_o curve and the deep resistivity curve (R_{ILd} or R_{LLd}) indicates the presence of hydrocarbons (see Fig. 56, Chapter VIII). By using values of R_o and deep resistivity (R_{ILd} or R_{LLd}), a value for water saturation can be calculated by the formula:

$$S_w = \sqrt{R_o/R_t}$$

Where:

S_w = water saturation of the uninvaded zone
R_o = wet resistivity
R_t = true resistivity (R_{ILd}) or (R_{LLd})
Note: when $R_o = R_t$ then S_w = 100%

An R_o curve is obtained in three ways: (1) a logging engineer can plot R_o as an overlay on the resistivity log, (2) a geologist can calculate and plot R_o on the resistivity log, and (3) some density logs (example Fig. 28) have an F (formation factor) curve plotted with the bulk density log. When a formation's R_w is known, an R_o curve can be created by overlaying and tracing an F curve on the resistivity log.

Pickett Crossplot Method

The Pickett crossplot (Pickett, 1972) is one of the simplest and most effective crossplot methods in use. This technique not only gives estimates of water saturation, but can also help determine: (1) formation water resistivity (R_w), (2) cementation factor (m), and (3) matrix parameters for porosity logs (Δt_{ma} and ρ_{ma}).

The Pickett method is based on the observation that true resistivity (R_t) is a function of porosity (ϕ), water saturation (S_w), and cementation factor (m). A Pickett crossplot is developed by plotting porosity values with deep resistivity (R_{ILd} or R_{LLd}) values on two-by-three cycle log-log paper (Fig. 40). On the plot, a zone with constant R_w, m, and S_w equal to 100% will have data points plotted along a single, straight-line-trend (Fig. 40). This straight-line-trend represents the R_o (wet resistivity) line. The slope of the R_o line is equal to cementation factor (m). Data plotted above the R_o line represent water saturation values less than 100%. The geologist *must* remember that data points plotted above the R_o line only represent water saturation less than 100% *when R_w and m are both constant.* A value for R_w can be obtained from a Pickett Crossplot (see Figure 40 for the procedure).

Water saturation (S_w can be quantified from the Pickett crossplot method by remembering that $S_w = (R_o / R_t)^{1/2}$. A porosity value of 10% (0.10) will have a wet resistivity (R_o) value of 5.6 ohms (Fig. 40). The values of various water saturation lines (Fig. 40), parallel to the R_o line, are determined as follows:

Porosity	R_o	R_t	$S_w = \sqrt{(R_o/R_t)}$
0.10	5.6	$2 \times R_o$ = 11.2	71%
0.10	5.6	$4 \times R_o$ = 22.4	50%
0.10	5.6	$6 \times R_o$ = 33.6	41%
0.10	5.6	$8 \times R_o$ = 44.8	35%
0.10	5.6	$14 \times R_o$ = 78.4	27%
0.10	5.6	$20 \times R_o$ = 112.0	22%

After you determine the R_o line (S_w = 100%), you can plot the lower water saturation values (see above listing) parallel to the R_o line. Your next step is plotting on the crossplot actual values from the zone you are interested in. This will give you a "quick look" assessment of a zone's water saturation.

As an example, given: $\phi = 0.21$, $R_o = 1.5$, $R_t = 40$. By the formula:

$$S_w = \sqrt{R_o/R_t}$$

Then a value for S_w = 19.4% is calculated. This is already plotted on Figure 40. As other points are added, you will have a better picture of the range of water saturations for the well.

In addition to plotting true porosity (ϕ) versus deep resistivity (R_{ILd} or R_{LLd}) on a Pickett crossplot, the following can also be plotted on the vertical (or y) axis:

$$\Delta t - \Delta t_{ma}$$

Where:
Δt = interval transit time of formation
Δt_{ma} = interval transit time of matrix

$$\rho_{ma} - \rho_b$$

Where:
ρ_{ma} = density of matrix
ρ_b = bulk density of formation

$$\phi_{snp} \text{ or } \phi_{cnl}$$

Where:
ϕ_{snp} = sidewall neutron porosity, limestone ϕ units
ϕ_{cnl} = compensated neutron porosity, limestone ϕ units

When $\Delta t - \Delta t_{ma}$ or $\rho_{ma} - \rho_b$ are plotted versus R_t (R_{ILd} or R_{LLd}), a value for formation matrix (Δt_{ma} or ρ_{ma}) must be used. Pickett (1972) suggests that whenever Δt_{ma} or ρ_{ma}, selected for the log-log crossplot, is incorrect, the R_o line for $\Delta t - \Delta t_{ma}$ or $\rho_{ma} - \rho_b$ versus R_t plot will not plot as a straight line (Fig. 40), but will curve. A geologist should try several matrix values (Δt_{ma} or ρ_{ma}) until the R_o line is straight. By such trial and error, a correct matrix parameter (Δt_{ma} or ρ_{ma}) for a formation is determined. Determining matrix parameters (Δt_{ma} or ρ_{ma}) is an additional benefit of the Pickett crossplot technique.

Hingle Crossplot

The oldest of the resistivity versus porosity crossplot methods, which can be used to determine water saturation (S_w), is the Hingle (1959) crossplot. As in other crossplot techniques, a significant benefit of Hingle's technique is that, even if matrix properties (ρ_{ma} or Δt_{ma}) of a reservoir are unknown, you can still determine a value for water saturation (S_w). This is also true if a reservoir's water resistivity (R_w) is unknown. The procedure for constructing a Hingle crossplot to determine water saturation is:

1. Select the correct crossplot graph paper (Fig. 41, sandstones; Fig. 42, carbonates).
2. Scale the x axis on a linear scale, using values taken from a porosity log (Δt, ρ_b, or ϕ_N; example, Table 9). Be sure to select the scale so that the maximum porosity log values will plot on the graph paper (Fig. 43).
3. Plot deep resistivity values (R_{ILd} or R_{LLd}; example, Table 9) on the y axis (Fig. 43) versus the porosity log data (Δt, ρ_b, or ϕ_N). The resistivity scale can be changed, by any order of magnitude, to fit the log data without changing the validity of the graph paper.

Table 9. Density - Resistivity Crossplot Data, Morrow Sandstones, Cimmarron County, Oklahoma.

No.	Depth (ft)	ρ_b(gm/cc)	R_t
1	4,400	2.38	1.7
2	4,402	2.44	2.1
3	4,410	2.35	1.3
4	4,414	2.42	1.6
5	4,426	2.42	1.8
6	4,430	2.33	1.0
7	4,438	2.30	0.9
8	4,536	2.30	40
9	4,540	2.30	45
10	4,546	2.30	40

4. Construct a straight line through the most northwesterly points (Fig. 43), and extrapolate this line until it intersects the x axis (Fig. 43; $\phi = 0$; and $R_t = \infty$). The straight line defines $S_w = 1.0$, and is called the R_o line.
5. At the intersection point of the x axis and the R_o line (where S_w = 100%; example Fig. 43), determine the matrix value (ρ_{ma} = 2.70 gm/cc) and scale the x axis in porosity units (Fig. 43).
6. Calculate a value for R_w from any corresponding set of ϕ and R_o values by the formula $R_w = R_o/F$. On Figure 43, $R_o = 6.0$, ϕ = 10%, and $F = 0.62/\phi^{2.15}$ (F = 87.6), Therefore:

$$R_w = R_o/F$$
$$R_w = 6.0/87.6$$
$$R_w = 0.068$$

7. Determine lines of constant S_w based on the formula: $S_w = \sqrt{(R_o/R_t)}$ for any given ϕ value. On Figure 43:

ϕ	R_o	R_t	$S_w = \sqrt{(R_o/R_t)}$
.10	6.0	$2 \times R_o$ = 12	71%
.10	6.0	$4 \times R_o$ = 24	50%
.10	6.0	$11 \times R_o$ = 66	30%
.10	6.0	$25 \times R_o$ = 150	20%

Remember that all lines of constant S_w must be constructed so that they converge at the matrix point ($\phi = 0$ and $R_t = \infty$; Fig. 43). The lines of constant S_w (Fig. 43) are only valid if the R_w is constant.

8. Evaluate S_w values for all the points plotted on the crossplot; make sure the plotted data points are numbered (Table 9 and Fig. 43) to avoid confusion. In Figure 43, the water-bearing Morrow sands from 4,400 to 4,438 ft (points

1 to 7) were used to establish the R_o line ($S_w = 1.0$). The hydrocarbon-bearing Morrow sand from 4,536 to 4,546 ft (points 8 to 10) plot below the 20% water saturation ($S_w = 0.2$) line indicating the sand is productive.

The limitation imposed by evaluating a log from a crossplot is that a relatively large range of porosity values in water zones is required to properly define the R_o line (Fig. 43) and determine resistivity of formation water (R_w). Also, the lithology and mud filtrate must stay fairly constant in the interval being evaluated.

Permeability From Logs

Log-derived permeability formulas are only valid for *estimating* permeability in formations at irreducible water saturation ($S_{w\,irr}$; Schlumberger, 1977). When a geologist evaluates a formation by using log-derived permeability formulas, the permeability values, if possible, should be compared with values of nearby producing wells from the same formation. You can make productivity estimates based on log derived permeabilities if the formation evaluated is compared with *both* good and poor production histories in these nearby wells. By using comparisons of log-derived permeabilities from several wells, a geologist is not using an absolute value for log derived permeability.

Two methods for calculating log-derived permeability are discussed here: the Wyllie and Rose (1950) formulas and the Coates and Dumanoir (1973) formula. Before these formulas can be applied, a geologist must first determine whether or not a formation is at irreducible water saturation.

Whether or not a formation is at irreducible water saturation depends upon bulk volume water ($BVW = S_w \times \phi$) values. When a formation's bulk volume water values are constant (Fig. 39), a zone is at irreducible water saturation. If the values are not constant, a zone is not at irreducible water saturation (Fig. 39).

The Wyllie and Rose (1950) method for determining permeability utilizes a chart (Fig. 44), or the following two formulas:

$$K^{1/2} = 250 \times \phi^3/S_{w\,irr} \text{ (medium gravity oils)}$$
$$K^{1/2} = 79 \times \phi^3/S_{w\,irr} \text{ (dry gas)}$$

Where:

$K^{1/2}$ = square root of permeability; therefore: K is equal to permeability in millidarcies
ϕ = porosity
$S_{w\,irr}$ = water saturation (S_w) of a zone at irreducible water saturation

A more modern, but also more complex, method for calculating permeability is the Coates and Dumanoir (1973) formula. Unlike the Wyllie and Rose (1950) formulas, hydrocarbon density is put into the equation, instead of adjusting by constants for the effect hydrocarbon density has on permeability (Wyllie and Rose, 1950; formulas). The following data are required to calculate permeability by the Coates and Dumanoir (1973) formula.

R_w = formation water resistivity at formation temperature
$R_{t\,irr}$ = true formation resistivity from a formation at irreducible water saturation
ρ_h = hydrocarbon density in gm/cc
ϕ = porosity

A first step in the Coates and Dumanoir (1973) permeability formula is calculation of values for two constants: C and W.

$$C = 23 + 465\rho_h - 188\rho_h^2$$

Where:

C = constant in Coates and Dumanoir (1973) permeability formula
ρ_h = hydrocarbon density in gm/cc

$$W^2 = (3.75 - \phi) + \left\{\frac{[\log(R_w/R_{t\,irr}) + 2.2]^2}{2.0}\right\}$$

Where:

W = constant in Coates and Dumanoir (1973) permeability formula
ϕ = porosity
R_w = formation water resistivity at formation temperature
$R_{t\,irr}$ = deep resistivity from a zone at irreducible water saturation ($S_{w\,irr}$)

Once determined, the constants C and W can be used to calculate permeability (Coates and Dumanoir, 1973).

$$K^{1/2} = \frac{C \times \phi^{2W}}{W^4 \times (R_w/R_{t\,irr})}$$

Where:

$K^{1/2}$ = square root of permeability; therefore: K equals permeability in millidarcies (md)
C = constant based on hydrocarbon density
W = constant
ϕ = porosity
$R_{t\,irr}$ = deep resistivity from a zone at irreducible water saturation ($S_{w\,irr}$)
R_w = formation water resistivity at formation temperature

Shaly Sand Analysis

The presence of shale (i.e. clay minerals) in a reservoir can cause erroneous values for water saturation and porosity derived from logs. These erroneous values are not limited to sandstones, but also occur in limestones and dolomites.

Whenever shale is present in a formation, all the porosity tools (sonic, density, and neutron) will record too high a porosity. The only *exception* to this is the density log. It will not record too high a porosity if the density of shale is equal to or greater than the reservoir's matrix density. Also, the presence of shale in a formation will cause the resistivity log to record too low a resistivity. Hilchie (1978) notes that the most significant effect of shale in a formation is to reduce the resistivity contrast between oil or gas, and water. The net result is that if enough shale is present in a reservoir, it may be very difficult, or perhaps impossible, to determine if a zone is productive. Hilchie (1978) suggests that for shale to significantly affect log-derived water saturations, shale content must be greater than 10 to 15%.

Remember that all shaly sandstone formulas reduce the water saturation value from the value that would be calculated if shale effect was ignored. However, this lowering of water saturation can be a problem in log evaluation, because, *if a geologist overestimates shale content, a water-bearing zone may calculate like a hydrocarbon zone*.

The first step in shaly sand analysis is to determine the volume of shale from a gamma ray log (see Chapter V). Volume of shale from a gamma ray log is determined by the chart (Fig. 38) or by the following formulas (Dresser Atlas, 1979):

Older rocks (consolidated):

$$V_{sh} = 0.33\,[2^{(2 \times I_{GR})} - 1.0]$$

Tertiary rocks (unconsolidated):

$$V_{sh} = 0.083\,[2^{(3.7 \times I_{GR})} - 1.0]$$

Where:

V_{sh} = volume of shale
I_{GR} = gamma ray index

$$I_{GR} = \frac{GR_{log} - GR_{min}}{GR_{max} - GR_{min}}$$

Where:

GR_{max} = gamma ray maximum (shale zone)
GR_{min} = gamma ray minimum (clean sand)
GR_{log} = gamma ray log (shaly sand)

After the volume of shale (V_{sh}) is determined, it can then be used to correct the porosity log for shale effect. The formulas for correcting the sonic, density, and Combination Neutron-Density logs for volume of shale are:

Sonic Log (Dresser Atlas, 1979):

$$\phi_{sonic} = \left(\frac{\Delta t_{log} - \Delta t_{ma}}{\Delta t_f - \Delta t_{ma}} \times \frac{100}{\Delta t_{sh}}\right) - V_{sh}\left(\frac{\Delta t_{sh} - \Delta t_{ma}}{\Delta t_f - \Delta t_{ma}}\right)$$

Where:

ϕ_{sonic} = sonic log derived porosity corrected for shale
Δt_{log} = interval transit time of formation
Δt_{ma} = interval transit time of the formation's matrix (Table 6)
Δt_f = interval transit time of fluid (189 for fresh mud and 185 for salt mud)
Δt_{sh} = interval transit time of adjacent shale
V_{sh} = volume of shale

Density Log (Dresser Atlas, 1979):

$$\phi_{Den} = \left(\frac{\rho_{ma} - \rho_b}{\rho_{ma} - \rho_f}\right) - V_{sh}\left(\frac{\rho_{ma} - \rho_{sh}}{\rho_{ma} - \rho_f}\right)$$

Where:

V_{sh} = volume of shale
ϕ_{Den} = density log derived porosity corrected for shale
ρ_{ma} = matrix density of formation
ρ_b = bulk density of formation
ρ_f = fluid density (1.0 for fresh mud and 1.1 for salt mud)
ρ_{sh} = bulk density of adjacent shale

Combination Neutron-Density Log (Schlumberger, 1975):

$$\phi_{N\,corr} = \phi_N - \left[\left(\frac{\phi_{N\,clay}}{0.45}\right) \times 0.30 \times V_{sh}\right]$$

$$\phi_{D\,corr} = \phi_D - \left[\left(\frac{\phi_{N\,clay}}{0.45}\right) \times 0.13 \times V_{sh}\right]$$

$$\phi_{N\text{-}D} = \sqrt{\frac{\phi_{N\,corr}^2 + \phi_{D\,corr}^2}{2.0}}$$

Where:

$\phi_{N\,corr}$ = neutron porosity corrected for shale
$\phi_{D\,corr}$ = density porosity corrected for shale
V_{sh} = volume of shale
$\phi_{N\,clay}$ = neutron porosity of adjacent shale
ϕ_N = neutron porosity uncorrected for shale
ϕ_D = density porosity uncorrected for shale
$\phi_{N\text{-}D}$ = neutron-density porosity corrected for shale

Next, after the volume of shale has been determined and the log derived porosity has been corrected for volume of shale, the water saturation can be calculated. Three of the more commonly used shaly sand equations are:

(Simandoux, 1963):

$$S_w = \left(\frac{0.4 \times R_w}{\phi^2}\right) \times \left[-\frac{V_{sh}}{R_{sh}} + \sqrt{\left(\frac{V_{sh}}{R_{sh}}\right)^2 + \frac{5\phi^2}{R_t \times R_w}}\right]$$

(Fertl, 1975; where a = 0.25 Gulf Coast; a = 0.35 Rocky Mountains):

$$S_w = \frac{1}{\phi} \times \left[\sqrt{\frac{R_w}{R_t} + \left(\frac{a \times V_{sh}}{2}\right)^2} - \frac{a \times V_{sh}}{2}\right]$$

FLOW CHART FOR LOG INTERPRETATION

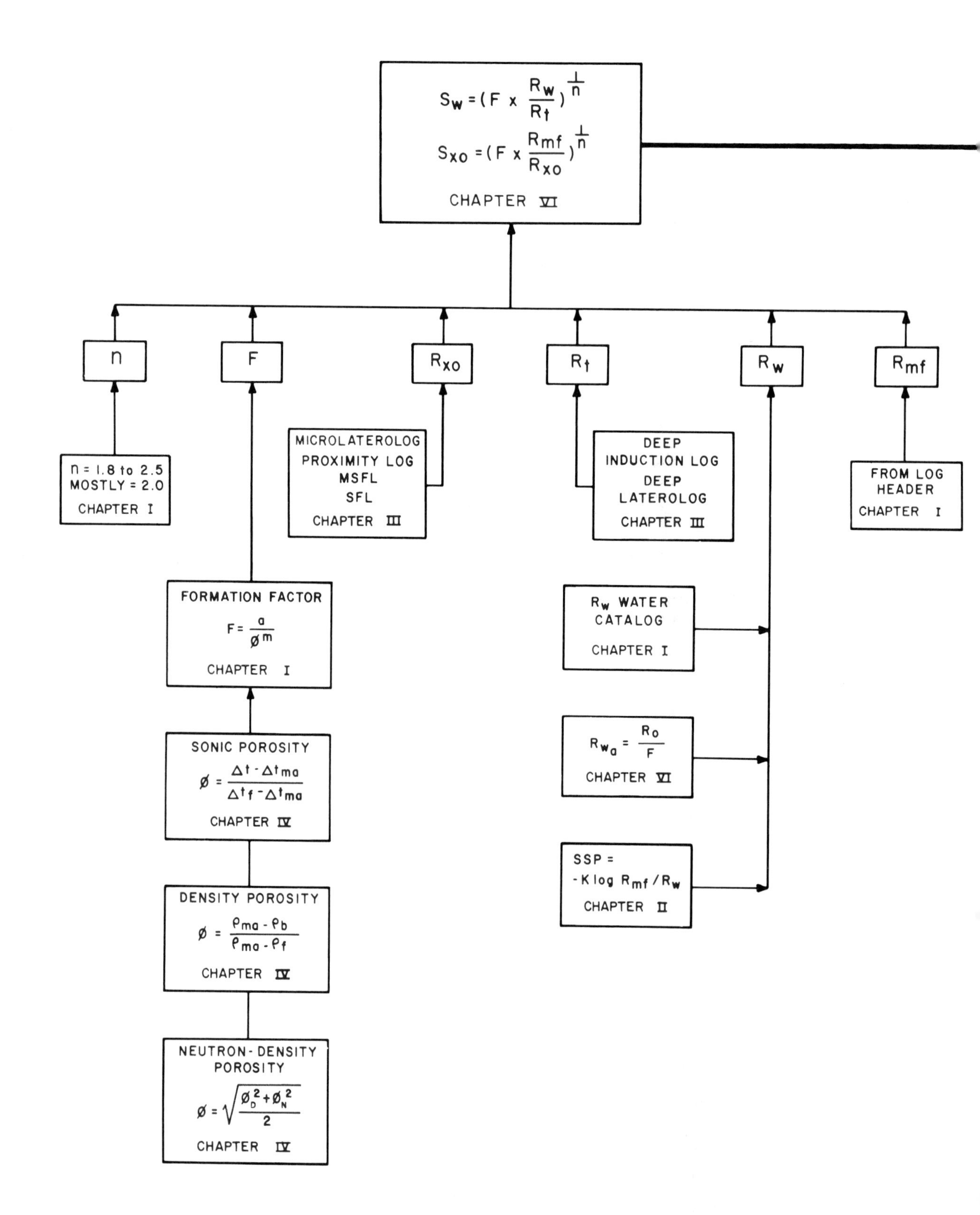
$S_w = (F \times \frac{R_w}{R_t})^{\frac{1}{n}}$
$S_{xo} = (F \times \frac{R_{mf}}{R_{xo}})^{\frac{1}{n}}$
CHAPTER VI
n
F
R_{xo}
R_t
R_w
R_{mf}
n = 1.8 to 2.5
MOSTLY = 2.0
CHAPTER I
MICROLATEROLOG
PROXIMITY LOG
MSFL
SFL
CHAPTER III
DEEP
INDUCTION LOG
DEEP
LATEROLOG
CHAPTER III
FROM LOG HEADER
CHAPTER I
FORMATION FACTOR
$F = \frac{a}{\phi^m}$
CHAPTER I
SONIC POROSITY
$\phi = \frac{\Delta t - \Delta t_{ma}}{\Delta t_f - \Delta t_{ma}}$
CHAPTER IV
DENSITY POROSITY
$\phi = \frac{\rho_{ma} - \rho_b}{\rho_{ma} - \rho_f}$
CHAPTER IV
NEUTRON-DENSITY POROSITY
$\phi = \sqrt{\frac{\phi_D^2 + \phi_N^2}{2}}$
CHAPTER IV
R_w WATER CATALOG
CHAPTER I
$R_{wa} = \frac{R_o}{F}$
CHAPTER VI
SSP =
$-K \log R_{mf}/R_w$
CHAPTER II

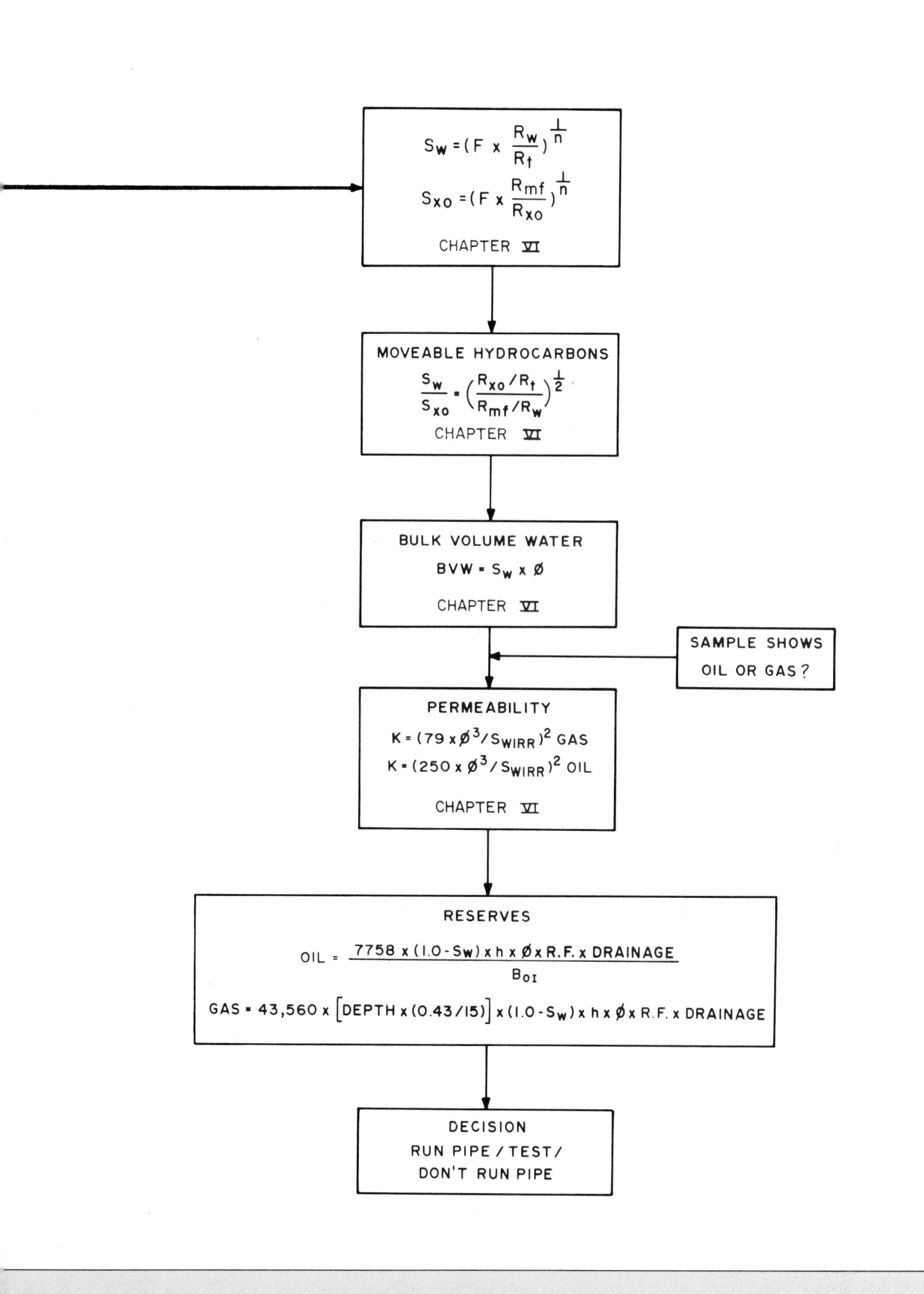
$S_w = (F \times \frac{R_w}{R_t})^{\frac{1}{n}}$
$S_{xo} = (F \times \frac{R_{mf}}{R_{xo}})^{\frac{1}{n}}$
CHAPTER VI
MOVEABLE HYDROCARBONS
$\frac{S_w}{S_{xo}} = \left(\frac{R_{xo}/R_t}{R_{mf}/R_w}\right)^{\frac{1}{2}}$
CHAPTER VI
BULK VOLUME WATER
$BVW = S_w \times \phi$
CHAPTER VI
SAMPLE SHOWS OIL OR GAS?
PERMEABILITY
$K = (79 \times \phi^3/S_{WIRR})^2$ GAS
$K = (250 \times \phi^3/S_{WIRR})^2$ OIL
CHAPTER VI
RESERVES
$OIL = \frac{7758 \times (1.0 - S_w) \times h \times \phi \times R.F. \times DRAINAGE}{B_{oi}}$
$GAS = 43{,}560 \times [DEPTH \times (0.43/15)] \times (1.0 - S_w) \times h \times \phi \times R.F. \times DRAINAGE$
DECISION
RUN PIPE / TEST / DON'T RUN PIPE

(Schlumberger, 1975)

$$S_w = \frac{-\frac{V_{sh}}{R_{sh}} + \sqrt{\left(\frac{V_{sh}}{R_{sh}}\right)^2 + \frac{\phi^2}{0.2 \times R_w \times (1.0 - V_{sh}) \times R_t}}}{\frac{\phi^2}{0.4 \times R_w \times (1.0 - V_{sh})}}$$

Where:

S_w = water saturation uninvaded zone corrected for volume of shale
R_w = formation water resistivity at formation temperature
R_t = true formation resistivity
ϕ = porosity corrected for volume of shale
V_{sh} = volume of shale
R_{sh} = resistivity of adjacent shale

A major problem encountered in shaly sand analysis is determining a resistivity value for shale in a formation. The percentage of shale is not the critical factor, rather, it is clay's cation exchange capacity (Hilchie, 1978), because cation exchange capacity greatly affects resistivity of the clay. Kaolinite and chlorite have extremely low cation exchanges; illite and montmorillonite have high cation exchanges. Therefore, montmorillonite and illite lower resistivity much more than kaolinite and chlorite. In shaly sand analysis, a geologist must make an assumption that resistivity of an adjacent shale (R_{sh}) is the same as resistivity of shale in the formation. *This assumption is not always correct.*

Most shaly sand interpretation problems occur in formations with R_w values which are not too salty (NaCL 20,000 ppm, or $R_w = 0.3$ at 80°; Hilchie, 1978). Where formation water is very salty, shale has less effect on the formation's resistivity. Therefore, calculated water saturations, without correction for shale, are close to true formation water saturation.

This chapter discussed several log interpretation techniques. These techniques are based on many of the formulas already presented in the text. What the formulas are, and where they are found is summarized on the log interpretation flow chart included at the end of this chapter.

Review - Chapter VI

1. The Archie equation is used to calculate a formation's water saturation in both the invaded (S_{xo}) and uninvaded (S_w) zones.
2. The Ratio Method for determining water saturation (S_w) does not require a value for porosity (ϕ).
3. The ratio between the water saturations in the invaded (S_{xo}) and uninvaded (S_w) zones (i.e. S_w/S_{xo}) can be used as an index for the degree of hydrocarbon moveability.
4. Bulk volume water (BVW) is important because it indicates when a reservior is at irreducible water saturation ($S_{w\,irr}$).
5. Quick look methods are important because they provide "flags" which indicate zones of potential interest.
6. Hingle and Pickett crossplot techniques are simple and rapid methods for determining: (1) a formation's matrix; (2) a formation's water resistivity (R_w); (3) a formation's water saturation (S_w); and (4) a formation's cementation factor (m).
7. Log derived data can be used to estimate permeability of a formation.
8. The flow chart included in this chapter will help you review the steps used in log interpretation and will also provide an index of where different pieces of information are located.

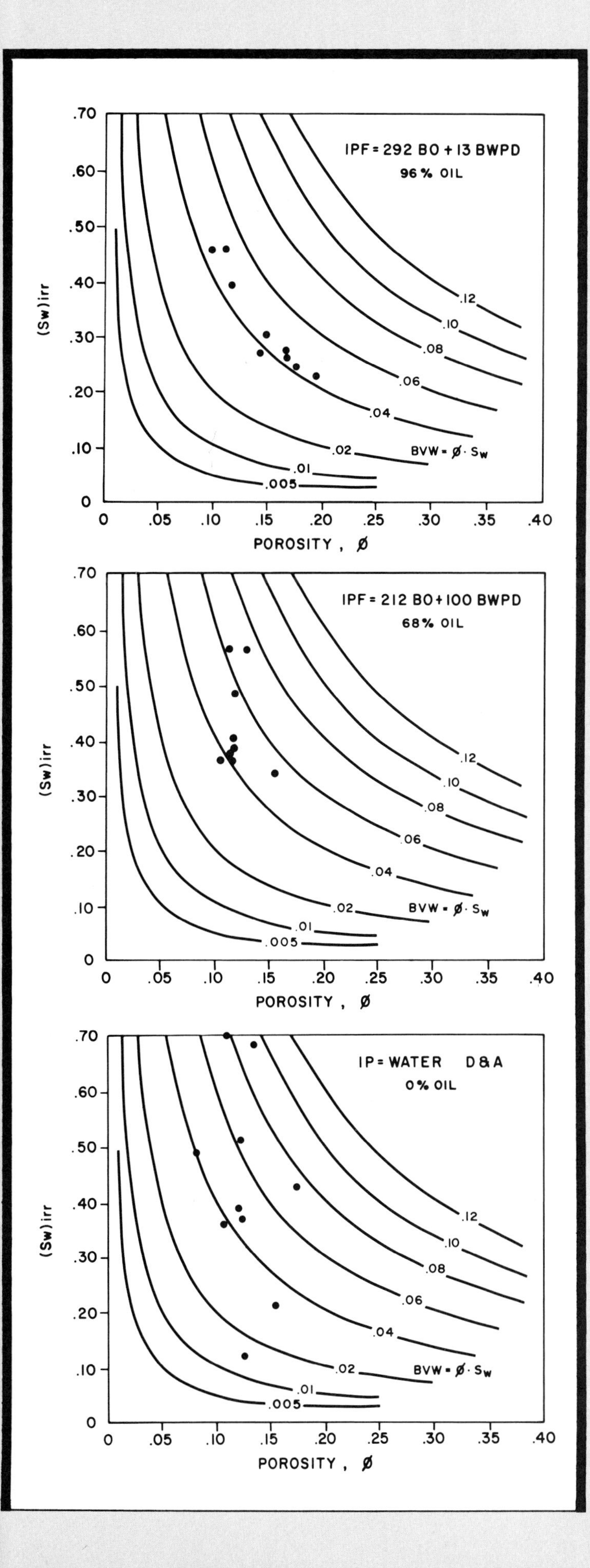

Figure 39. Crossplot of porosity versus water saturation used to determine bulk volume water. Because production of water in a well can affect a prospect's economics, it is important to know the bulk volume water and whether the formation is at irreducible water saturation ($S_{w\ irr}$).

This crossplot example is taken from the Ordovician Red River B-zone, Beaver Creek Field, Golden Valley County, North Dakota (after Jaafar, 1980).

When values of bulk volume water plot along hyperbolic lines or, in other words, are constant or close to constant, the formation is homogeneous and close to irreducible water saturation ($S_{w\ irr}$). Remember, a reservoir at $S_{w\ irr}$ will not produce water. Note that in Figure 39A (top diagram) the bulk volume water values are close to constant (parallel to the 0.04 hyperbolic line) and the formation produces 96% oil.

As the amount of formation water increases, the bulk volume water values become scattered from the hyperbolic lines and the formation has more water than it can hold by capillary pressure. Thus more water is produced relative to oil. Figure 39B (middle diagram) shows a well producing 68% oil, and Figure 39C (lower diagram) shows a well producing 0% oil (all water). Note the scatter of crossplot values from the hyperbolic lines in Figures 39B and 39C.

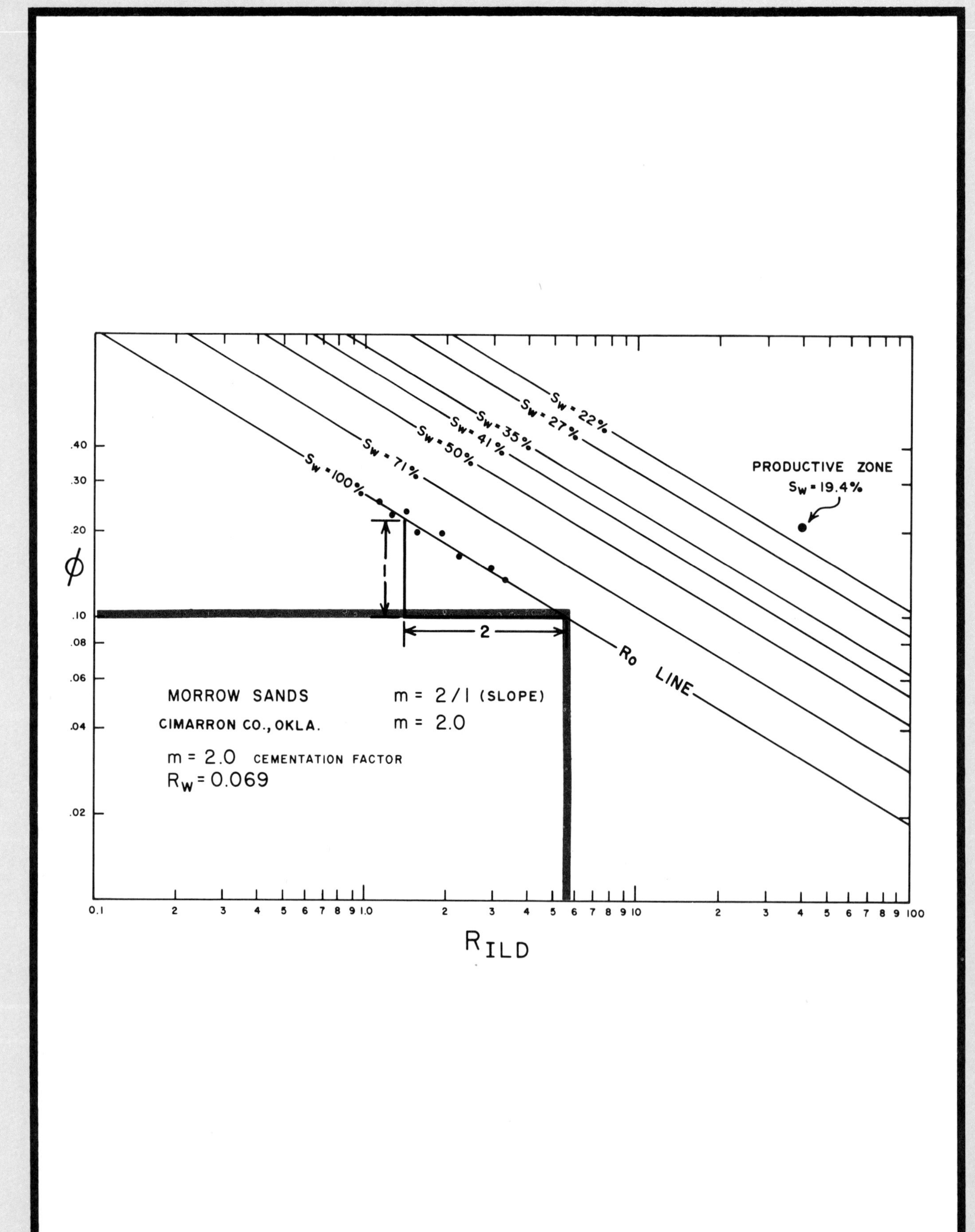

S_w = 22%
S_w = 27%
S_w = 35%
S_w = 41%
S_w = 50%
S_w = 71%
S_w = 100%
PRODUCTIVE ZONE
S_w = 19.4%
R_o LINE
2
.40
.30
.20
ϕ
.10
.08
.06
.04
.02
MORROW SANDS
CIMARRON CO., OKLA.
m = 2/1 (SLOPE)
m = 2.0
m = 2.0 CEMENTATION FACTOR
R_W = 0.069
0.1 2 3 4 5 6 7 8 9 1.0 2 3 4 5 6 7 8 9 10 2 3 4 5 6 7 8 9 100
R_{ILD}

Figure 40. Example of a resistivity versus porosity (Pickett) crossplot. Example taken from the Morrow sandstone, Cimarron County, Oklahoma.

Use the chart to find wet resistivity (R_o) which can be used to compute R_w.

Given: Porosity (ϕ) equals 10%; cementation factor (m) is determined by the slope of the R_o line (see chart) and is equal to 2; formation factor (F) is equal to $0.81/\phi^m$ (see Table 1).

Procedure:

1. Find the porosity value (10%) on the left-hand scale.
2. Follow the value horizontally until it intersects the sloping R_o line.
3. Follow the value vertically down from the intersection to the R_{ILd} scale at the bottom, and read the value of R_o. In this case, R_o equals 5.6 ohms.

In computing R_w from R_o, remember that:

$R_w = R_o/F$ (see text under heading R_{wa} Curve)

$R_w = 5.6/81$

$R_w = 0.069$ at formation temperature

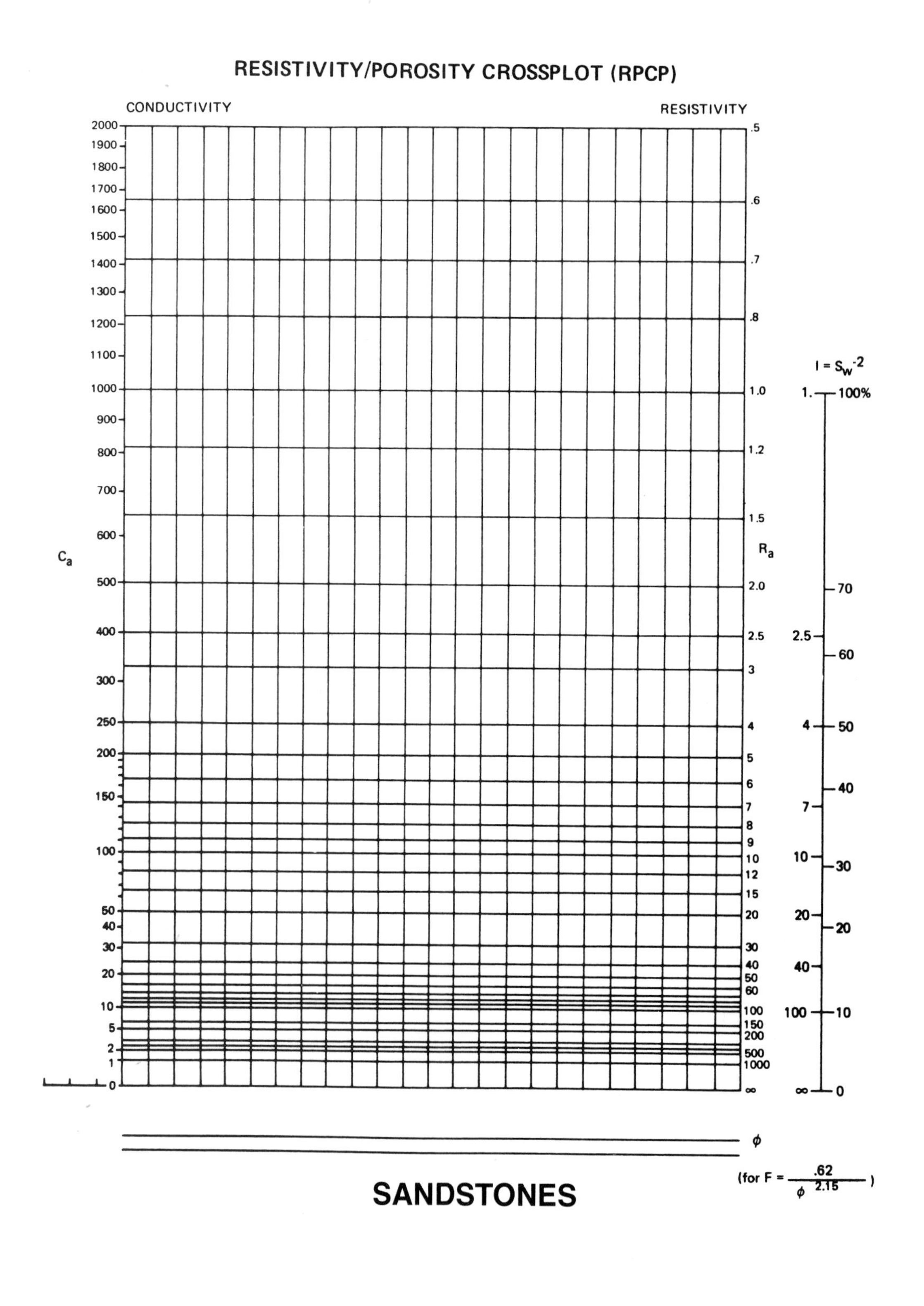
RESISTIVITY/POROSITY CROSSPLOT (RPCP)
CONDUCTIVITY
RESISTIVITY
C_a
R_a
2000
1900
1800
1700
1600
1500
1400
1300
1200
1100
1000
900
800
700
600
500
400
300
250
200
150
100
50
40
30
20
10
5
2
1
0
.5
.6
.7
.8
1.0
1.2
1.5
2.0
2.5
3
4
5
6
7
8
9
10
12
15
20
30
40
50
60
100
150
200
500
1000
∞
$I = S_w^{-2}$
1.
100%
70
2.5
60
4
50
40
7
10
30
20
20
40
100
10
∞
0
ϕ
SANDSTONES
(for $F = \frac{.62}{\phi^{2.15}}$)

Figure 41. Example of a resistivity-versus-porosity (Hingle) crossplot. Note that this crossplot is for use in plotting *sandstones*. A similar, but separate crossplot is used for plotting carbonates (Figure 42).

Courtesy, Dresser Industries.
Copyright 1979, Dresser Atlas.

This crossplot example was intentionally left blank so it can be used by the reader to construct a Hingle plot.

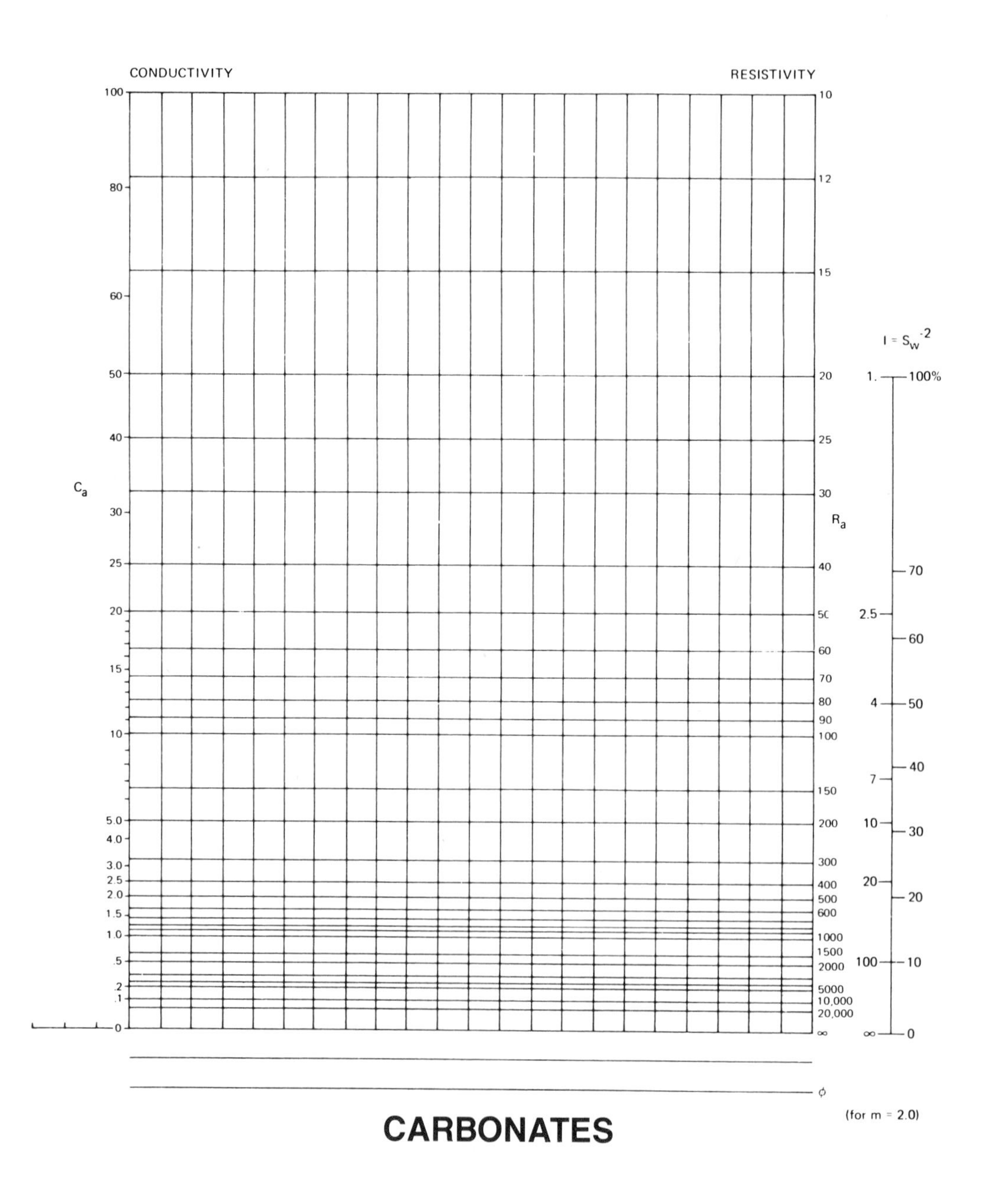
RESISTIVITY/POROSITY CROSSPLOT (RPCP)
CONDUCTIVITY
RESISTIVITY
C_a
R_a
100
80
60
50
40
30
25
20
15
10
5.0
4.0
3.0
2.5
2.0
1.5
1.0
.5
.2
.1
0
10
12
15
20
25
30
40
50
60
70
80
90
100
150
200
300
400
500
600
1000
1500
2000
5000
10,000
20,000
∞
$I = S_w^{-2}$
1.
2.5
4
7
10
20
100
∞
100%
70
60
50
40
30
20
10
0
φ
(for m = 2.0)
CARBONATES

Figure 42. Example of a resistivity-versus-porosity (Hingle) crossplot. Note that this crossplot is for use in plotting *carbonates*. A similar, but separate crossplot is used for plotting sandstone (Figure 41).

Courtesy, Dresser Industries.
Copyright 1979, Dresser Atlas.

This crossplot example was intentionally left blank so it can be used by the reader to construct a Hingle plot

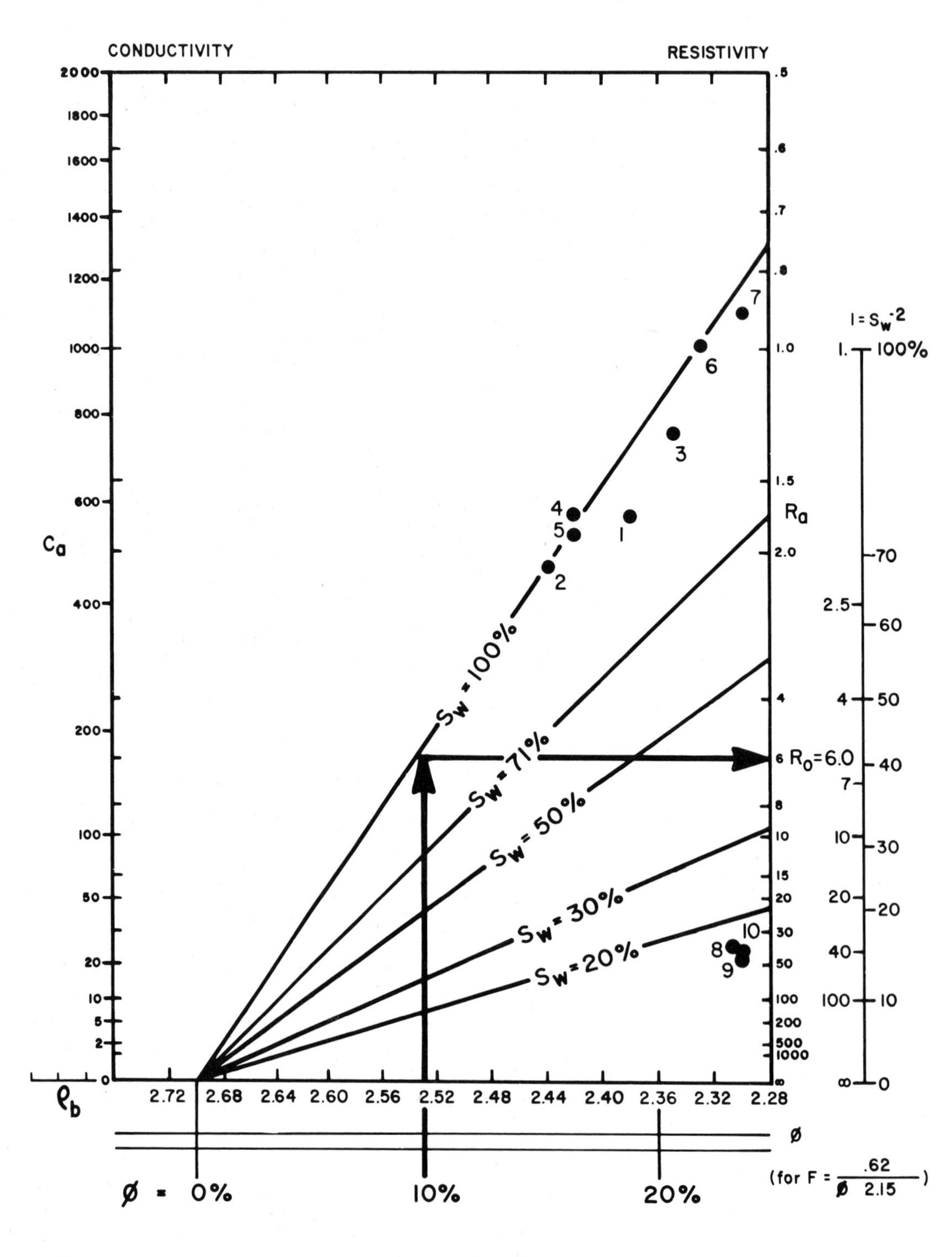
RESISTIVITY/POROSITY CROSSPLOT
CONDUCTIVITY
RESISTIVITY
C_a
R_a
ρ_b
$I = S_w^{-2}$
$S_w = 100\%$
$S_w = 71\%$
$S_w = 50\%$
$S_w = 30\%$
$S_w = 20\%$
$R_0 = 6.0$
ø
(for F = .62 / ø 2.15)
ø = 0%
10%
20%

ρ_{ma} = 2.70 gm/cc

R_w = 0.068

Figure 43. Example of a resistivity versus porosity (Hingle) crossplot. Morrow sandstone, Cimarron County, Oklahoma.

Before using the Hingle crossplot to determine water saturation (S_w) for a well-completion decision (see text, steps 1 through 8), you must first calibrate the x-axis scale for porosity (ϕ).

Given: Fluid density (ρ_f) = 1.0 gm/cc for freshwater mud; matrix density (ρ_{ma}) = 2.7 gm/cc (from Hingle crossplot); derived porosity is 10% (arbitrary).

Procedure:

Remember that the density of derived porosity (ϕ_{Den}) is calculated as follows (see text, under heading: Density Log; Chapter IV):

$$\phi_{Den} = \frac{\rho_{ma} - \rho_b}{\rho_{ma} - \rho_f}$$

Therefore:

$$0.10 = \frac{2.70 - \rho_b}{2.70 - 1.0} = \frac{2.70 - \rho_b}{1.7}$$

$$0.17 = 2.70 - \rho_b$$

$$\rho_b = 2.53 \text{ gm/cc bulk density at 10\% porosity when } \rho_{ma} = 2.7 \text{ gm/cc and } \rho_f = 1.0 \text{ gm/cc}$$

The values ρ_b = 2.53 gm/cc and ϕ = 10% should coincide on the x-axis. In step 2 of the text, you scaled the x-axis. This exercise (Fig. 43) gives you one point on your x-axis (ρ_b at 2.53; ϕ = 10%); steps 4 and 5 in the text give you the end-point of your scale (ρ_{ma} at 2.70; ϕ = 0%).

Scale the x-axis to cover values between 0% and 10% porosity, and continue above 10% to the end of the chart.

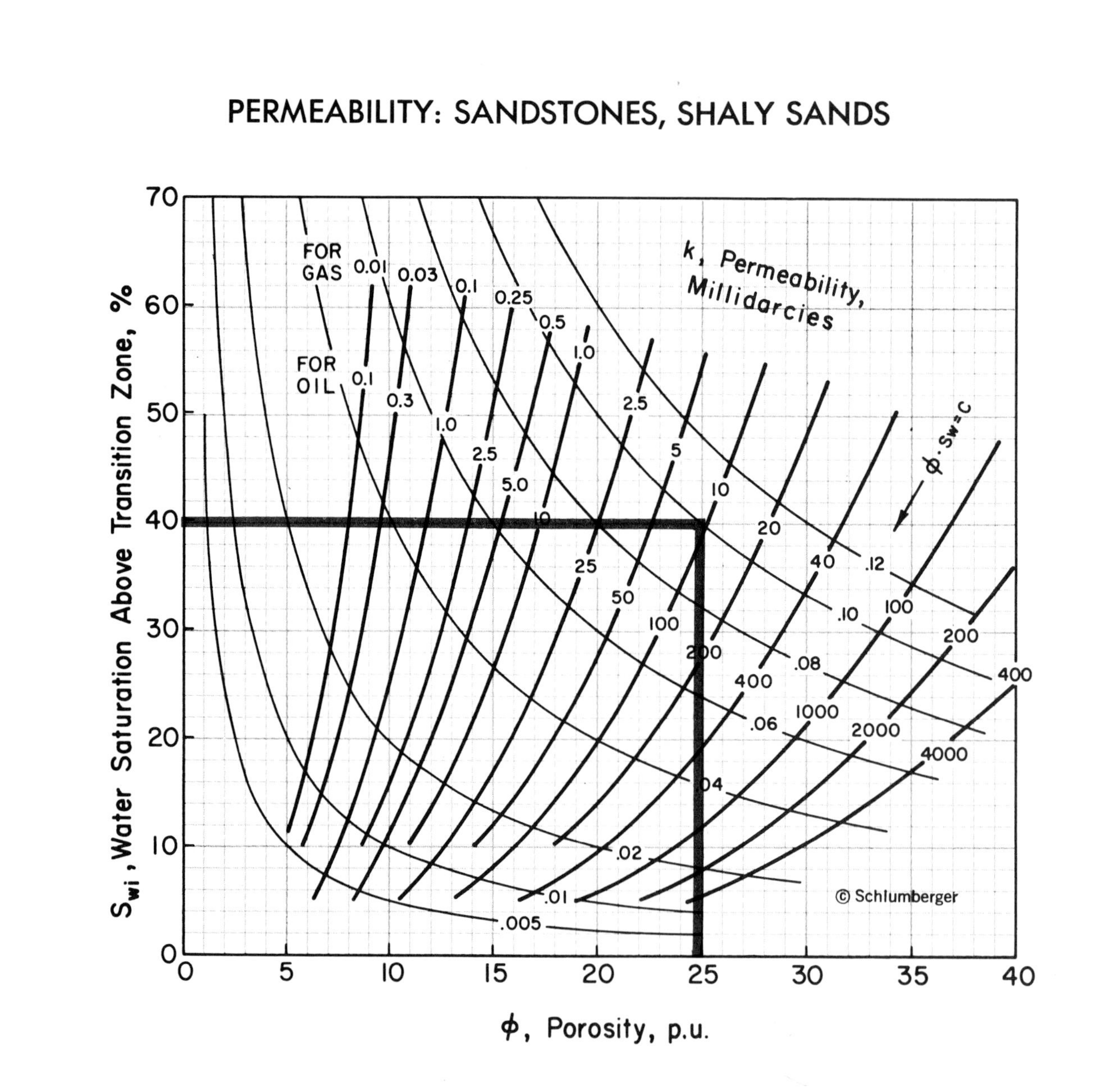
PERMEABILITY: SANDSTONES, SHALY SANDS
k, Permeability, Millidarcies
FOR GAS
FOR OIL
$\phi \cdot S_w = C$
S_{wi}, Water Saturation Above Transition Zone, %
ϕ, Porosity, p.u.
© Schlumberger

Figure 44. Chart of porosity (ϕ) versus irreducible water saturation ($S_{w\ irr}$) for estimating permeability and determining bulk volume water ($C = S_w \times \phi$).

Courtesy, Schlumberger Well Services.
Copyright 1969, Schlumberger.

Given: Porosity (ϕ) = 25% and irreducible water saturation ($S_{w\ irr}$) = 40% for an oil-bearing sandstone.

Procedure:

1. Find porosity (ϕ = 25%) on the bottom horizontal scale, and find irreducible water saturation ($S_{w\ irr}$ = 40%) on the left-hand vertical scale.
2. Follow the two values into the chart to the point where they intersect.
3. The vertically oriented curved line on which this point falls represents permeability. In this case, permeability (K) equals 100 md for oil (lower scale) and 10 md for gas (top scale).
4. The diagonally oriented, curved, hyperbolic lines ($C = S_w \times \phi$) represent lines of equal value for bulk volume water. In this example, the bulk volume water equals 0.10 (BVW = 0.10).

Note: It is important to remember that *this chart is only valid for estimating permeability* (K) *in zones at irreducible water saturation* ($S_{w\ irr}$).

Zones at irreducible water saturation ($S_{w\ irr}$) have bulk volume water values ($BVW = \phi \times S_w$) which are fairly constant. On the chart, data points from different intervals in a zone at irreducible saturation ($S_{w\ irr}$) will plot in a coherent pattern, on or parallel to one of the curved hyperbolic lines.

Data points from zones not at $S_{w\ irr}$ will scatter from this pattern.

CHAPTER VII

LITHOLOGY LOGGING AND MAPPING TECHNIQUES

General

Petrophysical logs provide most of the subsurface data available to an exploration geologist. Besides their importance in completion decisions, they are also invaluable tools for mapping and identifying lithologies. Six techniques are presented here which can assist geologists with lithologic determination and mapping. They are: (1) Gamma Ray Neutron-Density Log, (2) M-N* lithology plot, (3) MID* lithology plot, (4) Alpha mapping from SP log, (5) clean sand or carbonate maps from gamma ray log, and (6) rock typing and facies mapping. These techniques are especially important to a geologist when lithologic data from cores or samples are unavailable.

Combination Gamma Ray Neutron-Density Log

The gamma ray log (Chapter V) measures the natural radiation of a formation, and primarily functions as a lithology log. It helps differentiate shales (high radioactivity) from sands, carbonates, and anhydrites (low radioactivity). The neutron log is a porosity device that is used to measure the amount of hydrogen in a formation (Chapter IV). The density log is a porosity device that measures electron density (Chapter IV). When these three logs are used together (i.e. Combination Gamma Ray Neutron-Density Log), lithologies can be determined. Figure 35 in Chapter IV is a schematic illustration of how Gamma Ray Neutron-Density Log responses are related to rock type. Figure 45 is a Gamma Ray Neutron-Density Log through the Ordovician Stony Mountain Shale and Red River Formation in Richland County, Montana. Note in Figure 45 how the different rock types are related to log responses. As a quick look method, where there are a limited number of lithologies, this log works well for basic lithologic and facies mapping. Whenever lithologies are more complex, additional logging devices and techniques, such as the M-N* and MID* plot, must be used. Of course, after you determine lithology, you can prepare lithology or facies maps.

M-N* Lithology Plot

The M-N* plot requires a sonic log along with neutron and density logs. The sonic log is a porosity log (Chapter IV) that measures interval transit time. Interval transit time (Δt) is the reciprocal of the velocity of a compressional sound wave through one foot of formation. A sonic log, neutron log, and density log are all necessary to calculate the lithology dependent variables M* and N*. M* and N* values are essentially independent of matrix porosity

Table 10. Matrix and Fluid Coefficients of Several Minerals and Types of Porosity (Liquid-filled Boreholes).

	Δt_{ma}	ρ_{ma}	$(\phi_{SNP*})_{ma}$	$(\phi_{CLN*})_{ma}$
Sandstone (1) (V_{ma} = 18,000) $\phi > 10\%$	55.5	2.65	−0.035†	−0.05†
Sandstone (2) (V_{ma} = 19,500) $\phi < 10\%$	51.2	2.65	−0.035†	−0.005
Limestone	47.5	2.71	0.00	0.00
Dolomite (1) (ϕ = 5.5 to 30%)	43.5	2.87	0.035†	0.085†
Dolomite (2) (ϕ = 1.5% to 5.5% & > 30%)	43.5	2.87	0.02†	0.065†
Dolomite (3) (ϕ = 0.0% to 1.5%)	43.5	2.87	0.005†	0.04†
Anhydrite	50.0	2.98	−0.005	−0.002
Gypsum	52.0	2.35	0.49††	
Salt	67.0	2.03	0.04	−0.01

From Schlumberger Log Interpretation Manual/Principles. Courtesy Schlumberger Well Services; Copyright 1972, Schlumberger.

†Average values

††Based on hydrogen-index computation

(sucrosic and intergranular). A crossplot of these two variables makes lithology more apparent. M* and N* values are calculated by the following equations (Schlumberger, 1972):

$$M^* = \frac{\Delta t_f - \Delta t}{\rho_b - \rho_f} \times 0.01$$

$$N^* = \frac{\phi_{Nf} - \phi_N}{\rho_b - \rho_f}$$

Where:

Δt_f = interval transit time of fluid (189 for fresh mud and 185 for salt mud)

Δt = interval transit time from the log

ρ_f = density of fluid (1.0 for fresh mud and 1.1 for salt mud)

ρ_b = bulk density of formation

ϕ_N = neutron porosity of the formation from Compensated Neutron or Sidewall Neutron Porosity Log

ϕ_{Nf} = neutron porosity of fluid (use 1.0)

When the matrix parameters (Δt_{ma}, ρ_{ma}, ϕ_{Nma}; Table 10) are used in the M* and N* equations instead of formation parameters, M* and N* values can be obtained for the various minerals (Table 11).

Figure 46 is a M-N* plot of data from the Ordovician Red River C-zone in the Alpar Resources Federal 1-10, Richland County, Montana at a depth of 11,870 to 11,900 ft (Fig. 45). Data from this interval, cluster together in the M-N* lithology triangle. Lithology is defined by the end-members: anhydrite, dolomite, and limestone. In this case, lithology is an anhydritic limey dolomite (Fig. 46). Notice that two of the data points are above the dolomite-limestone line, indicating secondary porosity from vugs and/or fractures.

MID* Lithology Plot

The MID* (Matrix Identification) plot, like the M-N* is a crossplot technique which helps identify lithology and secondary porosity. Also, like the M-N* plot, the MID* plot requires data from neutron, density, and sonic logs.

The first step in constructing a MID* plot is to determine values for the apparent matrix parameters $(\rho_{ma})_a$ and $(\Delta t_{ma})_a$. These values are determined from neutron (ϕ_N), density (ρ_b), and sonic (Δt) data obtained from the log. Next, these values (i.e. ϕ_N, ρ_b, and Δt) are crossplotted on appropriate neutron-density and sonic-density charts to obtain $(\rho_{ma})_a$ and $(\Delta t_{ma})_a$ values. Crossplot charts, along with instructions on how to use them, can be obtained from Schlumberger's Log Interpretation Manual/Applications, Volume II (1974). The method for obtaining apparent matrix parameters $(\rho_{ma})_a$ and $(\Delta t_{ma})_a$ is also illustrated in the case studies discussed in Chapter VIII.

Once obtained, apparent matrix parameters $(\rho_{ma})_a$ and $(\Delta t_{ma})_a$ are plotted on the MID* plot (Fig. 47). Data plotted in Figure 47 arerom the Alpar Resources Federal 1-10 well, Richland County, Montana (Fig. 45), and include the same Red River C-zone interval (11,870 to 11,900 ft) illustrated in the M-N* plot (Fig. 46). The data points form a cluster (Fig. 47) defined by the end-members: anhydrite, dolomite, and limestone. The lithology is an anhydritic limey dolomite. The three points that plot above the dolomite-limestone line indicate secondary porosity.

Alpha Mapping from SP Log

The spontaneous potential (SP) log (Chapter II) can be used to map clean sands (shale-free) versus shaly sands. The technique is called Alpha mapping (α; Dresser Atlas, 1974), and is based on the observation that the presence of

Table 11. Values of M* and N* constants, calculated for Common Minerals.

	Fresh Mud (ρ = 1.0)		Salt Mud (ρ = 1.1)	
	M*	N*	M*	N*
Sandstone (1) V_{ma} = 18,000	.810	.628	.835	.669
Sandstone (2) V_{ma} = 19,500	.835	.628	.862	.669
Limestone	.827	.585	.854	.621
Dolomite (1) $\phi = 5.5 - 30\%$	.778	.516	.800	.544
Dolomite (2) $\phi = 1.5 - 5.5\%$	.778	.524	.800	.554
Dolomite (3) $\phi = 0 - 1.5\%$	.778	.532	.800	.561
Anhydrite ρ_{ma} = 2.98	.702	.505	.718	.532
Gypsum	1.015	.378	1.064	.408
Salt			1.269	1.032

From Schlumberger Log Interpretation Manual/Principles. Courtesy Schlumberger Well Services; Copyright 1972, Schlumberger.

shale in a formation decreases the SP response.

The alpha method can be extremely valuable in mapping because it can allow you to more narrowly define desirable zones. Alpha values from nearby wells can be used to construct clean sand (high energy) maps (in effect, you are mapping iso-alpha values).

To construct an Alpha map, first calculate the static spontaneous potential (SSP) that a sand would have, if it was 100% shale-free and unaffected by bed thickness. The equation for SSP is:

$$SSP = -K \times \log(R_{mf}/R_w)$$

Where:
- SSP = static spontaneous potential
- K = 60 + (0.133 × formation temperature)
- R_{mf} = resistivity of mud filtrate at formation temperature
- R_w = resistivity of formation water at formation temperature

The SSP must be calculated for the formation in each well, so that variations in R_{mf} and R_w can be corrected. Next, determine alpha values by the method shown in Figure 48. The alpha cutoff (50%, 75%, whatever) is arbitrary, but should be based on production histories in the area.

The resulting alpha (α) map delineates clean sand environments. In the above example (in Fig. 48), the greater alpha thickness for a given alpha cut-off (i.e. 75% α, or 50% α) indicates a greater thickness of higher energy, low-shale sandstones. Also, because the presence of shale in a sandstone can cause a loss of permeability, an alpha map is indicative of better reservoir conditions.

The problem with alpha mapping from an SP log is that SP response is decreased, not only by shale, but also by thin beds (<10 feet) and the presence of hydrocarbons (Chapter II). Bed thickness problems are minimized by making an SP log bed thickness correction (Chapter II). But, the SP log can't be corrected for hydrocarbons.

Clean Sand or Carbonate Maps from Gamma Ray Log

The gamma ray log can be used to map clean (shale-free) sandstones or carbonates versus shaly sandstones and carbonates. Because shales are more radioactive than clean sandstones or carbonates (Chapter V), when the percentage of shale increases in these rock types, the gamma ray reading also increases.

Figure 49 is from a Gamma Ray Neutron-Density Log through the Mississippian, upper Mission Canyon Formation in Roosevelt County, Montana. In this interval of the Mission Canyon Formation, crinoid-fenestrate bryozoan bioherms are often developed. Because the bioherm facies is composed of clean carbonate relative to the non-bioherm facies, *the gamma ray log can be used to map the bioherm facies*. The procedure for obtaining a clean carbonate cut-off from a gamma ray log is described in Figure 49.

A gamma ray API value of 20 on the gamma ray log (Fig. 49) will represent clean carbonate with a volume of shale (V_{sh}) equal to or less than 5%. By drawing a vertical line on the gamma ray log equal to 20 API units (Fig. 49), the geologist can identify and map the clean carbonate (or sand).

Figure 50 is an isopach map of clean carbonate for the upper Mission Canyon Formation in Roosevelt County, Montana. Because the relationship between clean carbonate and the crinoid-fenestrate bryozoan bioherm facies is already established, the map (Fig. 50) delineates the distribution of the bioherm facies. Clean carbonate maps have also been used to map the Pennsylvanian banks (bioherms) of north central Texas (Wermund, 1975).

Rock Typing and Facies Mapping

An important contribution to subsurface analysis of carbonate rocks has been the attempt to establish relationships between log responses and carbonate facies. Pickett (1977), Asquith (1979), and Watney (1979; 1980) used crossplots to identify log response/rock type relationships. Table 12 is a list of the crossplots applied by these authors.

Table 12. Types of Carbonate Rock Type Identification Crossplots (after Pickett, 1977; Asquith, 1979; and Watney, 1979 and 1980).

Δt	(interval transit time) vs. ϕ_N (neutron porosity)
ρ_b	(bulk density) vs. ϕ_N (neutron porosity)
ρ_b	(bulk density) vs. Δt (interval transit time)
R_t	(deep resistivity) vs. ϕ_N (neutron porosity)
GR	(gamma ray) vs. ϕ_N (neutron porosity)†
R_t	(deep resistivity) vs. ϕ_s (sonic porosity)

†Watney (1979 and 1980) also uses neutron log readings measured in counts/sec.

To date, crossplots have been used to establish log versus lithology relationships only when petrographic data is available from cores or cuttings in selected wells. Petrographic analysis from selected wells is essential to firmly establish rock type.

When establishing log/lithology relationships, log responses from control wells (i.e. wells with petrographic analysis) are crossplotted. Next, areas that delineate rock-type clusters are outlined (see Fig. 51) on the crossplot. Finally, log responses from wells without cores or cuttings are added to the crossplot. The carbonate rock type and depositional environment of wells without petrographic analysis can then be determined by the cluster in which each occurs on the crossplot chart (see Fig. 51).

In Figure 51, the solid black circles and squares represent data from wells where petrographic analysis was used to determine carbonate rock type and depositional environment. The open circles represent data from a well without petrographic analysis. The carbonate rock types and depositional environments were determined by the cluster in which the open circles were plotted (Fig. 51).

Figure 52 is a crossplot of deep resistivity (R_t) versus sonic porosity (ϕ_s) for the Lower Permian, Council Grove B-zone in Ochiltree County, Texas. Clusters for the three carbonate rock types (oolite grainstone, oolitic wackestone, and argillaceous bioclastic wackestone) were established by petrographic analysis of cores and cuttings (open circles). The solid circles represent data from wells with only log control. Figure 53 is a facies map of the Council Grove B-zone based on the percentage distribution of the three carbonate rock types established by the resistivity/sonic porosity crossplot (Fig. 52).

The advantage of log crossplot techniques is that they maximize use of available information. Cores and cuttings are required from only a few control wells rather than all wells. This is very important in subsurface facies mapping because of the difficulty in obtaining cores and cuttings from every well in an area. Also, because petrographic analysis of every well is unnecessary, you can save a great deal of time.

However, it should be emphasized that petrographic analysis of cores or cuttings from control wells is an essential first step to firmly establish the rock-type cluster used in the crossplots.

Review - Chapter VII

1. A Combination Gamma Ray Neutron-Density Log can be used to determine lithology when a limited number of rock types are present.

2. Where lithology is more complex, a sonic log plus a Combination Neutron-Density Log are both necessary to construct M-N* or MID* lithology identification crossplots.

3. The spontaneous potential (SP) and gamma ray logs can be used to map shaly versus non-shaly carbonates or sandstones.

4. Crossplotting of multiple log reponses can be used to establish relationships between log responses and rock types, provided some petrographic data from cores or cuttings is available.

FED. 1—10
SW/SE SEC. 10 26N—55E
Richland County, Montana

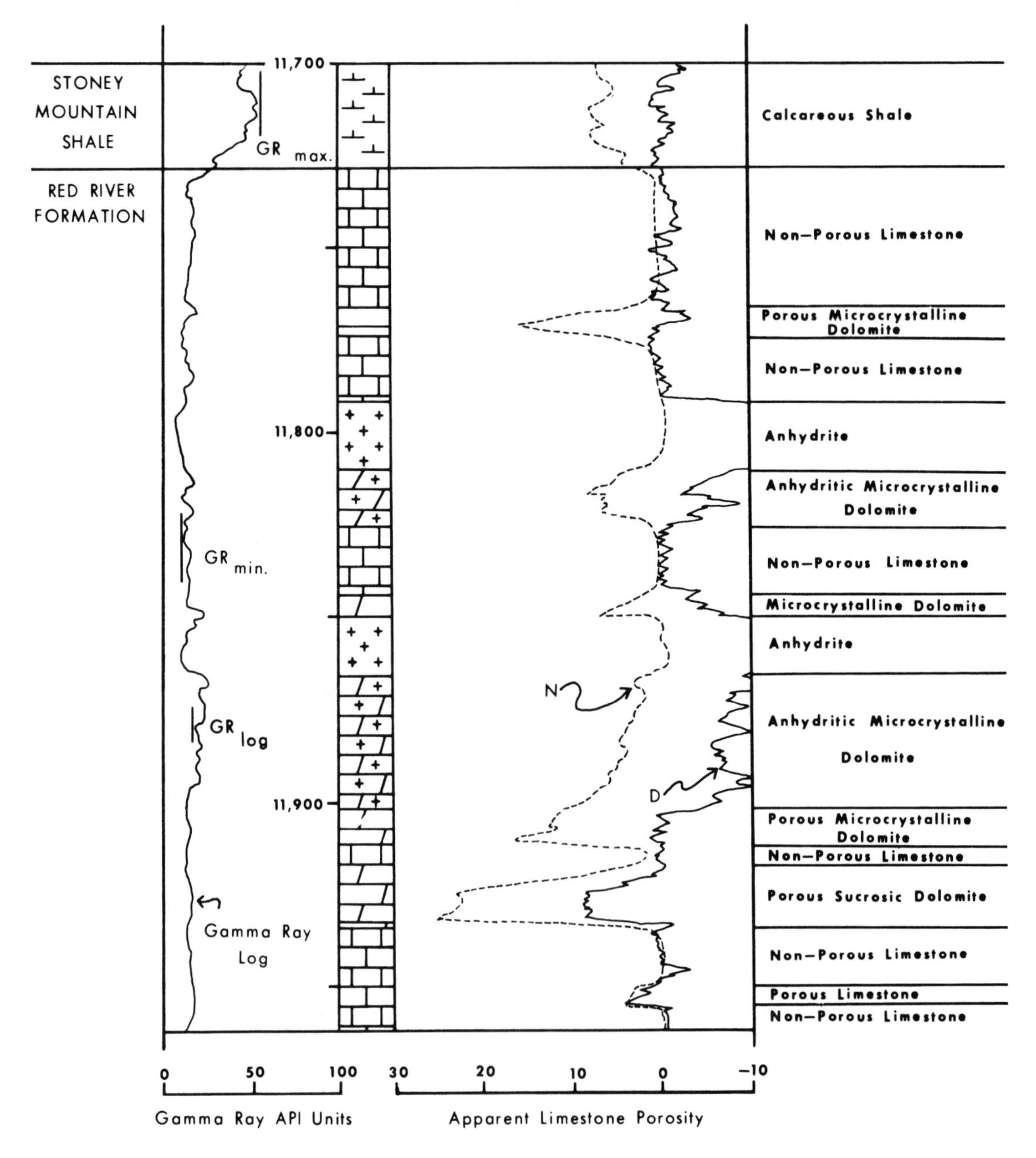

Figure 45. Example of a Combination Gamma Ray Neutron-Density Log showing corresponding lithologies. See Chapters IV and V about log interpretation.

Example taken from the Ordovician Red River Formation, Richland County, Montana. After Asquith (1979).

Note in log tracks #2 and #3 that the neutron log is represented by a dashed line and the density log is represented by a solid line (see Chapter IV).

The gamma ray log is in track #1 (see Chapter V).

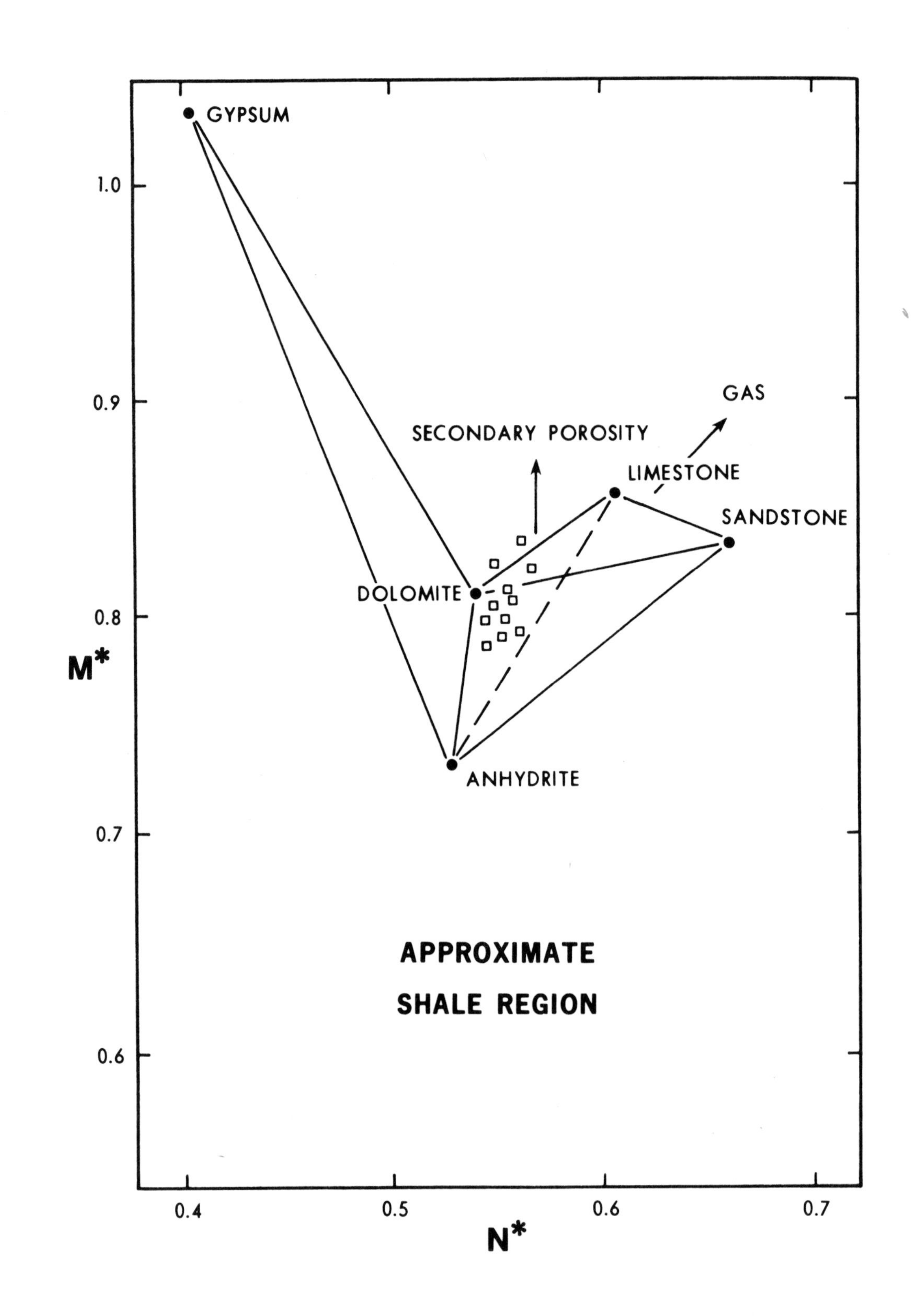
GYPSUM
1.0
0.9
0.8
0.7
0.6
M*
GAS
SECONDARY POROSITY
LIMESTONE
SANDSTONE
DOLOMITE
ANHYDRITE
APPROXIMATE
SHALE REGION
0.4
0.5
0.6
0.7
N*

Figure 46. Example of an M-N* crossplot of data from the well illustrated in Figure 45; interval 11,870 to 11,900 ft. After Asquith (1979).

This crossplot helps to determine lithology. Note how data points are clustered within a lithology triangle bounded by three corners: dolomite, anhydrite, and limestone. In this case, the rock is identified as anhydritic limey dolomite (see text).

Note that two points plot above the limestone-dolomite line, and into the zone of secondary porosity. This indicates secondary porosity from vugs and/or fractures.

The triangle end-members are plotted from common matrix values for M* and N* found in Table 11.

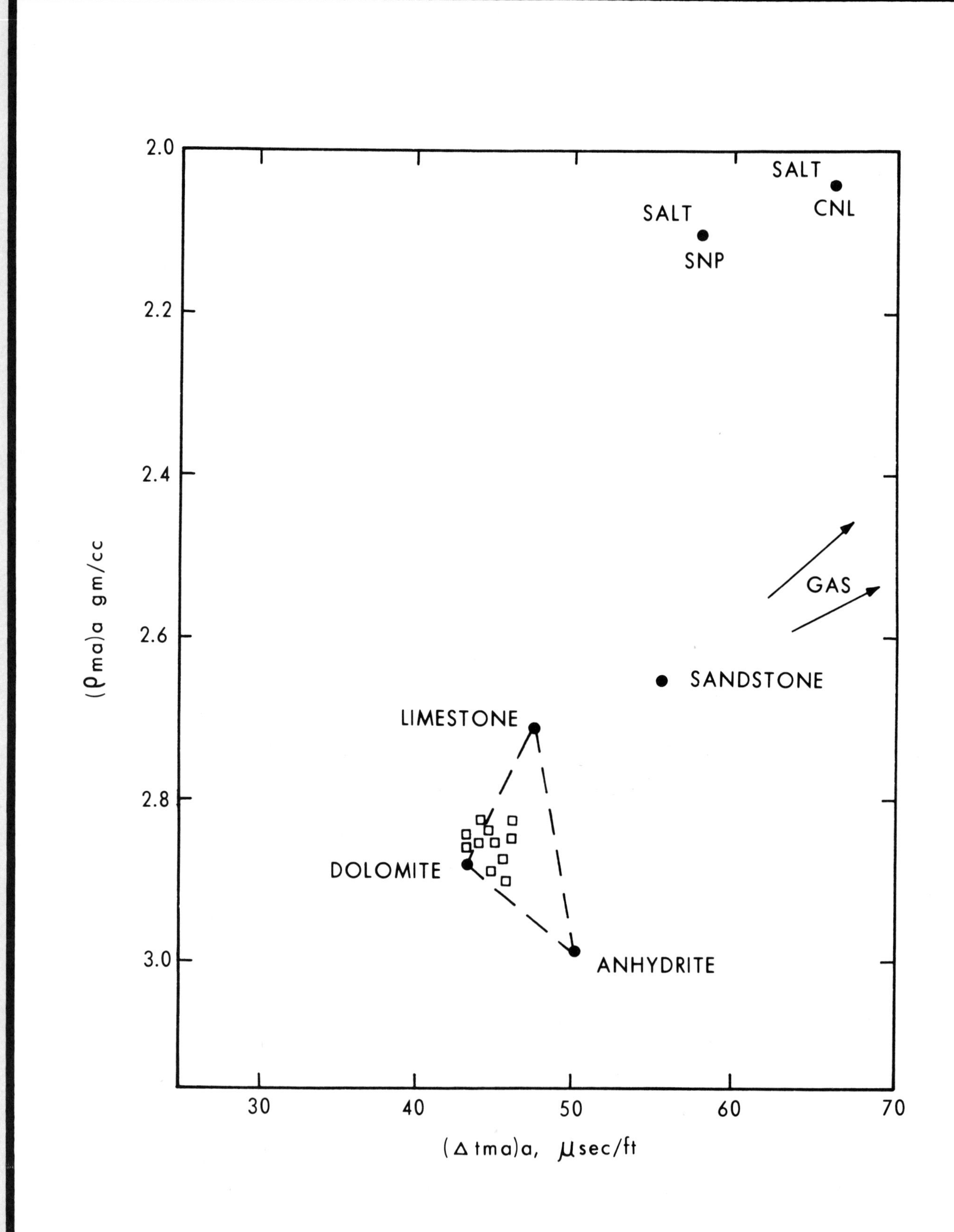
SALT
CNL
SALT
SNP
GAS
SANDSTONE
LIMESTONE
DOLOMITE
ANHYDRITE
2.0
2.2
2.4
2.6
2.8
3.0
$(\rho_{ma})a$ gm/cc
30
40
50
60
70
$(\Delta tma)a$, μsec/ft

Figure 47. Example of a MID* crossplot of data from the log illustrated in Figure 45; interval is 11,870 to 11,900 ft. After Asquith (1979). The lithologic determination is the same as in Figure 46 M-N* plot.

Note that the data points cluster in a triangle defined by the end-members—limestone, dolomite, and anhydrite—indicating the lithology is an anhydritic limey dolomite.

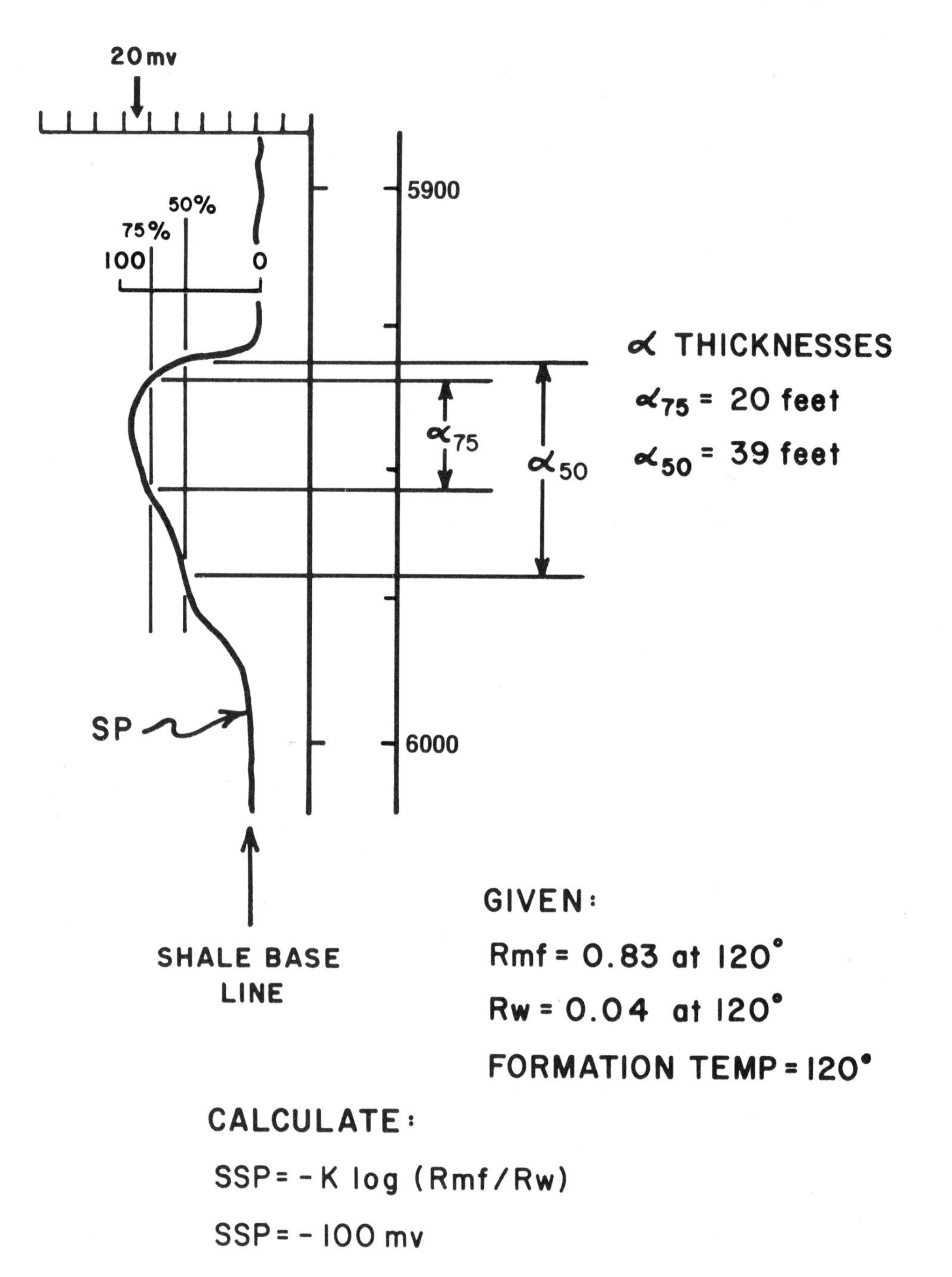
SP Scale 20 mv / div.
20 mv
50%
75%
100
0
5900
6000
SP
SHALE BASE LINE
α75
α50
α THICKNESSES
α75 = 20 feet
α50 = 39 feet
GIVEN:
Rmf = 0.83 at 120°
Rw = 0.04 at 120°
FORMATION TEMP = 120°
CALCULATE:
SSP = - K log (Rmf / Rw)
SSP = - 100 mv

Figure 48. Determining Alpha (α) from an SP log. Two different cutoffs are demonstrated: 50% alpha (α_{50}) and 75% alpha (α_{75}). The alpha percentage is determined as an inverse function of shaliness (100% alpha is shale-free; 50% alpha is 50% shaly).

Given: You must first determine SSP (see text for formula). R_{mf} = 0.83 at 120°F, R_w = 0.04 at 120°F, T_f = 120°F.

Procedure:

1. By formula, we determine that SSP = −100mv. Plot a scale of 100mv on the SP log, using the shale baseline as the zero point; then use the SP scale to establish the value of your scale increments (in this case, each increment is 20mv).
2. In this exercise, α thickness and depth will be determined for both α_{50} and α_{75}, so draw vertical lines through your SSP scale approximately halfway (50%) and three-quarters of the way (75%) across, and drop your vertical lines to intersect the SP curve at the desired depth range.
3. From the intersections, follow the values horizontally to the depth scale on the log. From this log-depth scale you can count depth-increments to determine alpha thickness, as well as the top and bottom boundaries of the given alpha zone.

In this example α_{75} is the thinner of the two, and measures 20 ft, from approximately 5,935 to 5,955 ft; whereas α_{50} is thicker and measures 39 ft, from approximately 5,931 to 5,970 ft.

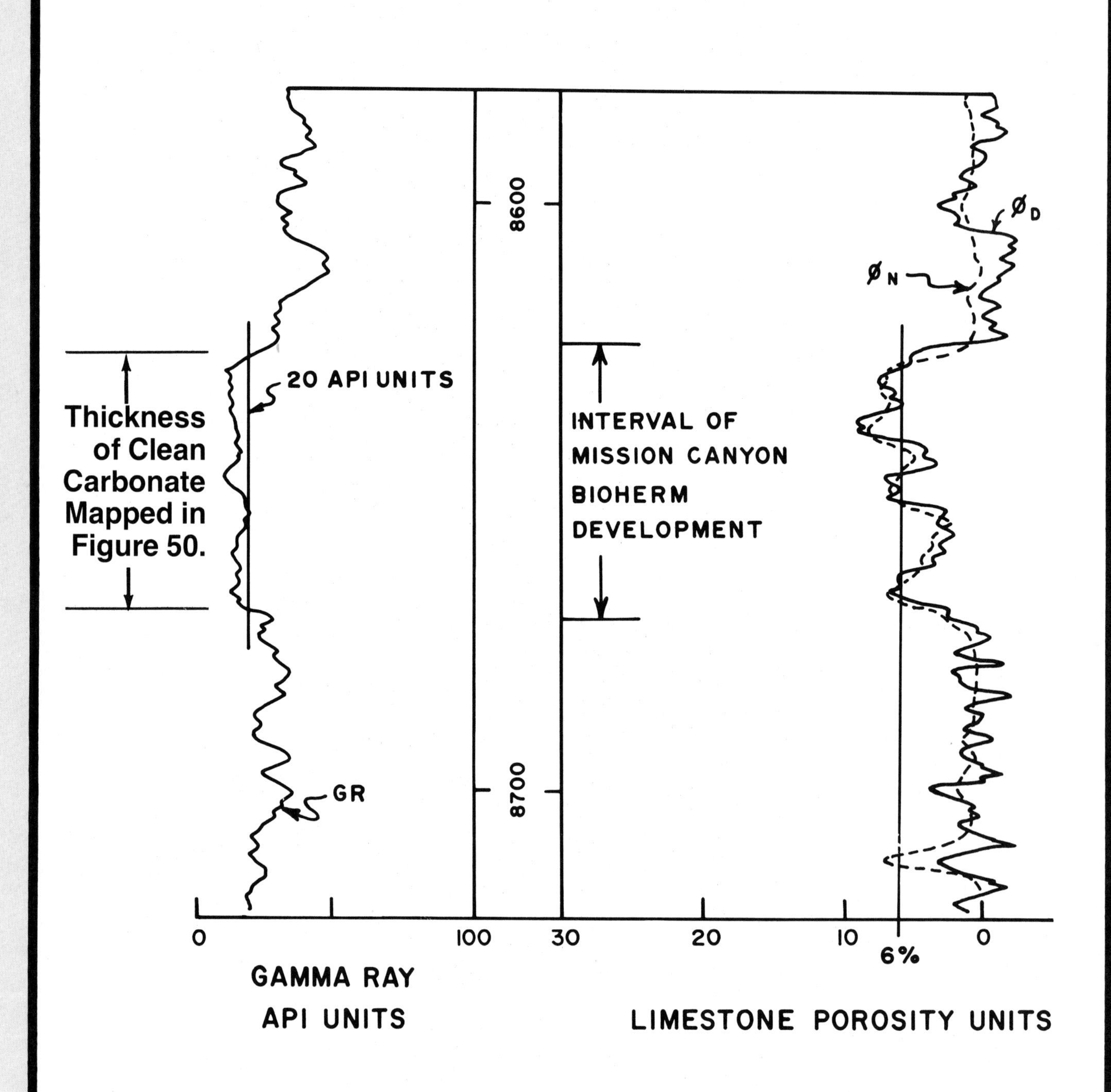
Thickness of Clean Carbonate Mapped in Figure 50.
20 API UNITS
GR
8600
8700
INTERVAL OF MISSION CANYON BIOHERM DEVELOPMENT
ØD
ØN
0
100
30
20
10
6%
0
GAMMA RAY API UNITS
LIMESTONE POROSITY UNITS

Figure 49. Example gamma ray log depicting a 20 API unit gamma ray cutoff used to determine a clean carbonate interval.

Given: The volume of shale (V_{sh}) cutoff is arbitrarily set at 5% ($V_{sh} = 0.05$). Next, determine the gamma ray index from the chart in Figure 38; Chapter V (gamma ray index where $V_{sh} = 5\%$, is $I_{GR} = 0.10$).

Procedure:

1. Determine gamma ray cutoff (see log; and Fig. 38).

   Remember:

$$I_{GR} = \frac{GR_{log} - GR_{min}}{GR_{max} - GR_{min}}$$

   Where:

   GR_{log} = gamma ray log
   GR_{max} = gamma ray maximum (shale)
   GR_{min} = gamma ray minimum (shale-free sandstone or carbonate)

   From Log:

   $GR_{max} = 90$ (from shale zone on log)
   $GR_{min} = 12$ (from clean carbonate zone on log)
   $I_{GR} = 0.10$ (I_{GR} for $V_{sh} = 0.05$; given)

   Then:

$$I_{GR} = \frac{GR_{log} - GR_{min}}{GR_{max} - GR_{min}}$$

   Or:

$$0.10 = \frac{GR_{log} - 12}{90 - 12}$$

   Therefore:

   $GR_{log} = 19.8$ (round off to 20 API units)

2. 20 API represents clean carbonate where the volume of shale is equal to (or less than) 5%.

   Draw a vertical line from the scale value of 20 API units and determine the thickness and limits of the clean carbonate formation (bioherm) much as you determined alpha values in Figure 48.

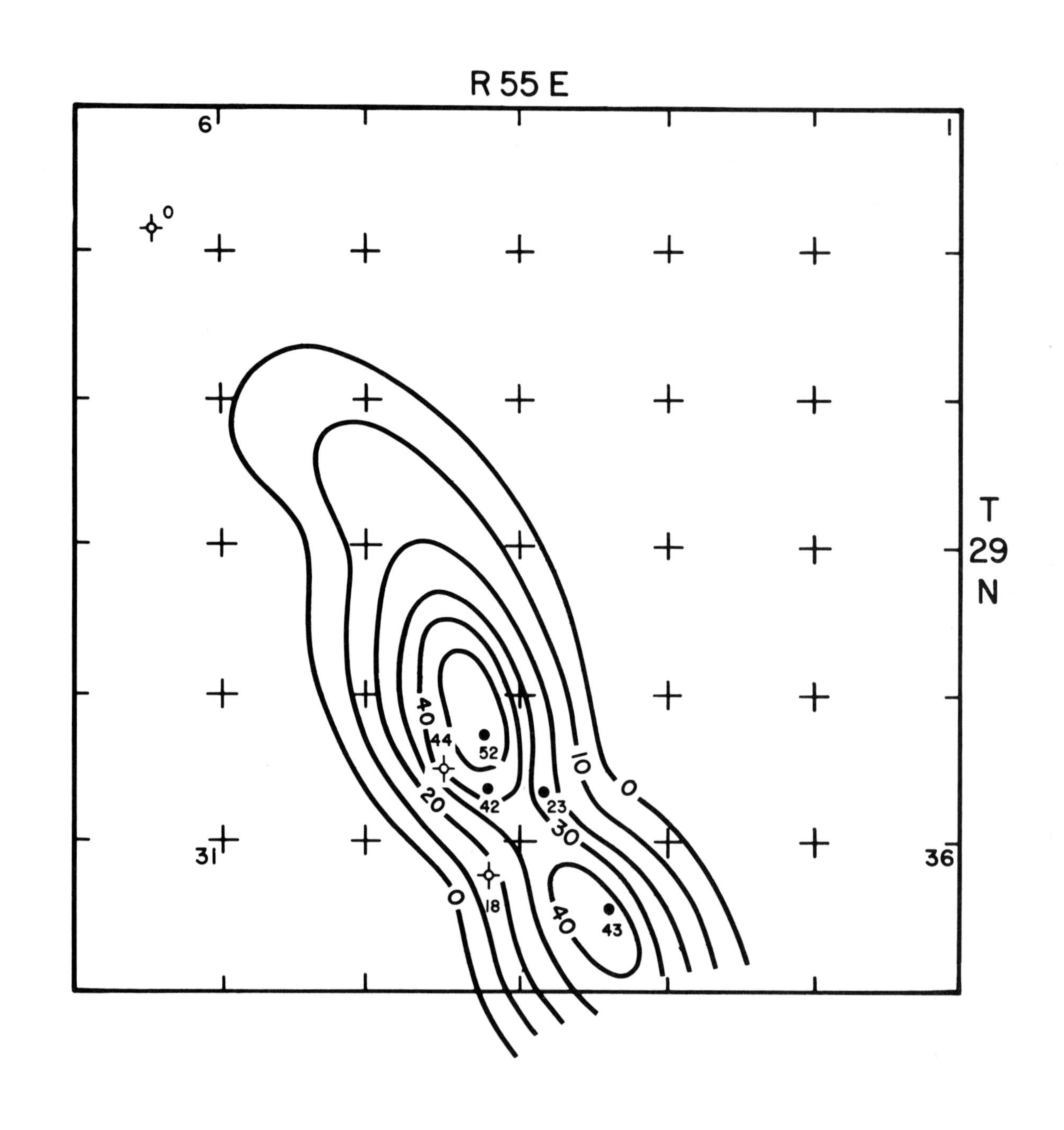
R 55 E
6
1
T
29
N
31
36
0
40
44
52
20
42
10
0
23
30
0
18
40
43

Figure 50. Example isopach map of clean carbonates from the Mississippian Mission Canyon Formation, Roosevelt County, Montana, described in the text and in Figure 49.

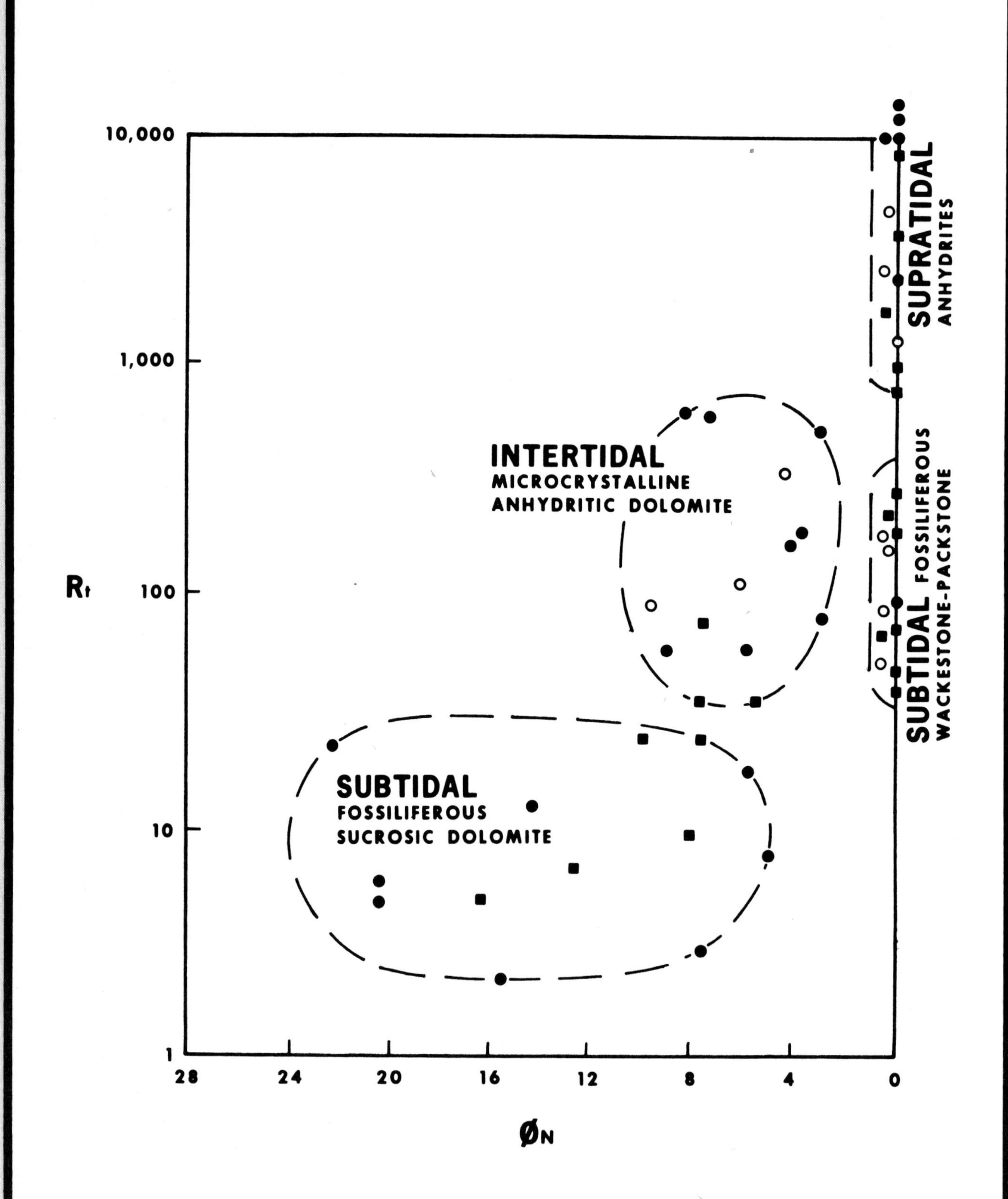
10,000
1,000
100
10
1
Rt
28
24
20
16
12
8
4
0
ØN
INTERTIDAL
MICROCRYSTALLINE
ANHYDRITIC DOLOMITE
SUBTIDAL
FOSSILIFEROUS
SUCROSIC DOLOMITE
SUPRATIDAL
ANHYDRITES
SUBTIDAL FOSSILIFEROUS
WACKESTONE-PACKSTONE

Figure 51. Example crossplot of formation resistivity (R_t; in this case deep Laterolog*) versus neutron porosity (ϕ_N). This comparison of log response to facies helps the geologist develop rock type clusters.

This example is from the Ordovician Red River C and D zones in Richland and Roosevelt counties, Montana. After Asquith (1979).

Solid squares and circles represent wells with core or cuttings available, in addition to log response. Open circles represent wells with log control only.

Facies classifications are first confirmed by core/cuttings analysis, but once clusters are established then, only log control is necessary.

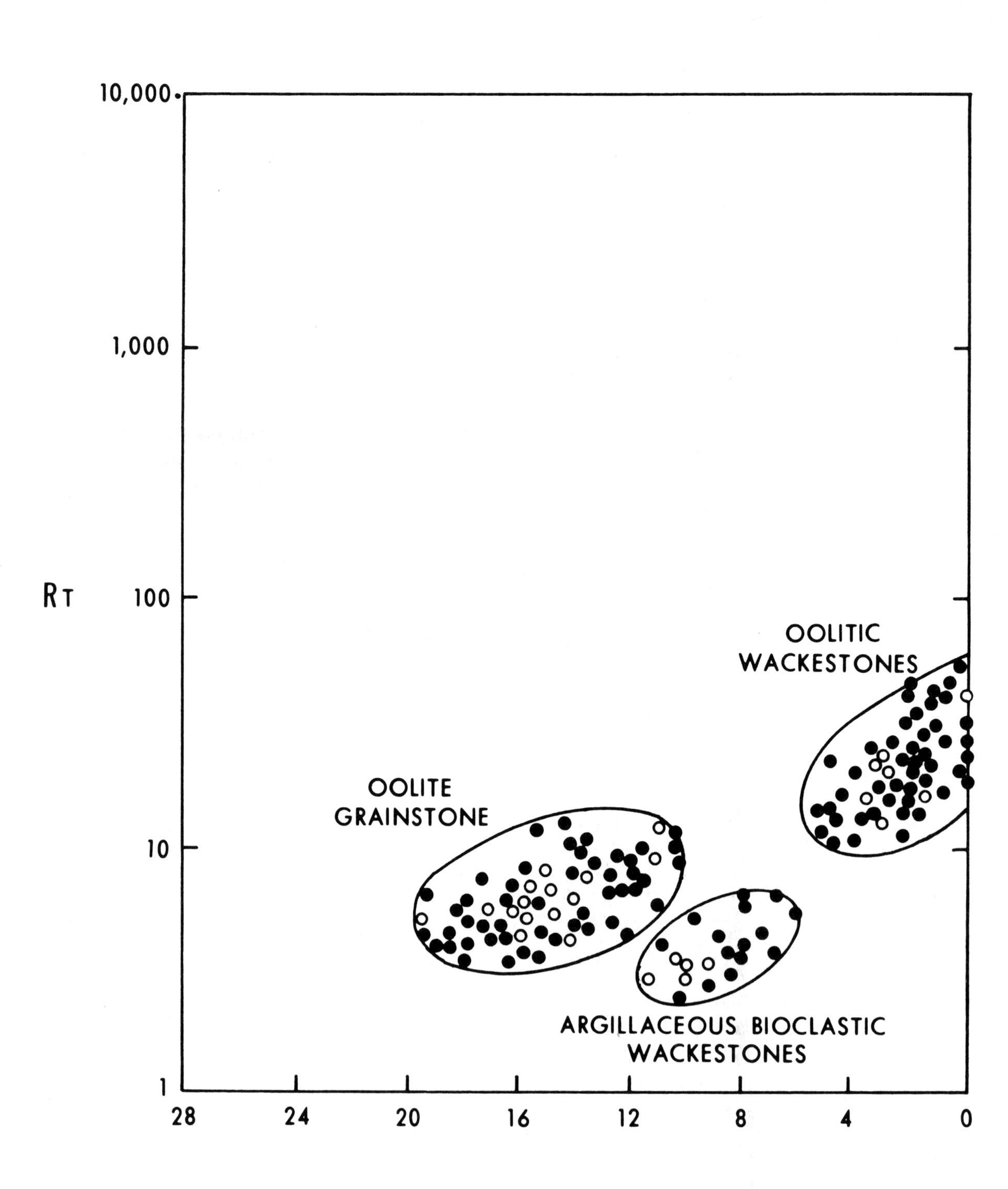
10,000
1,000
RT
100
10
1
28
24
20
16
12
8
4
0
ØS
OOLITIC
WACKESTONES
OOLITE
GRAINSTONE
ARGILLACEOUS BIOCLASTIC
WACKESTONES

Figure 52. Example crossplot of formation resistivity (R_t; in this case from deep induction) with sonic porosity (ϕ_s). As with Figure 51, the rock type clusters are developed by core or cuttings analysis, but logging control is all that's necessary once the relationship is defined.

This example comes from the Lower Permian Council Grove B-zone, Ochiltree County, Texas. After Asquith (1979).

Open circles represent wells with both core/cuttings analysis and log control. Solid black circles represent wells with only log control.

(In this example, sonic porosity (ϕ_s) is based on a limestone matrix, where Δt_{ma} = 47.6 μsec/ft.)

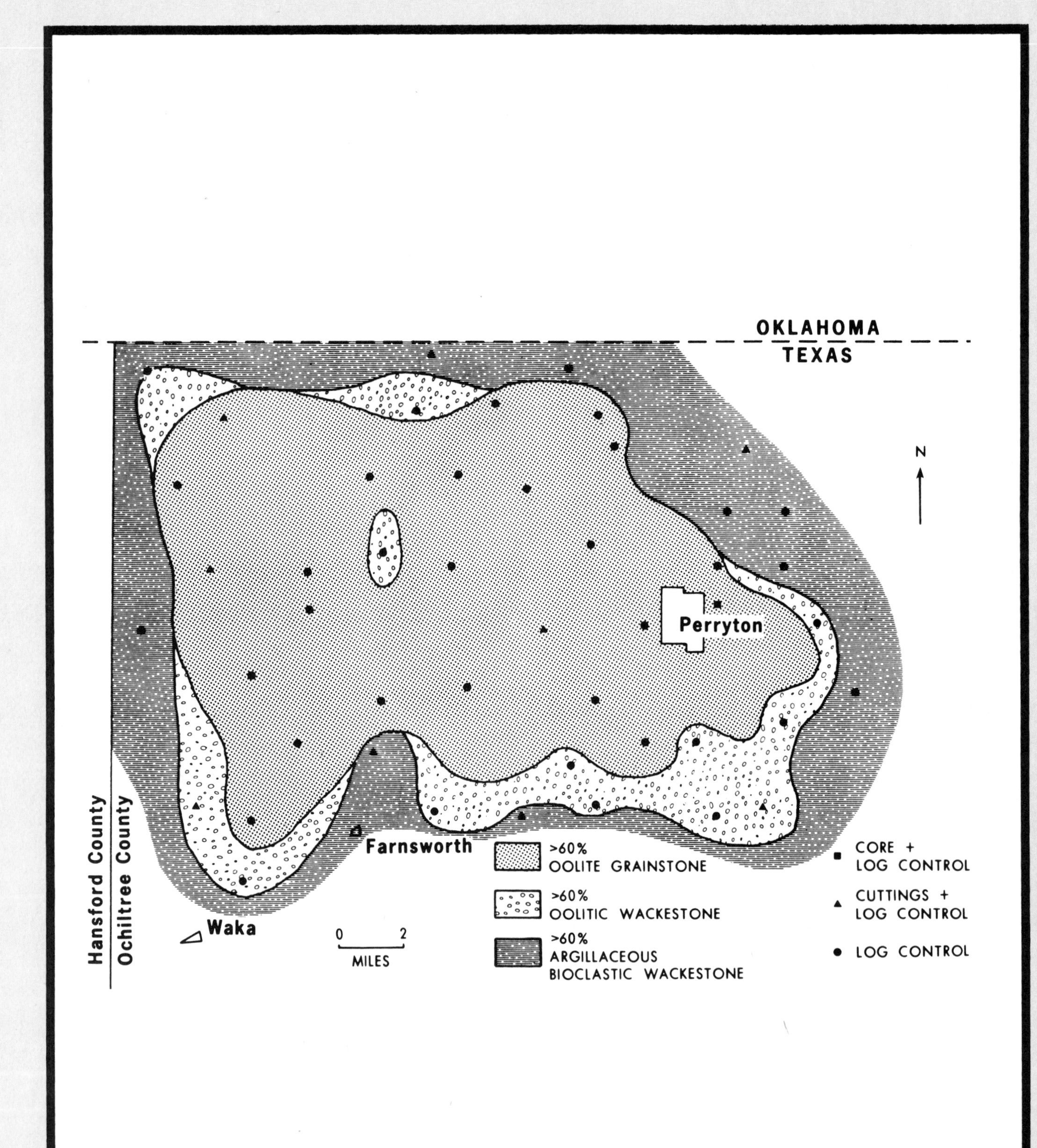
OKLAHOMA
TEXAS
N
Perryton
Farnsworth
Waka
Hansford County
Ochiltree County
0 2
MILES
>60%
OOLITE GRAINSTONE
>60%
OOLITIC WACKESTONE
>60%
ARGILLACEOUS
BIOCLASTIC WACKESTONE
CORE +
LOG CONTROL
CUTTINGS +
LOG CONTROL
LOG CONTROL

Figure 53. Example facies map of the Lower Permian Council Grove B-zone, Ochiltree County, Texas, prepared from the facies clusters established by crossplotted log data in Figure 52. After Asquith (1979).

The legend defines the position of core, cuttings, and log-only control.

CHAPTER VIII

LOG INTERPRETATION CASE STUDIES

Introduction

A major hurdle for geologists is learning how the theory and the many formulas of log interpretation are translated into practice. The learning process is slow, and only takes place after a good deal of dreary effort is supported by actual experience with logs. There are absolutely no short-cuts. Memorizing log patterns and curve values just doesn't work, and can, in fact, prove disastrous. However, to assist the process of changing what's done into how it's done, case studies can be an invaluable asset. The ones presented here cover a variety of geographical areas, geologic ages, lithologies and different log packages. These are not necessarily a classic representation of any of the variations. The reader will need to actively engage his or her intellect in finding appropriate answers for each problem, and will also need to apply material already learned from a thorough study of the preceding text.

Although all the case studies are based on actual field examples, in some studies the amount of log data analyzed would be superfluous in real-life. A pipe-setting decision doesn't always require a full-blown log analysis. *The unusual amount of detail is offered here as a learning experience*. Nevertheless, careful study of a log can always yield information which will further the development of an expertise.

These case studies are offered as a progressive series of problems. In the first example the reader is asked to do very little, but in the final example the reader should be prepared to proceed most of the way alone. An effort is made to define variables and terminology appearing in the case studies, even though they may already have been defined earlier in the text.

In each of the case studies the moveable hydrocarbon index (S_w/S_{xo}) is calculated. Sometimes, the water saturations of the flushed zone (S_{xo}) may exceed 100%. Whenever this happens, a value of 100% is then used to calculate the moveable hydrocarbon index.

No matter how significant log interpretation is to a geologist as an exploration tool, it must also be viewed in the larger context of exploration decision making. That's why each case study includes a volumetric calculation of reserves. And, in several studies, the reader is asked to determine a rate of return on investment. As everyone intimately connected with petroleum exploration knows, wells can simultaneously be geologic successes and economic failures. A successful exploration program will use petrophysical log interpretation as a way to tip the risk scale toward economically successful wells.

The list of formulas which follows is an important part of petroleum exploration. A derivation of variables is included in this listing, although in most instances, the reader will not be asked to solve them. Instead, the values will be given and will need to be "plugged" into their position in the formulas.

Formulas Used for Calculating Volumetric Producible Hydrocarbon Reserves

OIL

$$N_r = \frac{7758 \times DA \times h \times \phi \times S_h \times RF}{BOI}$$

Where:

N_r = volumetric recoverable oil reserves in stock tank barrels (STB)
DA = drainage area in acres
h = reservoir thickness in feet
ϕ = porosity
S_h = hydrocarbon saturation ($1.0 - S_w$)
RF = recovery factor
BOI = oil volume factor or reservoir barrels per stock tank barrel

$$BOI = 1.05 + 0.5 \times \left(\frac{GOR}{100}\right)$$

Where:

$$\text{GOR (gas oil ratio)} = \frac{\text{gas in cubic feet}}{\text{oil in barrels}}$$

GAS

Most geologists use formula I; but an alternative, Formula II, is offered.

$$\textbf{I.}\ G_r = 43{,}560 \times DA \times h \times \phi \times S_h \times \left(\frac{Pf_2}{Pf_1}\right) \times RF$$

Where:

G_r = volumetric recoverable gas reserves in standard cubic feet (SCF)
DA = drainage area in acres
h = reservoir thickness in feet

ϕ = porosity
S_h = hydrocarbon saturation $(1.0 - S_w)$
RF = recovery factor
Pf_1 = surface pressure
Pf_2 = reservoir pressure

$$\frac{Pf_2}{Pf_1} = \frac{0.43\dagger \times \text{depth}}{15}$$

†0.43 is a universal average pressure gradient which may need to be adjusted for local conditions.

II. $G_r = 43{,}560 \times DA \times h \times \phi \times S_h \times B_{gi} \times RF$

Where:

G_r = volumetric recoverable gas reserves in standard cubic feet (SCF)
DA = drainage area in acres
h = reservoir thickness in feet
ϕ = porosity
S_h = hydrocarbon saturation $(1.0 - S_w)$
RF = recovery factor
B_{gi} = gas volume factor in SCF/cu ft

$$B_{gi} = \frac{T_{sc}}{P_{sc}} \times \frac{P}{Z \times T_f}$$

$$= \left[\frac{(459.7 + 60)}{15} \times \frac{P}{Z \times (459.7 + T_f)}\right]$$

Where:

T_{sc} = temperature at standard conditions
P_{sc} = surface pressure at standard conditions
P = reservoir pressure
Z = gas compressability factor
T_f = formation temperature (°F)

FORMATION PRESSURE ESTIMATION

1. Static mud column pressure = depth × mud weight × .052
2. Rule of thumb for static bottom hole pressure:

$$P_{ws} = P_{wh} + 0.25 \times (P_{wh}/100) \times (\text{depth}/100)$$

Where:

P_{ws} = static bottom hole pressure
P_{wh} = well head pressure

GEOTHERMAL GRADIENT ESTIMATION

$$g = \frac{(T_f - T_s) \times 100}{\text{depth}}$$

Where:

g = temperature gradient in °F/100 ft
T_f = formation temperature in °F

$$T_f = T_s + g \times (\text{depth}/100)$$

T_s = mean surface temperature
depth = formation depth in feet

As you begin trying out the various formulas in the case studies, keep in mind that many values such as water saturation (S_w) and porosity (ϕ) are given in percent. So, even though this will be immediately obvious to you, remember to change percentages to decimals before entering the numbers into your calculator.

Case Study 1
Pennsylvanian Atoka Sandstone
Permian Basin

Your company has just finished drilling a 15,900 ft wildcat well in the Permian basin. The primary target, Morrowan sandstones, are not sufficiently developed for commercial production and you are facing a decision of declaring the well dry and abandoned (D&A) unless you can find another zone which will produce hydrocarbons.

Sample cuttings from the well indicate that at 14,600 ft the Atoka Sandstone is predominantly loose, subrounded, coarse to very coarse, quartz sandstone with minor, tan, arkosic sandstone—medium-grained to coarse-grained, unsorted and friable. This information indicates that the Atoka Sandstone may have permeability because of its larger pore space developed from the coarse grain size. Better permeability is also indicated by poor cementation because the sands are friable.

Other information from the Atoka which you consider favorable is the emission of a few gas bubbles from sample cuttings and also gas on the mud logger's chromatograph, an instrument designed to measure the amount and type of gas in drilling muds. Total gas background on the chromatograph increased from 10 units to 40 units of gas with a trace of C-2 and C-3 hydrocarbons during drilling of the sandstone's bottom four feet.

Even though you are optimistic about the productive potential of the Atoka Sandstone because of the nature of the sample cuttings and the gas shows, you are concerned about a reverse drilling break which occurred as the Atoka was penetrated. The drill penetration rate was 8 to 10 min/ft before, and again after, the Atoka Sandstone. However, the penetration rate through the Atoka was 15.5 min/ft and decreased to 20 min/ft through the bottom 10 ft. The slower drilling times may mean that the sandstone doesn't have the porosity and permeability suggested by sample examination.

You are now preparing to calculate, and then assess, the log parameters needed to determine whether the Atoka Sandstone might produce.

You have the following information: (1) resistivity of formation water (R_w) = 0.065 at T_f from your logging engineer; (2) resistivity of mud filtrate (R_{mf}) = 0.65 at T_f from the log header and corrected to formation temperature by the Arps equation (see Chapter I); (3) temperature of the formation (T_f) = 187°F from the estimation of formation temperature chart (see Chapter I); and (4) the surface temperature = 75°F from an estimation by the well site geologist.

Complete the following Pennsylvanian Atoka Sandstone Log Evaluation Table (work Table A). Five depths were selected for your convenience and are listed in the table.

You may find this list of formulas helpful as you pursue your calculations:

Neutron Density Porosity—Values for neutron (N) and density (D) porosity read on neutron-density log (Fig. 55).

$$\phi_{N\text{-}D} = \sqrt{\frac{\phi_D^2 + \phi_N^2}{2}}$$

R_t Minimum—Value for LL-8* from log reading (Fig. 54) used to correct R_{ILd}* to R_t in thin, resistive zones (Chapter III).

$$R_{t\,min} = (\text{LL-8* or SFL*}) \times (R_w/R_{mf})$$

Water Saturation Archie—Here, ϕ is the value of $\phi_{N\text{-}D}$ from the neutron-density formula. Also, in this example, R_t is equal to either $R_{t\,min}$ or to R_t from the tornado chart in Appendix 6 (page 212).

$$S_{wa} = \sqrt{\frac{0.81}{\phi^2} \times \frac{R_w}{R_t}}$$

Water Saturation Ratio—Value for R_{xo} is from the shallow Laterolog* (Fig. 54). R_t value is the larger value when $R_{t\,min}$ formula and tornado chart R_t are compared.

$$S_{wr} = [(R_{xo}/R_t)/(R_{mf}/R_w)]^{0.625}$$

Moveable Hydrocarbon Index:

$$S_w/S_{xo} = [(R_{xo}/R_t)/(R_{mf}/R_w)]^{1/2}$$

Water Saturation Corrected:

$$S_{w\,corr} = S_{wa} \times \left(\frac{S_{wa}}{S_{wr}}\right)^{0.25}$$

Bulk Volume Water—Porosity (ϕ) in this example is neutron-density porosity ($\phi_{N\text{-}D}$).

$$BVW = \phi \times S_{wa}$$

Residual Oil Saturation—S_{xo} is water saturation of flushed zone calculated by: $S_{xo} = [0.81/\phi^2 \times (R_{mf}/R_{xo})]^{1/2}$

$$ROS = 1.0 - S_{xo}$$

Moveable Oil Saturation:

$$MOS = S_{xo} - S_w$$

Irreducible Water Saturation—Formation factor (F) equal to $0.81/\phi^2$ in consolidated sands (see Table 1). This formula calculates an approximate value for irreducible water saturation. It should be used only in crossplots where you are trying to determine the relative permeabilities: K_{ro}, K_{rg}, and K_{rw}.

$$S_{w\,irr} = \sqrt{F/2000}$$

Volumetric Recoverable Gas Reserves—Variables defined in introduction to case studies.

G_r = 43,560 × drainage area (DA) × reservior thickness (h) × porosity (ϕ) × hydrocarbon saturation (S_h) × gas volume factor (B_{gi}) × recovery factor (RF)

Work Table A:
Pennsylvanian Atoka Sandstone Log Evaluation Table

Depth	ILd	ILm	LL8	GR	Φ_N	Φ_D	R_t†	R_{tmin}	$\Phi_{N\text{-}D}$	S_{wa}	S_{wr}	S_w/S_{xo}	S_{wcorr}	BVW
14604	120	220	1000	17.5	7	22	75.9	100	16	14	100	1.00	9	.022
14608	160	200	1150	21	6.5	17.5								
14612	270	350	1700	16	9	24.5								
14616	160	190	450	19	10.5	24								
14620	90	130	350	17.5	9.5	22.5								

†$R_t = R_{ILd}$ corrected for invasion by tornado chart (Chapter III).

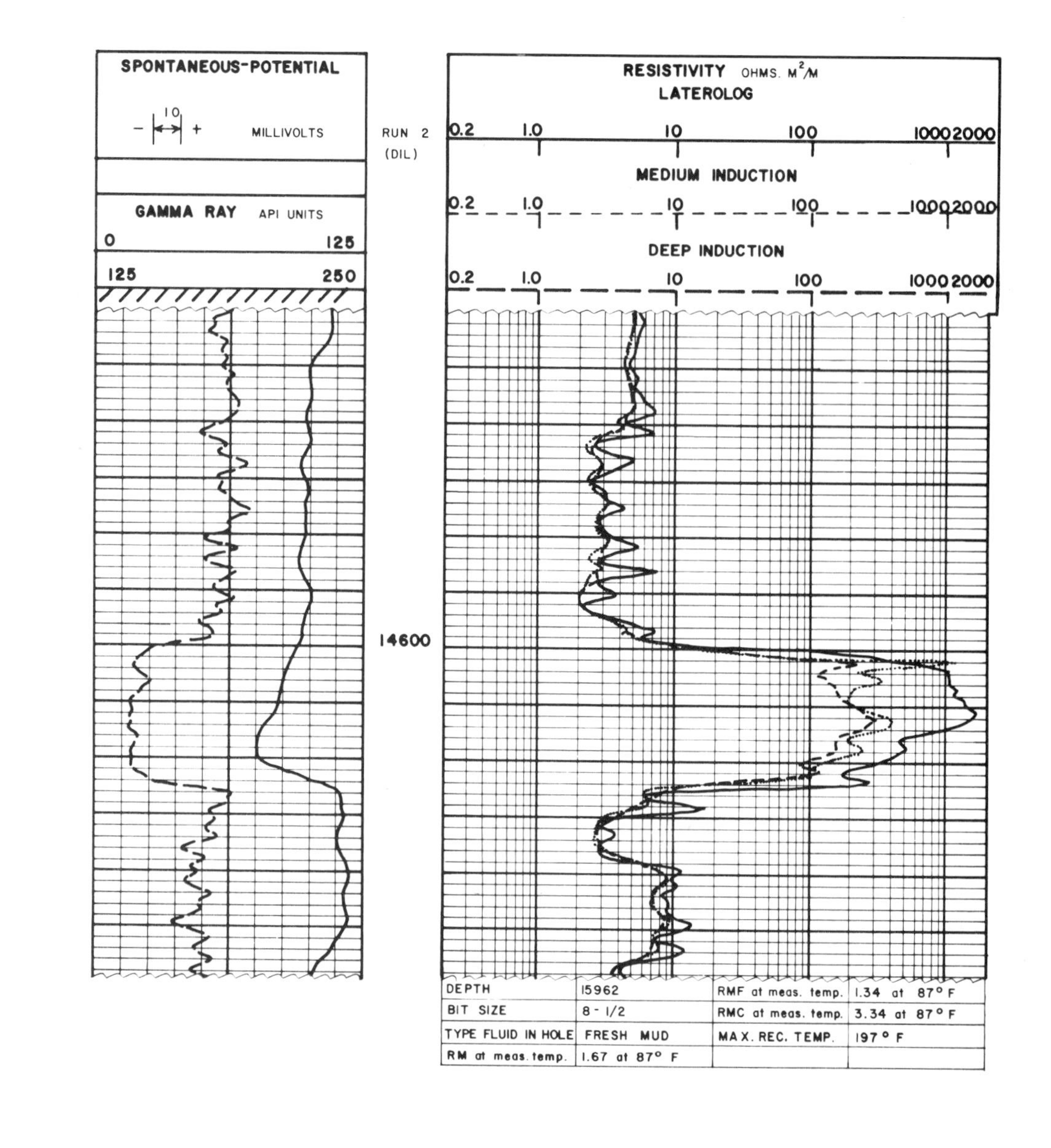

Figure 54. Dual Induction Laterolog* with Spontaneous Potential and Gamma Ray Log, Pennsylvanian Atoka Sandstone, Permian basin. Values from this log are used in work Table A.

At a depth of 14,600 to 14,625 ft, note:

1. High resistivities on resistivity logs in tracks #2 and #3.
2. Hydrocarbon suppression of SP log in track #1 (solid line).
3. Gamma ray log (track #1 dashed line) has lower gamma ray count in Atoka Sandstone interval because sands have lower radioactivity than shales.

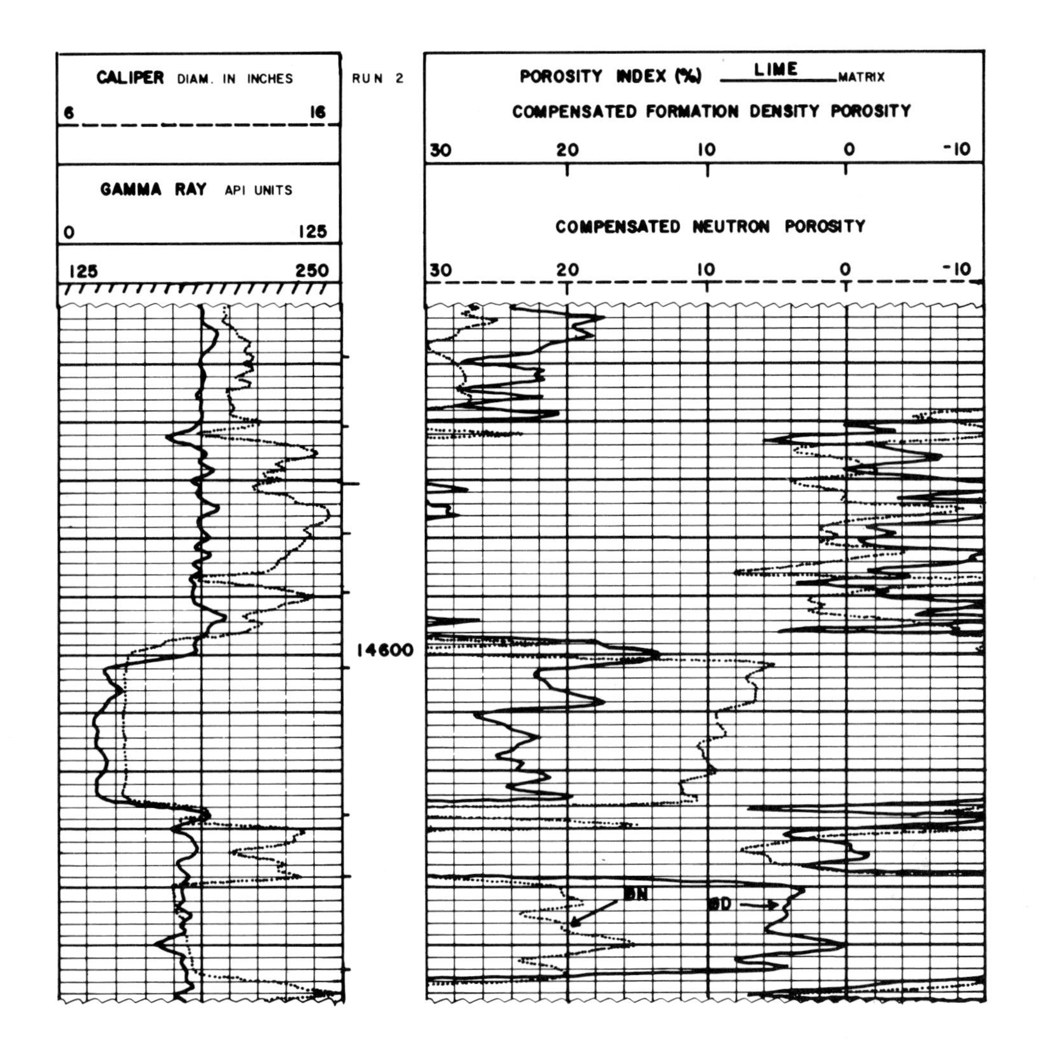

Figure 55. Combination Neutron-Density Log with Gamma Ray Log and caliper, Pennsylvanian Atoka Sandstone, Permian basin. Values from this log are used in work Table A.

From a depth of 14,600 to 14,625 ft, note:

1. Strong gas effect, tracks #2 and #3 (i.e. density log reads much higher porosities than the neutron log) and high porosity values on neutron-density logs in tracks #2 and #3.
2. Mudcake on caliper log in track #1 (dashed line). Mudcake is indicated because the hole diameter, as shown on the caliper log, is getting smaller.

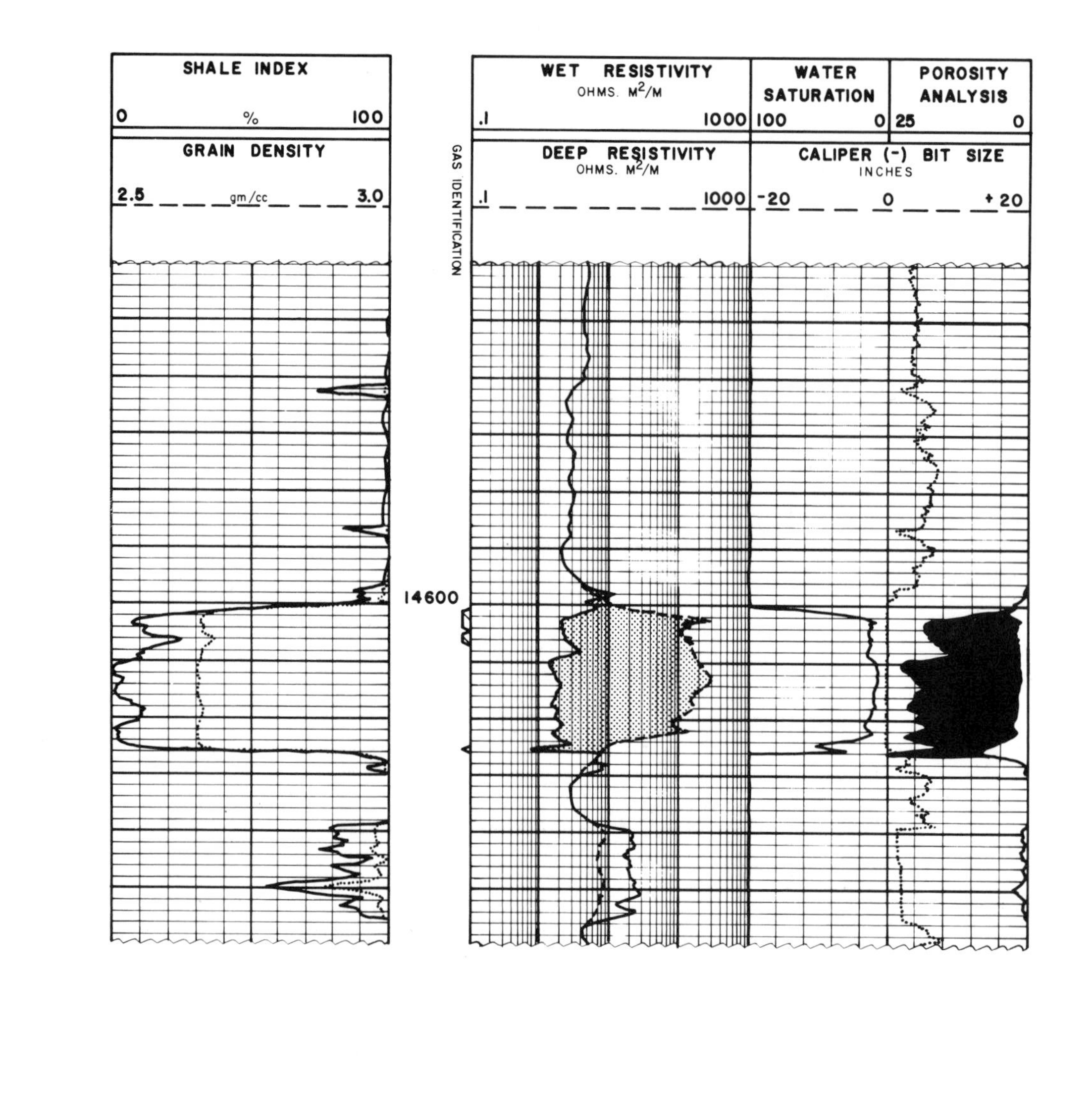

Figure 56. Computer processed Cyberlook* Log, Pennsylvanian Atoka Sandstone, Permian basin.

From a depth of 14,600 to 14,625 ft, note:

1. Low water saturations (indicated on left side of track #3) and high porosities (indicated by curve on right-hand side of track #3) calculated by Cyberlook* Log.
2. Shading in track #2 indicates separation between wet resistivity (R_o) and deep resistivity (R_{ILd}). The greater the separation, the greater the hydrocarbon saturation.

Case Study 1 Answer

A careful examination of the logs recorded through the Atoka Sandstone provides the following information. On the caliper log (Fig. 55), mudcake development is shown by the decreasing hole size. This mudcake development is important because it indicates a permeable zone. Mudcake forms by the accumulation of solid particles from drilling muds on the borehole walls when a porous and permeable zone is invaded with drilling fluids. The SP log (Fig. 54) shows hydrocarbon suppression because the SP reading is less than SSP. Such suppression suggests the presence of hydrocarbons; sample shows and the neutron-density log response (Fig. 55) through the Atoka tell you that in this case the hydrocarbons will probably be gas. The density log (Fig. 55) reads a much higher porosity than the neutron log. When this occurs, it is because of gas effect.

If water saturation Archie (S_{wa}) and porosity (ϕ) values are crossplotted (Fig. 57), a grain size variation of medium-grained to coarse-grained is shown on the plot. This supports sample cutting examination which indicates a medium-grained to coarse-grained sandstone.

Crossplotting irreducible water saturation versus porosity (see Fig. 58) reveals that the Atoka zone has good permeability with values ranging from 10 to over 100 millidarcies (md). Another crossplot in Figure 59, comparing a calculated irreducible water saturation ($S_{w\,irr} = \sqrt{F/2000}$) with water saturation Archie, shows a high relative permeability to gas (K_{rg}). The high relative permeability to gas (K_{rg}) means there is a correspondingly low permeability to water (K_{rw}). The reservoir, therefore, should not produce water.

The values derived for both water saturation and bulk volume water are low. These low values also indicate that the Atoka Sandstone is a reservoir with a high gas saturation at irreducible water saturation. Because water saturation values are low, it follows that the reservoir must have high gas saturation (remember: $1.0 - S_w$ = hydrocarbon saturation).

Other evidence that the reservoir is at irreducible water saturation and has high gas saturations can be interpreted from a crossplot (Fig. 60) of bulk volume water. This crossplot is created by plotting water saturation Archie versus porosity (ϕ). The data points cluster along a hyperbolic line and have, then, approximately equal values for bulk volume water (BVW). Bulk volume water values plotted along the hyperbolic line range from 0.015 to 0.027. Their clustering also supports the conclusion that the reservoir has high gas saturation at irreducible water saturation.

Values calculated for residual oil saturation (ROS) are high. This is anomalous because other evidence supports the conclusion that the reservoir has high porosity and permeability. Under these conditions, it could ordinarily be expected that hydrocarbons would be moved out instead of left behind in the rock. Furthermore, calculations of moveable oil saturation (MOS) values are also low. Again, this is initially puzzling because, in a good reservoir with high porosity and permeability, these values should be high.

Additional anomalous information comes from the calculation of the moveable hydrocarbon index. The values are greater than 0.7, but (usually) favorable moveability values are less than 0.7. Analyses of ROS, MOS, and the moveable hydrocarbon index provide negative evidence that hydrocarbons will move. Indeed, all of these factors suggest that most of the hydrocarbons will remain in place in the reservoir.

While information relating to inadequate hydrocarbon moveability should not be ignored in initial log evaluation, it can sometimes be explained in the following way. In this case, what is significant is the high residual gas saturation in the flushed zone ($1.0 - S_{xo}$ = residual hydrocarbon saturation). This high residual gas saturation is the result of the bypassing of gas by drilling fluids invading a reservoir. The high residual gas saturation left behind after invasion can be erroneously read by the logs as unmoved hydrocarbons ($S_{xo} < S_w^{1/5}$; Chapter VI).

Even though you are concerned about negative information from the drill penetration rate through the Atoka Sandstone, and about the pessimistic moveability data, you decide that the other evidence from sample examination, gas shows, and log interpretation supports a decision to set pipe. Log interpretation information especially significant to your decision is: high porosities on the neutron-density logs, strong gas effect on the neutron-density logs, low water saturation calculated by the Archie equation, high log-derived permeabilities from various permeability crossplots, and the low bulk volume water values. While you believe the well contains a gas-filled Atoka Sandstone reservoir, you want to determine whether or not the well will be a commercial success. To do so, you calculate volumetric gas reserves before making your final pipe setting decision.

An estimation of Atoka Sandstone gas recovery of 11.0 BCF is calculated by using the following parameters: geothermal gradient = 0.014 × formation depth; pressure gradient = 0.35 × formation depth; drainage area = 560 acres; reservoir thickness = 15 ft; effective porosity = 15%; water saturation = 13%; gas gravity = 0.65; recovery factor = 0.85; formation temperature = 205°F; initial bottom hole pressure = 5,117 PSI; Z factor = 0.988.

The Atoka Sandstone was perforated from 14,610 to 14,615 ft. The calculated absolute open flow (CAOF) was 21,900,000 cu ft of gas per day (21,900 mcfgpd) with a high shut-in tubing pressure (SITP) of 3,758 pounds per square inch (PSI) and a high initial bottom hole pressure (IBHP) of 5,556 PSI. The gas gravity was 0.599 at a bottom hole temperature of 219°F. The well's first year cumulative production was 3,268,129 mcf plus 95,175 barrels of condensate.

Answer Table A:
Pennsylvanian Atoka Sandstone Log Evaluation Table

Depth	ILd	ILm	LL8	GR	Φ_N	Φ_D	R_t	R_{tmin}	$\Phi_{N\text{-}D}$	S_{wa}	S_{wr}	S_w/S_{xo}	S_{wcorr}	BVW	S_{xo}	MOS†	ROS	S_{wirr}††
14602	170	600	450	20	5.5	20	68	45	15	19	77	.81	13	.029	23	4	77	13
14604	120	220	1000	17.5	7	22	75.9	100	16	14	100	1.00	9	.022	14	0	86	13
14606	150	300	1050	25	6.5	20.5	60	105	14.5	15	100	1.00	9	.022	15	0	85	14
14608	160	200	1150	21	6.5	17.5	157.3	115	13	14	82	.87	9	.018	17	3	83	16
14610	210	190	1400	16	9.5	26.5	210	140	20	8	78	.80	5	.016	10	2	90	11
14612	270	350	1700	16	9	24.5	261.7	170	18.5	8	76	.90	5	.015	10	2	90	11
14614	250	380	1200	16	9	22.5	207.3	120	17	9	71	.77	5	.015	12	3	88	12
14616	160	190	450	19	10.5	24	157.9	45	18.5	10	46	.52	7	.019	19	9	81	11
14618	155	250	500	17.5	10	23.5	97.1	50	18.0	13	66	.70	9	.023	18	5	82	11
14620	90	130	350	17.5	9.5	22.5	77.6	35	17	15	61	.67	11	.026	23	8	77	12
14622	120	105	180	17.5	12	23.5	120	18	19	11	31	.38	9	.021	29	18	71	11

†MOS = moveable oil saturation, or $S_{xo} - S_w$

†† value for $S_{w\,irr}$ used on relative permeability crossplot charts, is based on: $S_{w\,irr} \simeq \sqrt{F/2000}$

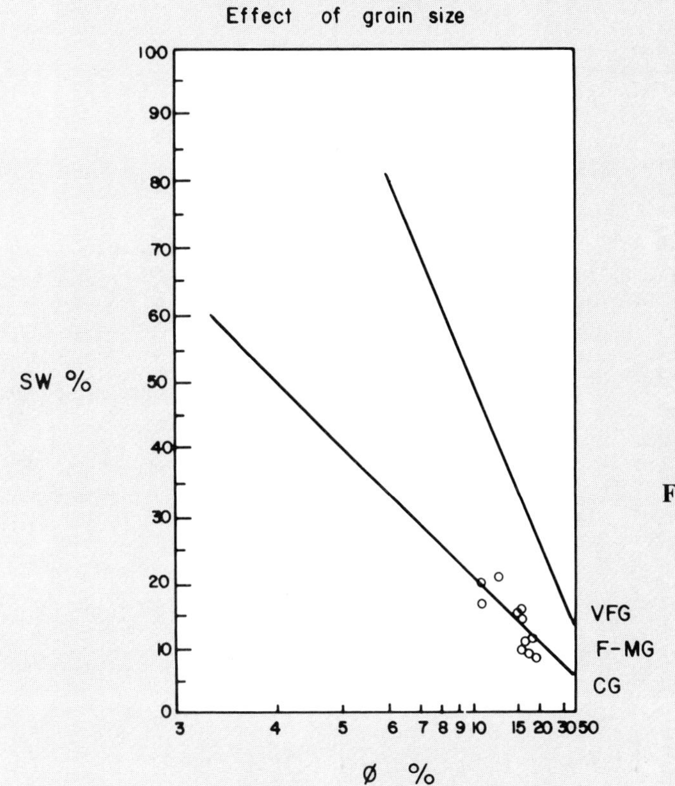

Figure 57. Grain size determination by water saturation (S_w) versus porosity (ϕ) crossplot, Pennsylvanian Atoka Sandstone, Permian basin.

Note:

Remember that water saturation Archie (S_{wa}) is equal to $S_{w\,irr}$ in zones at irreducible water saturation.

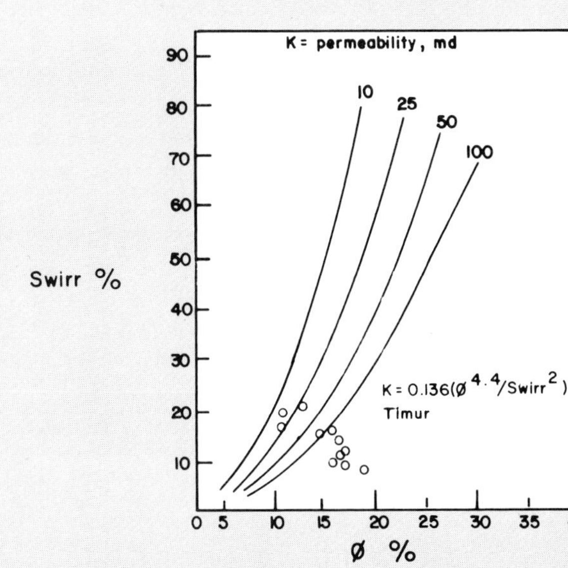

Figure 58. Irreducible water saturation ($S_{w\,irr}$) versus porosity (ϕ) crossplot for determining permeability, Pennsylvanian Atoka Sandstone, Permian basin.

Note:

Remember that water saturation Archie (S_{wa}) is equal to $S_{w\,irr}$ in zones at irreducible water saturation.

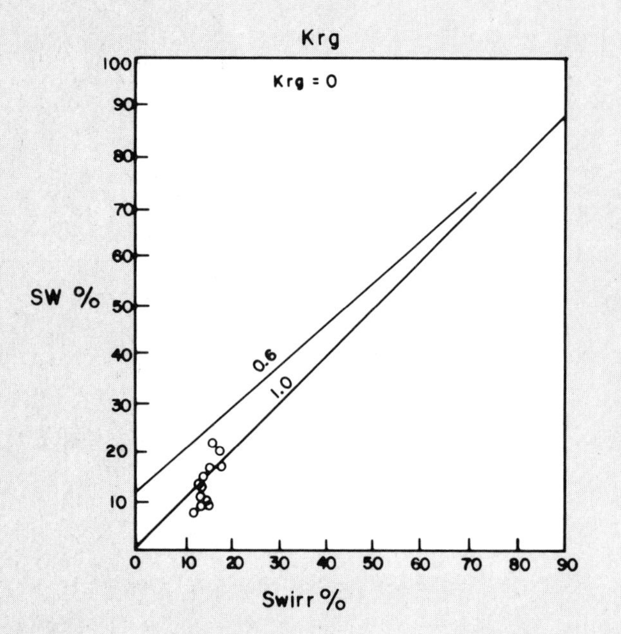

Figure 59. Irreducible water saturation ($S_{w\,irr}$)† versus water saturation (S_w)†† crossplot for determining relative permeability to gas (K_{rg}), Pennsylvanian Atoka Sandstone, Permian basin.

Note:

†$S_{w\,irr} = \sqrt{F/2000}$ This formula calculates an approximate, theoretical value for $S_{w\,irr}$. Values calculated for $S_{w\,irr}$ by this formula should only be used in crossplots to determine *relative* permeabilities: K_{ro}, K_{rg}, and K_{rw}.

††In this example, S_w is equal to water saturation Archie (S_{wa}).

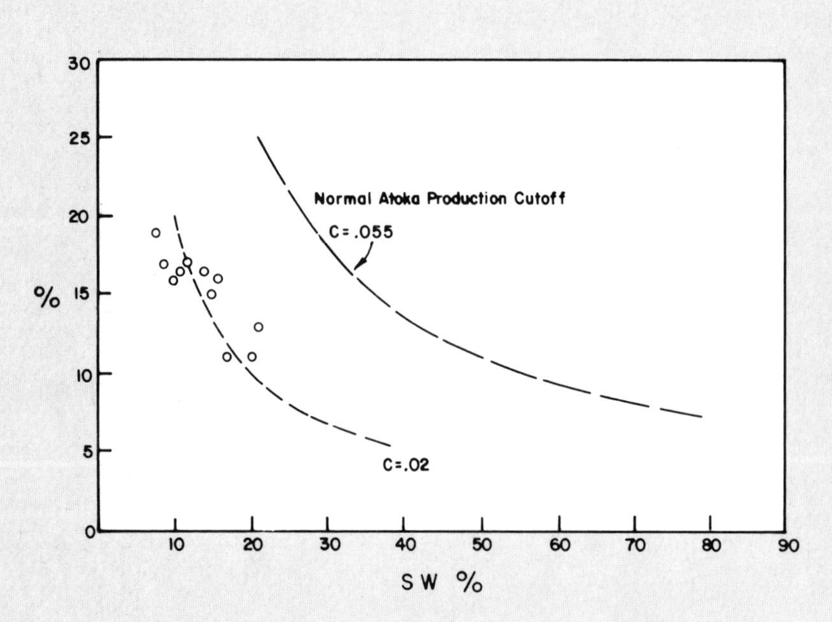

Figure 60. Bulk volume water crossplot (ϕ vs. S_w), Pennsylvanian Atoka Sandstone, Permian basin.

C stands for bulk volume water.

Note:

Data points cluster along the 0.02 hyperbolic line; the closer they are to the line, the closer you are to irreducible water saturation ($S_{w\,irr}$).

Case Study 2
Mississippian Mission Canyon Formation
Williston Basin

An Ordovician Red River wildcat well is in progress in the Williston basin. As drilling proceeds, two zones—the Mississippian Mission Canyon and Devonian Duperow—are encountered which have hydrocarbon shows. The well has just penetrated the Devonian Duperow, the second "show" zone. A decision is made to stop and drill stem test (DST) the Duperow. Unfortunately, the drill stem test tool fails and becomes firmly stuck. Several attempts are made to remove the tool but none succeed. Meanwhile, wall collapse (caving) begins to develop within the well bore.

Now you face the choice of either halting drilling with the hope that the first show zone, the Mississippian Mission Canyon Formation, might be a productive reservoir or, of continuing efforts to remove the DST tool. If the fish (the stuck DST tool) is successfully removed from the hole, the Duperow can be tested and drilling continued to the Red River Formation. However, continued attempts to recover the fish will undoubtedly lead to further deterioration of the hole, and logging measurements taken when hole conditions are poor, may be unreliable. Also, it is possible the DST tool cannot be dislodged.

Because of the exploratory nature of the well, you made a decision several weeks ago to core the Mission Canyon at a depth of 9,302 to 9,358 ft. Twenty-five feet of oil-stained, fractured, microcrystalline dolomite was recovered, and the remaining core consisted of microcrystalline limestone and anhydrite. After coring, se eral DST's were tried and when none were successful, the well was drilled ahead.

You decide the information from coring is favorable enough to halt drilling, to log to the top of the fish while hole conditions appear still reasonably good, and then to assess the potential of the Mission Canyon. Your company agrees with your suggestion that a very complete log package be run in the hole, since the well is exploratory and carbonates can often be harder to evaluate than sandstones.

The following data are available to you: (1) resistivity of the formation water (R_w) at $T_f = 0.023$; (2) resistivity of the mud filtrate (R_{mf}) at $T_f = 0.017$; (3) temperature of the formation (T_f) = 207°F; (4) $\Delta t_{ma} = 44.4$ μsec/ft, which is the interval transit time for a dolomite matrix (Table 6); (5) $\Delta t_f = 185$ μsec/ft, which is the interval transit time of fluid for saltwater mud (Chapter IV); (6) $\rho_{ma} = 2.82$ gm/cc, which is the matrix density for dolomite (Table 7); (7) ρ_f = 1.1 gm/cc, which is the fluid density for saltwater mud (Chapter IV); and (8) the surface temperature = 60°F.

Preliminary examination of a Cyberlook* Log reveals higher water saturations in the lower part of the Mission Canyon Porosity zone (Fig. 66). Consequently, you decide to begin your evaluation of the Mission Canyon by developing a Pickett crossplot which will give you a quick analysis of the distribution of the different water saturations. Values from the log (Fig. 62) for neutron porosity (ϕ_N) and from the log (Fig. 61) for deep resistivity (R_{LLd}*), when crossplotted on two-by-three cycle log-log paper, will show these water saturation distributions.

In order to establish an R_o line in the Pickett crossplot, you use the following information:

$$R_o = R_w \times F$$

Where 10% porosity, $F = 1/\phi^2$ or 100

$$R_o = 0.023 \times 100$$

or

$$R_o = 2.3 \text{ ohm/meters}$$

The slope of the R_o line equals 2 (average slope for carbonates).

Choose any five points (depths) in the following Mission Canyon Log Evaluation Table (work Table B) and calculate the values for the blank spaces (complete the Table for those five depths).

After determining values for the different log parameters, you can use the formula for volumetric calculation of recoverable oil to evaluate the productive potential of the Mississippian Mission Canyon Formation. The formula for volumetric calculation of recoverable oil is:

$$N_r = \frac{7758 \times DA \times h \times \phi \times S_h \times RF}{BOI}$$

Parameters used in your calculations of recoverable oil are: drainage area (DA) = 150 acres; reservoir thickness (h) = 28 ft; porosity (ϕ) = 11%; water saturation (S_w) = 33.5%; recovery factor = 20%; and reservoir barrels per stock tank barrel (BOI) = 1.35. When the equation is solved for N_r the resulting value for N_r represents stock tank barrels. A stock tank barrel is oil recovered at the surface after shrinkage has occurred as gas separates.

The Mississippian Mission Canyon Formation Log Evaluation Table (work Table B) has already been partly completed for you. Fill in the rest of the table as you proceed with your calculations.

Work Table B:
Mississippian Mission Canyon Formation Log Evaluation Table

Depth	LLd	LLs	MSFL*	Φ_N	Φ_D	R_t	Φ_{N-D}	S_{wa}	S_{wr}	S_{xo}	S_w/S_{xo}	MOS	ROS	BVW
9308	35	13	4	14.5	6	48.2								
9310	27	22	1	24	9	37.6								
9312	60	20	15	20	0	84.0								
9326	14	9	3.5	10.5	2	18.3								
9332	8.6	6	2.2	15	5	11.2								
9336	16	4.8	1.1	20	9	22.7								
9338	18	8	2.4	23.5	9	24.4								
9340	23	6.5	9	17	5	32.9								
9346	4	2	.75	21	6	5.3								
9352	5	2.2	1.6	21	2.5	6.7								
9354	5	2.8	1.1	23.5	6	6.6								
9376	4.6	3.2	1.9	17	7.5	5.7								
9386	3.5	1.7	1.6	13	8	4.6								

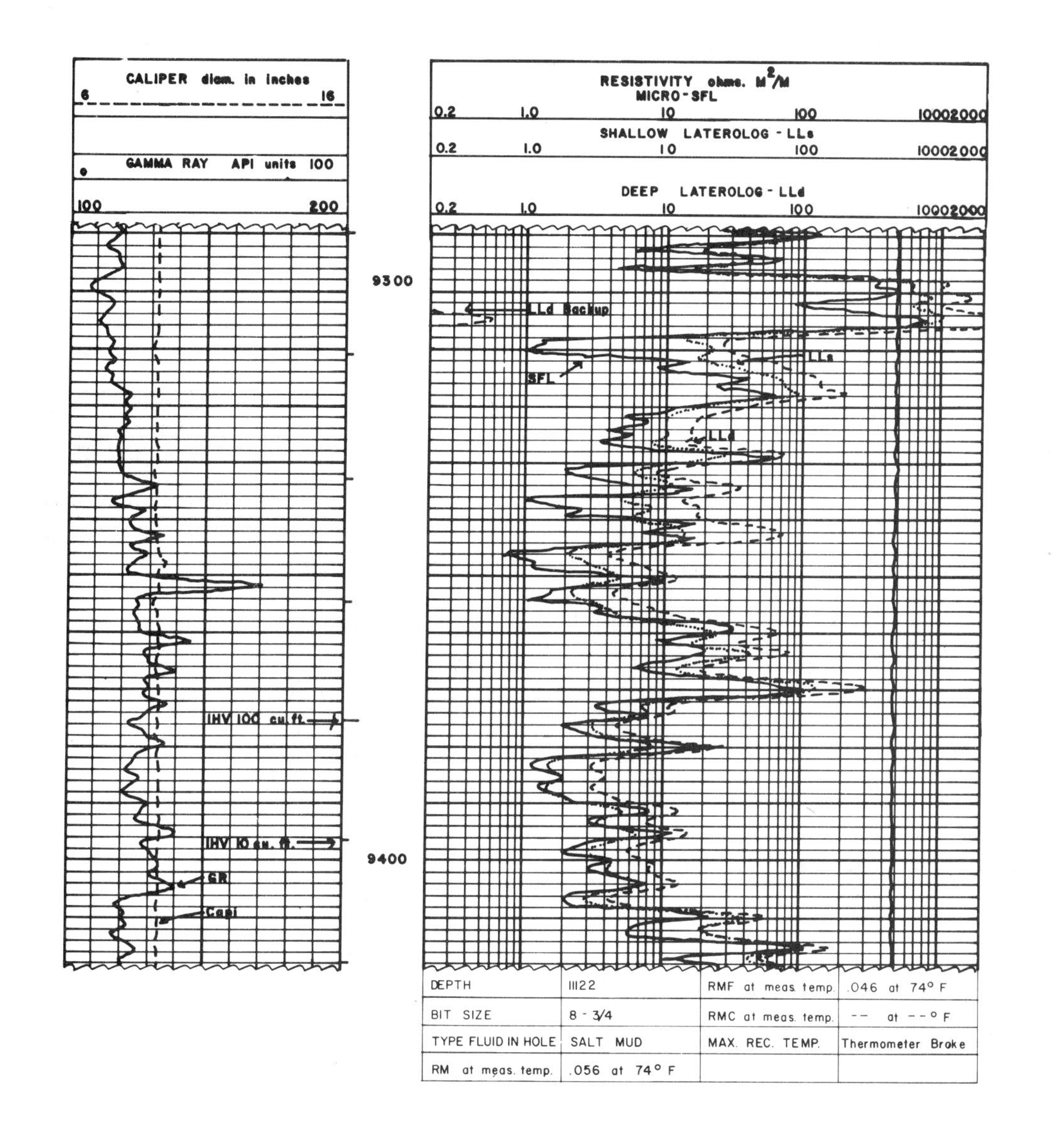

Figure 61. Dual Laterolog*—MSFL* with gamma ray log and caliper, Mississippian Mission Canyon Formation, Williston basin.

From a depth of 9,308 to 9,408 ft, note:

1. The separation of the three resistivity logs which read the following resistivities:

   R_{MSFL}* (R_{xo})—low resistivity
   R_{LLS} (R_i)—intermediate resistivity
   R_{LLd} (R_t)—high resistivity

2. This type of resistivity profile on a Dual Laterolog*—MSFL* indicates the presence of hydrocarbons (see Chapter I; Fig. 7B).

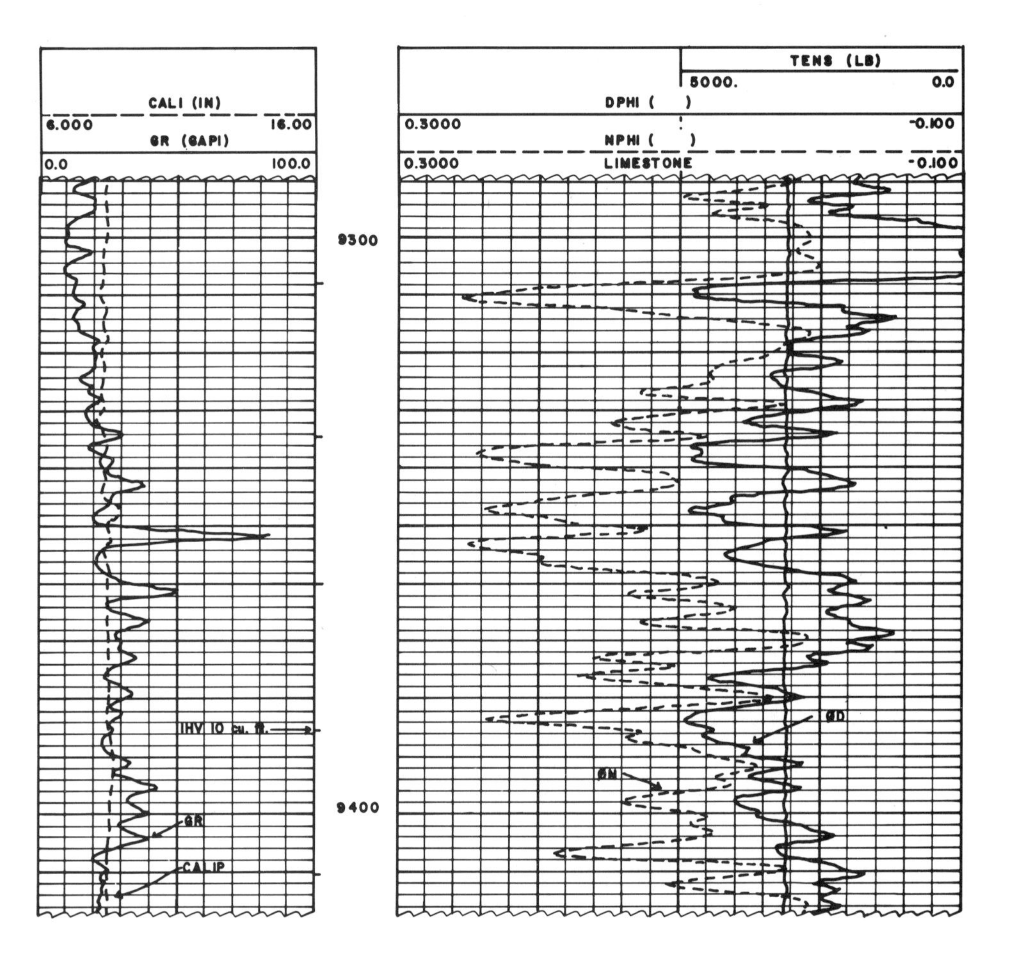

Figure 62. Combination Neutron-Density Log with gamma ray log and caliper, Mississippian Mission Canyon Formation, Williston basin.

From a depth of 9,308 to 9,408 ft, note:

1. The high porosities on the neutron and density logs (tracks #2 and #3).
2. The neutron log reads higher porosity than the density log indicating the lithology is dolomite (tracks #2 and #3).

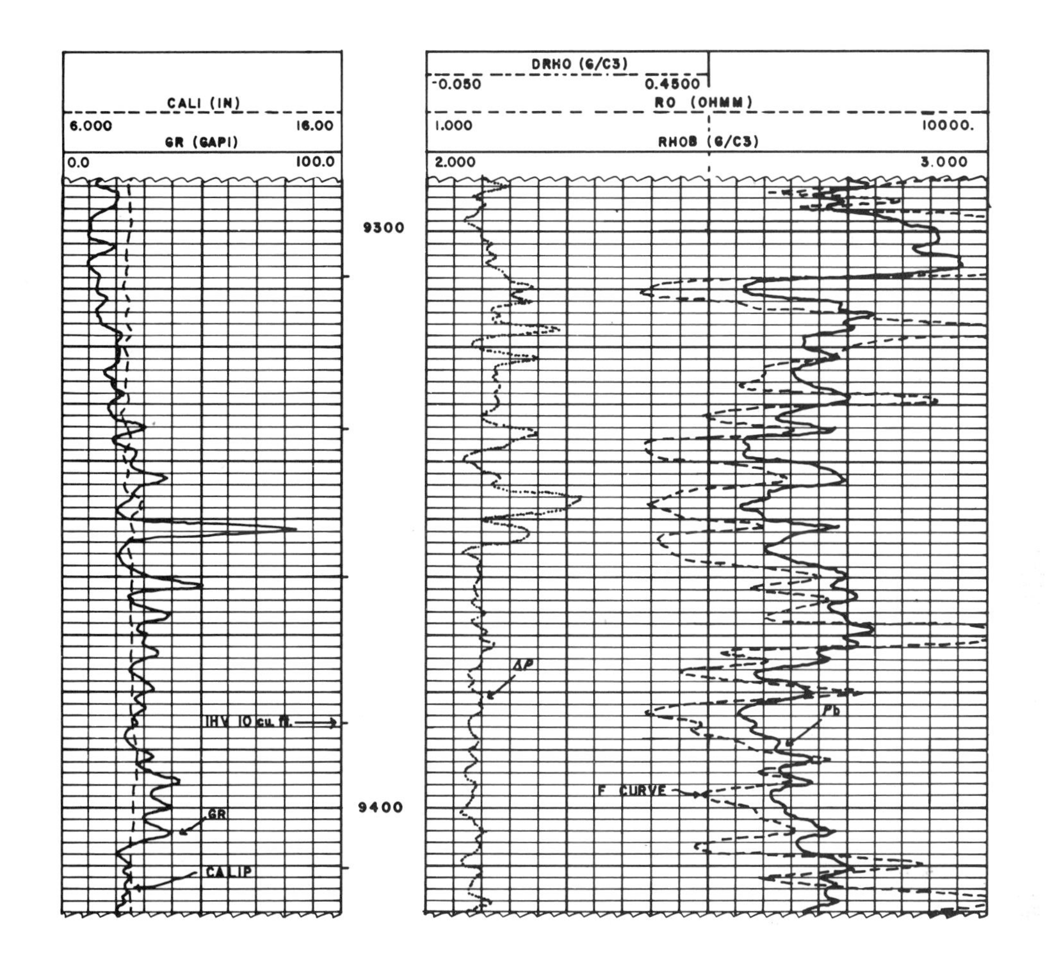

Figure 63. Density log with F curve, gamma ray log and caliper, Mississippian Mission Canyon Formation, Williston basin.

From a depth of 9,308 to 9,354 ft, note:

The increased amount of correction on the bulk density correction curve (Δ_ρ). This increase on the correction curve may indicate the presence of fractures in the Mission Canyon (tracks #2 and #3).

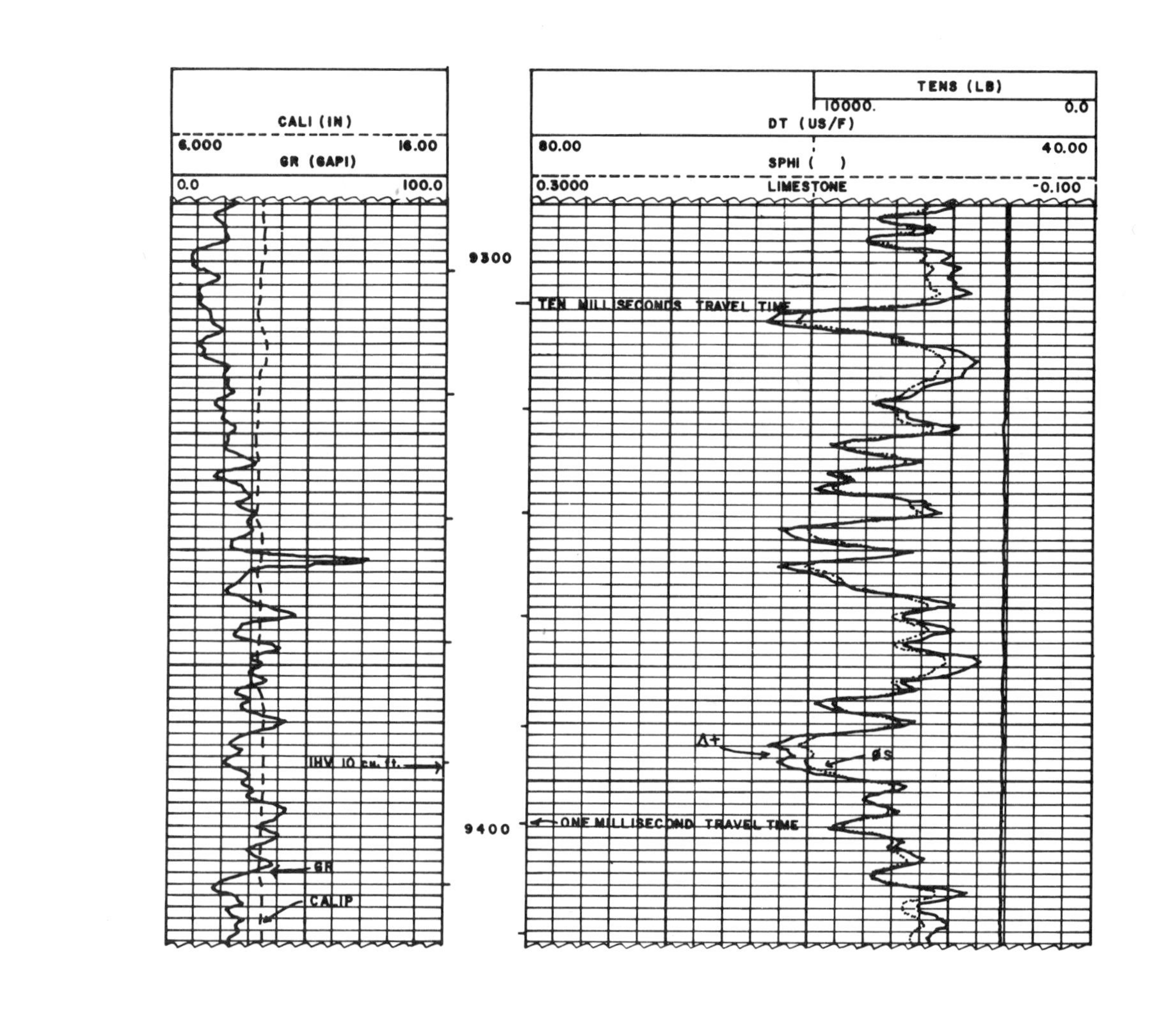

Figure 64. Sonic log with gamma ray log and caliper, Mississippian Mission Canyon Formation, Williston basin.

From a depth of 9,308 to 9,408, note:

The numerous porosity zones indicated by the increasing interval transit time (Δt) on tracks #2 and #3.

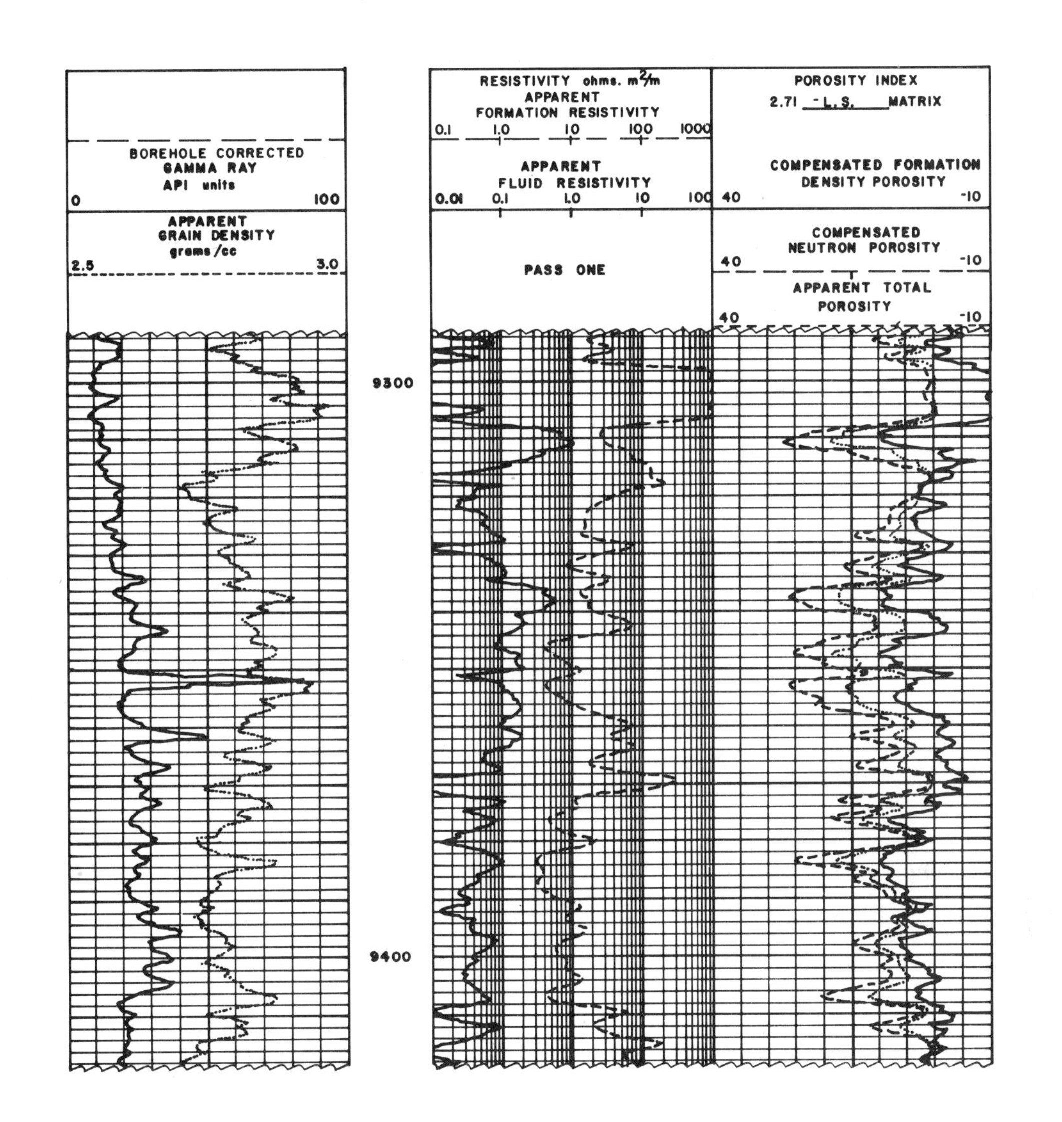

Figure 65. Computer processed Cyberlook* Log (pass-one), Mississippian Mission Canyon Formation, Williston basin.

Note:

1. The data on this log is used to generate the pass-two Cyberlook* Log (Fig. 66).
2. Pass-one is used as a check of the computer's input of logging parameters before the pass-two log is generated (Fig. 66).

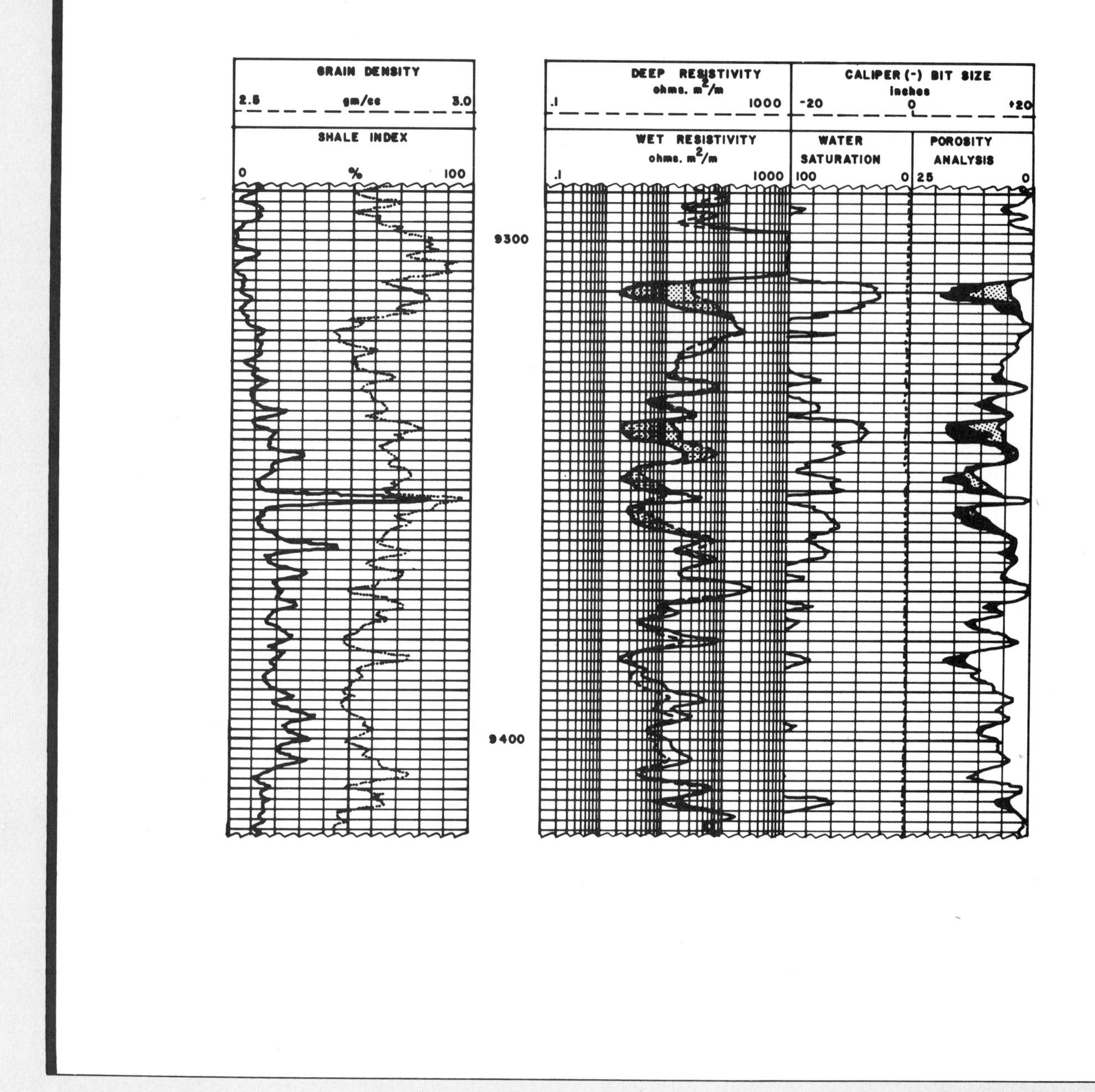

Figure 66. Computer processed Cyberlook* Log (pass-two), Mississippian Mission Canyon Formation, Williston basin.

From a depth of 9,360 to 9,408 ft, note:

The higher water saturations over this interval, indicating the lower part of the Mission Canyon Formation is probably water productive (wet).

Case Study 2 Answer

The very complete log package run in this well includes an electric log suite for resistivity measurements, and both a Combination Neutron-Density Log and a sonic log for porosity measurements. A Cyberlook* or computer processed log is also part of the package. It is used for a quick look examination of the well to identify zones with low water saturations (i.e. possible productive zones).

Because the borehole has experienced some caving problems, your assessment begins with a careful check of the caliper log (Fig. 61). The caliper log shows a relatively constant hole diameter and no intervals of significant hole enlargement due to washout. The constancy of the hole diameter means log measurements should be reliable.

Your next evaluation step includes an examination of the resistivity logs (Fig. 61). The salt saturated drilling mud ($R_{mf} \simeq R_w$) in the well has necessitated using a Dual Laterolog* with a Microspherically Focused Log (MSFL*).

The MSFL* measures the resistivity of the flushed zone (R_{xo}), while the Laterolog* shallow (LLS) and Laterolog* deep (LLD) measure the resistivities of the invaded (R_i) and uninvaded (R_t) zones, respectively.

You begin scrutinizing these resistivity logs to identify invasion profiles. Invasion profiles help you locate zones which merit a more detailed analysis.

Between depths of 9,308 and 9,415 ft, the resistivity logs—MSFL*, LLS, and LLD—read different values for R_{xo}, R_i, and R_t; and the curves separate. The curve separation suggests that invasion has taken place and hydrocarbons are present in porous and permeable zones occurring intermittently over the interval.† However, you note that the lower porosity zones from 9,370 to 9,415 ft have less separation between the Microspherically Focused Log* (MSFL* reading R_{xo}; see Fig. 61) and the deep Laterolog* (LLD reading R_t: see Fig. 61). The lessening of separation in the lower zones indicates higher water saturations (remember: higher water saturations mean lower hydrocarbon saturations).

Porous and permeable zones, which occur intermittently over the interval from 9,308 to 9,415 ft, are identified by analyzing the Combination Neutron-Density Log (Fig. 62), the density log (Fig. 63), and the sonic log (Fig. 64). Approximately eleven different porous and permeable zones can be identified on these logs from a depth of 9,308 to 9,415 ft. On the Combination Neutron-Density Log (Fig. 62), the zones of porosity and permeability are seen by an increase in both neutron and density porosity. They appear on the density log (Fig. 63) as a decrease in bulk density (ρ_b). Finally, they are identified on the sonic log by an increase in the interval transit time (Δt).

A check of the Cyberlook* Log (Fig. 66) verifies that the lower Mission Canyon zones from 9,370 to 9,415 ft have higher water saturations. You note in track #3 of the Cyberlook* Log an increase in water saturations with increasing depth.

Further verification of water problems from a depth of 9,370 to 9,415 ft comes from a Pickett crossplot (Fig. 67). On the plot, data points with water saturations above 35% are mostly from lower zones.

As you continue your evaluation of the Mission Canyon, you decide to compare your observations of the core lithologies with lithologies derived from log data. Your study of the core indicates it is microcrystalline dolomite, limestone and anhydrite. To compare this information with log data, you construct a MID* plot, a neutron-density porosity crossplot, and an M-N* plot. (Ordinarily you would probably construct only one of these lithology plots when evaluating a well, but all are presented here as a learning experience.)

By crossplotting $(\Delta t_{ma})_a$† versus $(\rho_{ma})_a$† on a MID* plot (Fig. 68), you determine that the interval you are studying has a matrix which varies from dolomite to dolomitic limestone. And, because the average $(\Delta t_{ma})_a$ is 44.4 μsec/ft (Fig. 69) and the average $(\rho_{ma})_a$ is 2.82 gm/cc (Fig. 70), the interval has an average lithology of limey dolomite.

The neutron-density crossplot (Fig. 71) shows porosities varying from 4 to 17%. The clustering of points between dolomite and limestone supports a judgement that lithology is a limey dolomite.

An M-N* plot (Fig. 72) suggests the presence of secondary porosity because many data points are plotted above the calcite-dolomite lithology tie-line. Once more, like the MID* and the neutron-density crossplots, the M-N* plot indicates a lithology varying from dolomite to dolomitic limestone.

Another crossplot (Fig. 73) is useful for establishing grain size. A plot of water saturation (S_w) versus porosity (ϕ) shows grain size variations from coarse-grained to fine-grained. However, data which cluster in the area of coarse or larger grain sizes probably don't reflect the grain size of the intercrystalline porosity. Rather, this data clustering in the larger size areas may, instead, reflect vuggy porosity. Data which cluster above very fine-grained

†Of course, the reverse would be true in a salt saturated mud system if all three resistivity curves—MSFL*, LLS, and LLD—had essentially the same values and separation did not occur. You would then conclude either invasion hadn't occurred or hydrocarbons weren't present.

†The value for $(\Delta t_{ma})_a$ is obtained by crossplotting (Fig. 69) interval transit time (Δt) with neutron porosity (ϕ_N). A crossplot (Fig. 70) of bulk density (ρ_b) versus neutron porosity (ϕ_N) provides a value of $(\rho_{ma})_a$. See Chapter VII, or in the book, *Log Interpretation Manual/Applications*, (Schlumberger, 1974).

are from lower porosity zones in the Mission Canyon interval. These data points above very fine-grained are probably also above irreducible water saturation ($S_{w\,irr}$) and, therefore, cannot be used for determining grain size.

At this juncture in your log evaluation of the Mississippian Mission Canyon from 9,308 to 9,415 ft, you are optimistic about the productive potential of the well. On logs, the interval shows invasion, it has intermittent permeable and porous zones, and its rock type—dolomite—is usually a good reservoir rock. But, you are concerned about whether or not completion should be attempted from the lower zones of the interval, especially from 9,370 to 9,415 ft. The Cyberlook* Log examination, the diminishing separation of the resistivity curves with increasing depth, and the high water saturations on the Pickett crossplot all strongly support your judgement that production from the lower porosity zones in the Mission Canyon interval will not be water-free. You don't know, though, how much water these zones will produce relative to oil.

Because you are primarily concerned about the water saturations in the lower Mission Canyon interval, you continue your log evaluation by comparing the relative permeability to water (K_{rw}), relative permeability to oil (K_{ro}), and percent water-cut.

Relative permeabilities to water (K_{rw}) of different zones are shown on a crossplot of $S_{w\,irr}$† vs. S_w (Fig. 74). Data points, clustering on or below the zero permeability to water line, represent zones from which water-free production can be expected. Data points above the zero line represent zones which will produce some water; the amount of water produced will increase as the points are further away from the zero K_{rw} line.

Relative permeabilities to oil (K_{ro}) of different zones are shown on a crossplot of $S_{w\,irr}$ vs. S_w (Fig. 75). Data points, clustering around the 100% (K_{ro} = 1.0) line, represent zones which should produce 100% oil. Data points, with increasing distance from the 100% line, indicate zones which will produce increasing amounts of water.

The relative permeability to water (K_{rw}) and oil (K_{ro}) plots illustrate that some of the zones in the Mission Canyon will produce water. However, neither plot gives information about the *amount* of water each zone will produce. To determine the amount or percent of water which can be expected from each zone, you construct a water-cut crossplot.

The water-cut crossplot (Fig. 76) reveals a percent water-cut variation from 0 to a high of 50%. The percent of water produced, however, shouldn't exceed 30% and will generally be less than 15%. Higher water-cut values are from lower porosity zones; therefore, the lower porosity zones should not be perforated.

†Remember to use the formula $S_{wirr} = \sqrt{F/2{,}000}$ in crossplots of: K_{ro}, K_{rw}, K_{rg}.

In order to support your decision to avoid perforating the lower zones, you construct a bulk volume water (BVW) crossplot (Fig. 77). On the bulk volume water plot, data points above 0.035 are from lower porosity zones which are not at irreducible water saturation, and so these zones will produce some water.

The position of data points on a bulk volume water crossplot can indicate changes in types of carbonate porosity. Points which are below 0.035 represent zones with vuggy porosity, along with intercrystalline porosity (Table 8).

Even though you have examined crossplots of relative permeabilities which gave you information about the relationship between fluids in porous zones, you want more specific information about each zones's permeability. This information is provided by a permeability plot of $S_{w\,irr}$ vs. ϕ (Fig. 78). Most of the data points plot with permeability values which are considered favorable in your area. Values range from 0.1 to over 100 millidarcies, but generally indicate a good reservoir.

One of your last log evaluation procedures is finding values for the moveable hydrocarbon index (S_w/S_{xo}), for moveable oil saturation (MOS), and for residual oil saturation (ROS). The moveable hydrocarbon index value is less than 0.7, and so the oil is moveable. Oil moveability is also apparent from the high moveable oil saturation and low residual oil saturation values.

Your log evaluation of this particular well has been unusually complete. The extensive evaluation has, in part, been necessitated by the exploratory nature of the well and also by the water problems presented in the lower porosity zones of the Mississippian Mission Canyon Formation. Furthermore, because of the log package used in the well, a large amount of data was available for analysis.

It was apparent rather early in the log evaluation process that the data seemed to support a decision to set pipe. Nevertheless, it was important to know the correct interval for perforating so that water production could be kept as low as possible.

The estimated oil recovery from the Mission Canyon Formation for a gross interval of 9,308 to 9,357 ft is 353,110 stock tank barrels (STB). This oil recovery figure is based on the following parameters: drainage area = 150 acres; reservoir thickness = 28 ft; porosity = 11%; water saturation = 33.5%; recovery factor = 20%; and BOI (estimated) = 1.35.

The Mission Canyon Formation was selectively perforated from 9,308 to 9,357 ft. After a light acid clean-up, the well potential was 569 barrels of oil per day (BOPD), 31 barrels of water a day (BWPD), and 700,000 cubic feet of gas per day (700 mcfgpd) with a gas/oil ratio of 1,230/1. During the first five months, the well produced 56,495 barrels of oil and 5,802 barrels of water.

Answer Table B:
Mississippian Mission Canyon Formation Log Evaluation Table

Depth	LLd	LLs	MSFL*	Φ_N	Φ_D	R_t	$\Phi_{N\text{-}D}$	S_{wa}	S_{wr}	S_{xo}	S_w/S_{xo}	MOS	ROS	BVW	Δt	ρ_b	Φ_{den}	Φ_s	$\Phi_s \times .9$	M	N	Δtma_a	ρma_a	S_{wirr}
9308	35	13	4	14.5	6	48.2	11	20	25	60	.33	40	40	.022	60	2.60	13	11	10	.833	.570	45.5	2.79	20
9310	27	22	1	24	9	37.6	18	14	12	74	.19	60	27	.025	63	2.57	15	12	11	.836	.519	42.3	2.86	12
9312	60	20	15	20	0	84.0	14	12	40	24	.48	13	76	.017	56	2.70	7	8	7	.809	.500	41.8	2.90	16
9314	100	50	14	10	-4	135.1	6.5	20	29	54	.37	34	46	.013	52	2.78	2	5	5	.792	.536	44.7	2.85	34
9326	14	9	3.5	10.5	2	18.3	8	44	42	88	.50	44	12	.035	53	2.68	8	6	5	.835	.566	44.8	2.79	28
9332	8.6	6	2.2	15	5	11.2	11	41	43	81	.51	40	19	.045	58	2.63	11	9	8	.833	.557	44.6	2.81	20
9336	16	4.8	1.1	20	9	22.7	15.5	21	18	81	.25	61	19	.031	58	2.57	15	9	8	.867	.546	42.7	2.82	14
9338	18	8	2.4	23.5	9	24.4	18	17	28	47	.36	30	53	.031	58	2.57	15	9	8	.867	.522	41.3	2.86	12
9340	23	6.5	9	17	5	32.9	12.5	21	53	35	.60	14	65	.026	58	2.65	10	9	8	.819	.535	43.8	2.84	18
9342	70	14	7	10.5	-1	103.6	7	21	22	71	.30	50	29	.015	52	2.73	5	5	5	.816	.549	44.5	2.83	32
9346	4	2	.75	21	6	5.3	15	44	35	100	.44	56	0	.066	62	2.61	12	12	11	.815	.523	43.4	2.85	15
9352	5	2.2	1.6	21	2.5	6.7	15	39	49	70	.56	31	30	.059	61	2.67	9	11	10	.790	.503	43.1	2.89	15
9354	5	2.8	1.1	23.5	6	6.6	17	35	39	74	.47	39	26	.059	59	2.62	12	10	9	.829	.503	41.5	2.89	13
9362	35	16	14	13	2	46.4	9	25	56	39	.63	15	61	.022	53	2.74	5	6	5	.808	.530	43.6	2.85	25
9366	19	13	6.3	11	2	24.2	8	39	51	66	.59	27	34	.031	53	2.74	5	6	5	.805	.543	44.7	2.83	28
9372	15	9	4.2	12	1	19.4	8.5	41	46	76	.53	35	24	.034	53	2.72	6	6	5	.818	.543	43.9	2.83	26
9376	4.6	3.2	1.9	17	7.5	5.7	13	49	60	74	.66	25	26	.064	59	2.59	13	10	9	.846	.557	44.0	2.81	17
9382	4	2	1.5	16	8	5.2	13	51	55	83	.62	32	17	.067	61	2.58	14	11	10	.838	.568	45.1	2.79	17
9386	3.5	1.7	1.6	13	8	4.6	11	64	62	95	.68	31	5	.071	62	2.58	14	12	11	.834	.588	46.5	2.76	20
																			mean	.824	.548	44.4	2.82	

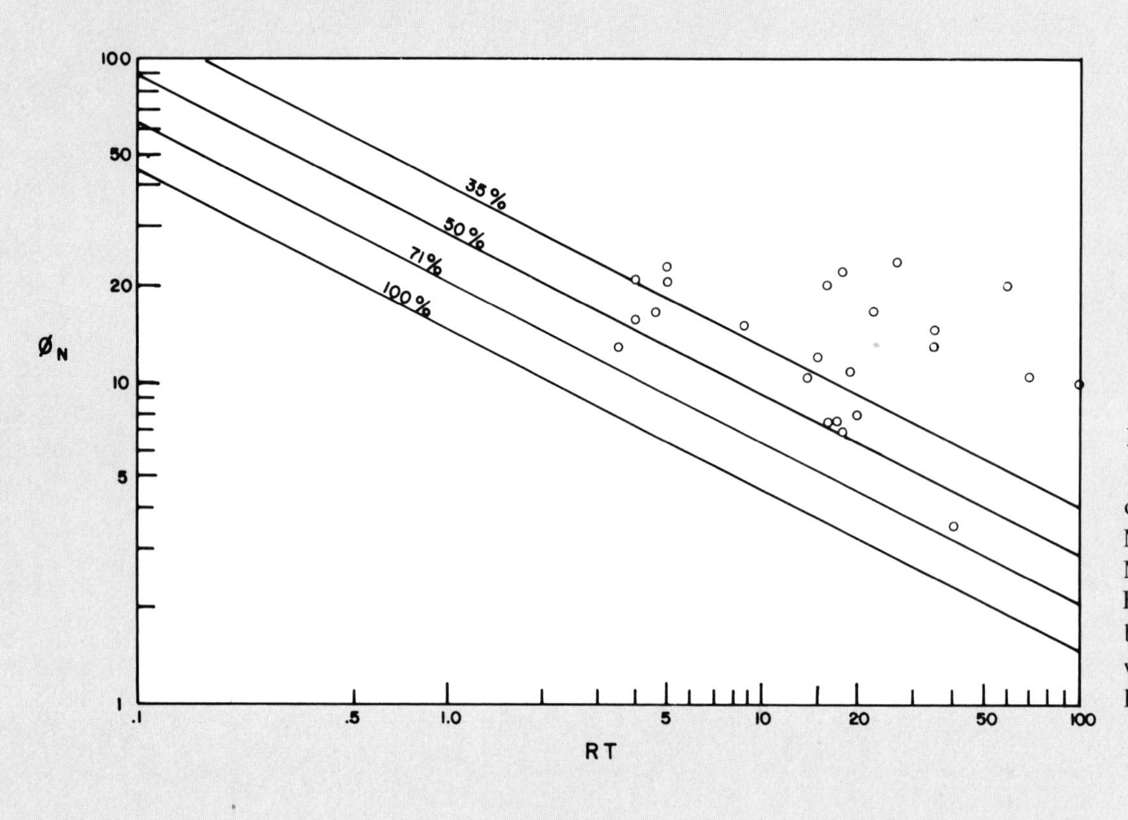

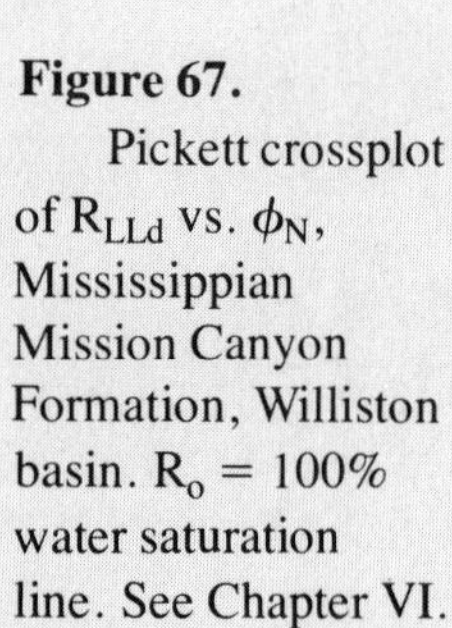

Figure 67.
Pickett crossplot of R_{LLd} vs. ϕ_N, Mississippian Mission Canyon Formation, Williston basin. R_o = 100% water saturation line. See Chapter VI.

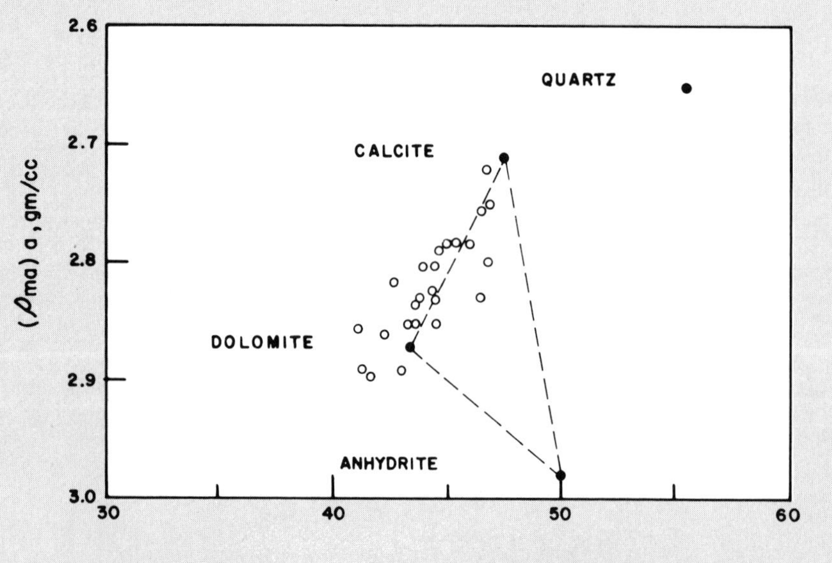

Figure 68.
MID* lithology plot of Mississippian Mission Canyon Formation, Williston basin. The solid circles represent matrix parameters for anhydrite, calcite, dolomite, and quartz.

Values for $(\rho_{ma})_a$ and $(\Delta t_{ma})_a$ are from Figures 69 and 70. See Chapter VII.

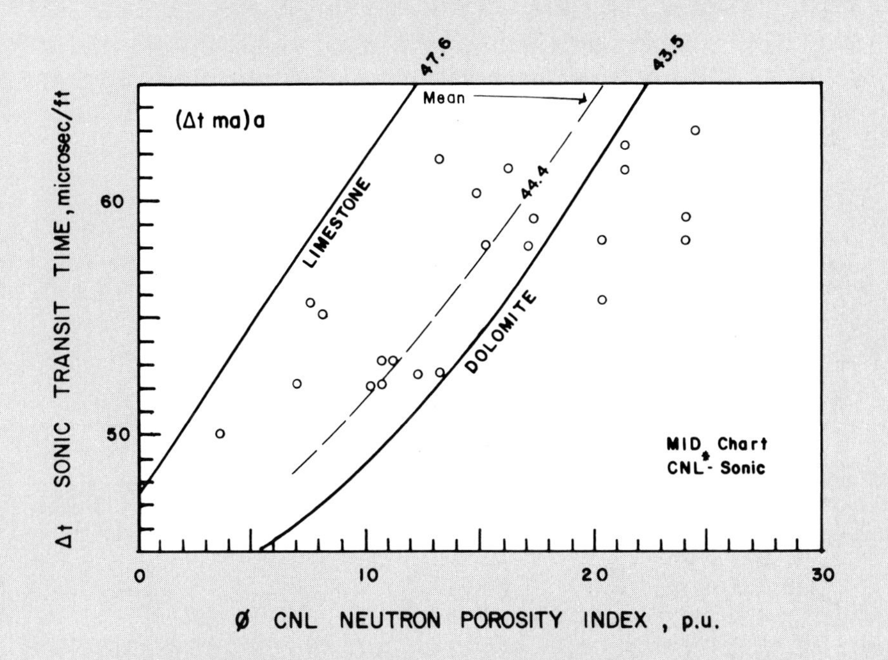

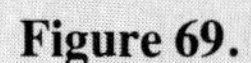

Figure 69.

Interval transit time (Δt) versus neutron porosity (ϕ_N) crossplot for determining $(\Delta t_{ma})_a$ for the MID* plot (Fig. 68), Mississippian Mission Canyon Formation, Williston basin. See Chapter VII.

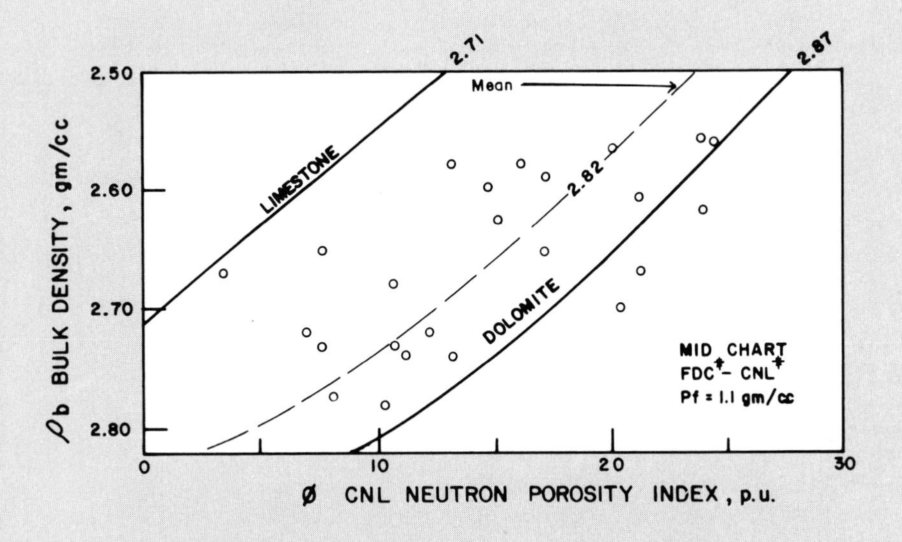

Figure 70.

Bulk density (ρ_b) versus neutron porosity (ϕ_N) crossplot for determining $(\rho_{ma})_a$ for the MID* plot (Fig. 68). Mississippian Mission Canyon Formation, Williston basin. See Chapter VII.

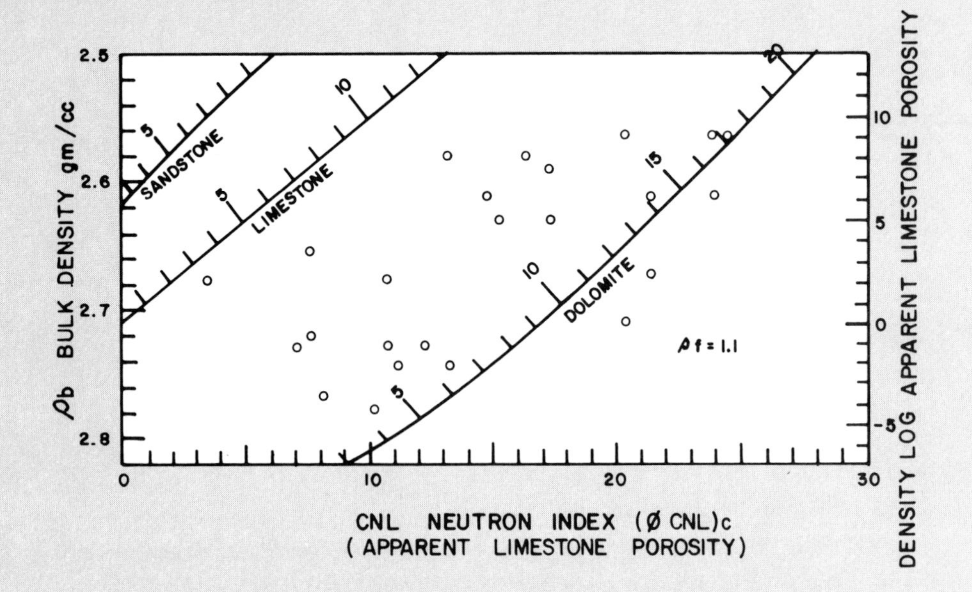

Figure 71.
Neutron-density crossplot for lithology and porosity identification, Mississippian Mission Canyon Formation, Williston basin. See Chapter IV.

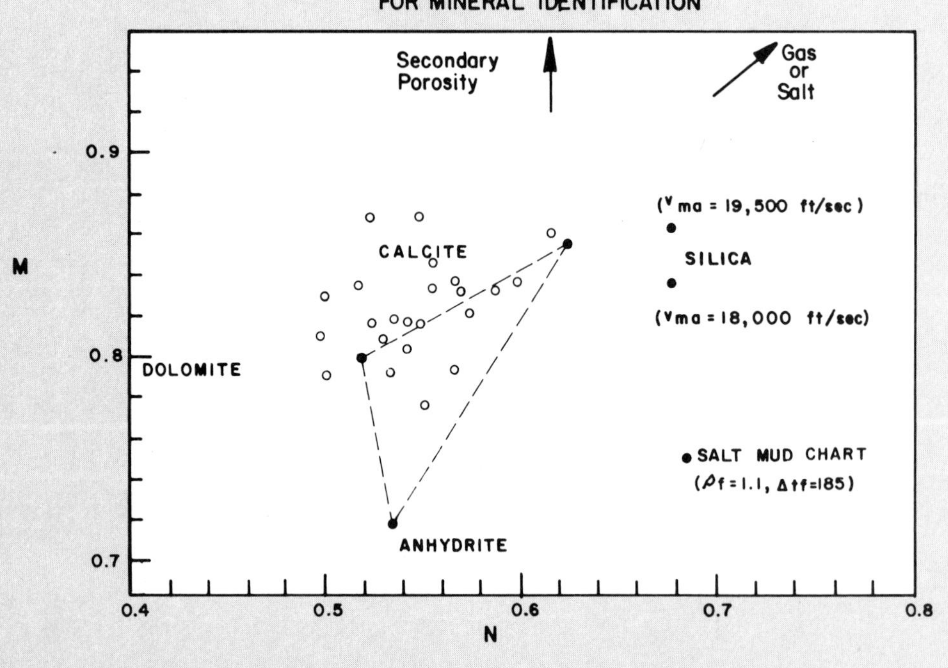

Figure 72.
M-N* lithology crossplot for matrix and secondary porosity identification, Mississippian Mission Canyon Formation, Williston basin. Solid circles represent matrix parameters for anhydrite, dolomite, calcite, and silica (quartz). See Chapter VII.

$$M^* = \frac{\Delta t_f - \Delta t}{\rho_b - \rho_f} \times 0.01$$

$$N^* = \frac{\phi_{Nf} - \phi_N}{\rho_b - \rho_f}$$

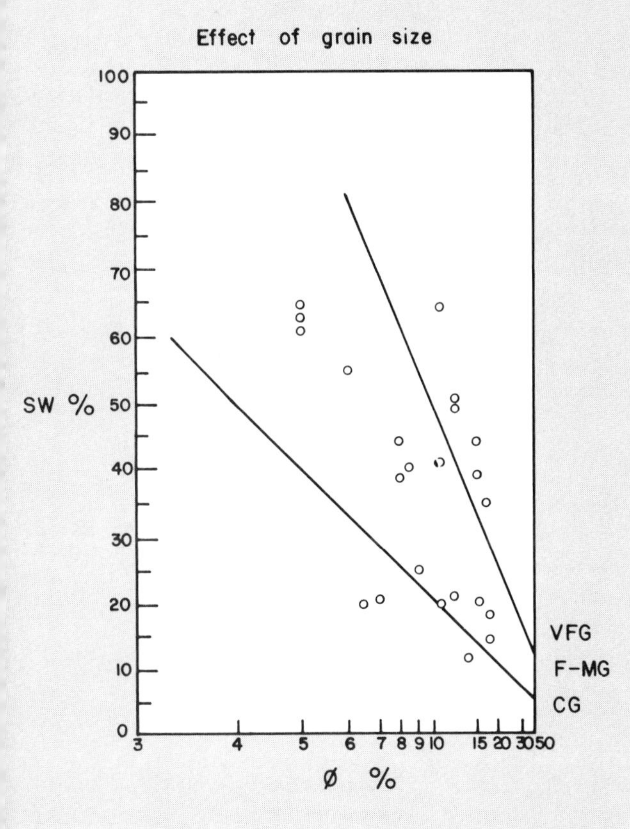

Figure 73.
Grain size determination by water saturation (S_w) versus porosity (ϕ) crossplot, Mississippian Mission Canyon Formation, Williston basin.

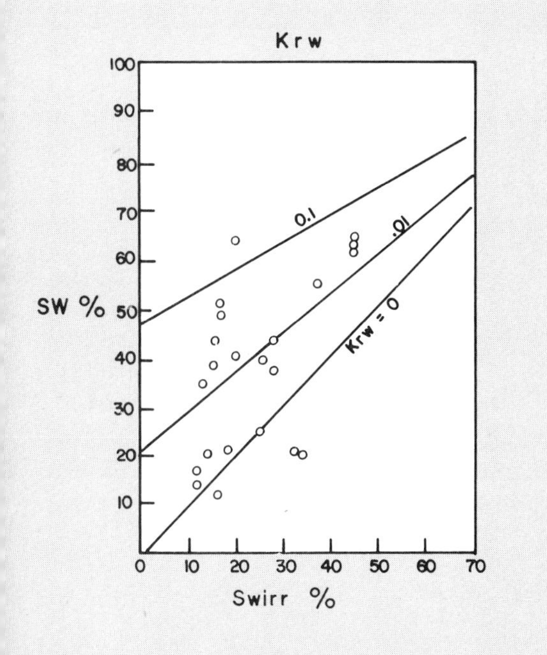

Figure 74.
Irreducible water saturation ($S_{w\,irr}$) versus water saturation (S_w) crossplot for determining relative permeability to water (K_{rw}), Mississippian Mission Canyon Formation, Williston basin.

$S_{w\,irr} = \sqrt{F/2000}$ This formula calculates an approximate, theoretical value for $S_{w\,irr}$. Values calculated for $S_{w\,irr}$ by this formula should only be used in crossplots where you are trying to determine K_{ro}, K_{rg}, K_{rw}, or percentage water-cut.

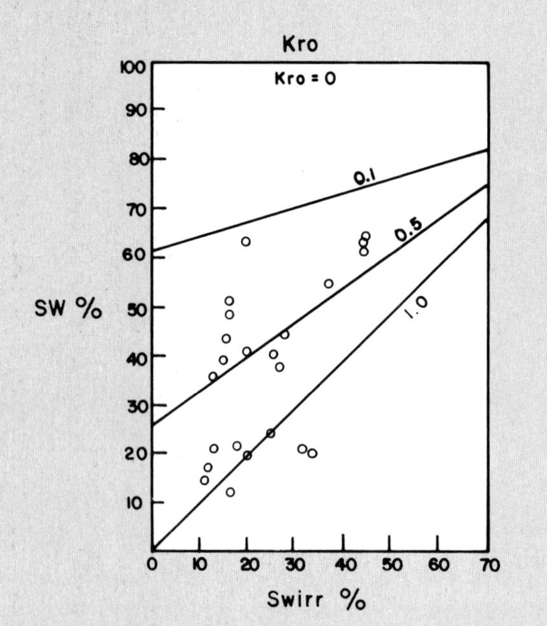

Figure 75.
Irreducible water saturation ($S_{w\,irr}$) versus water saturation (S_w) crossplot for determining relative permeability to oil (K_{ro}), Mississippian Mission Canyon Formation, Williston basin.

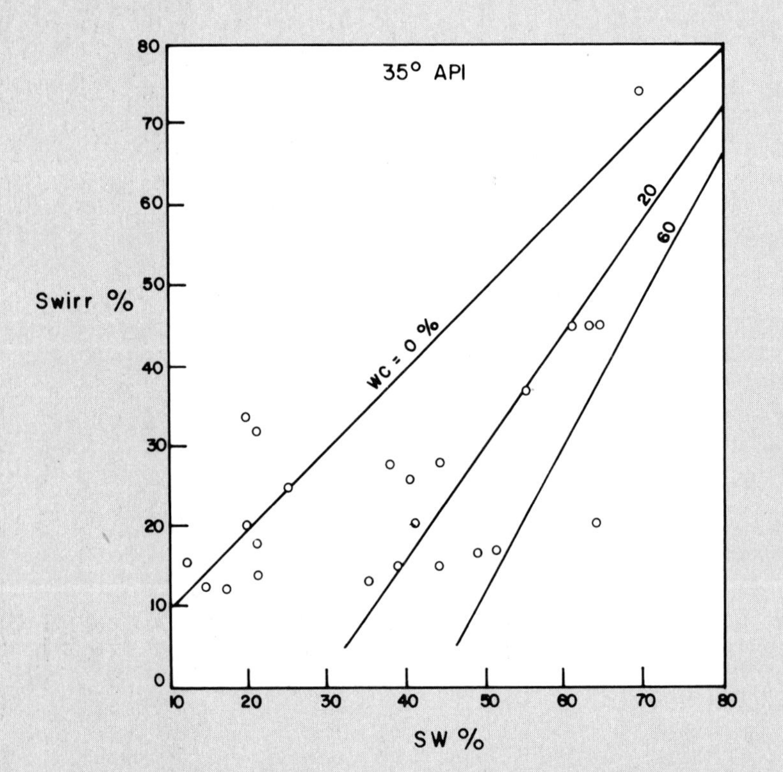

Figure 76.
Irreducible water saturation ($S_{w\,irr}$) versus water saturation (S_w) crossplot for determining percent water-cut, Mississippian Mission Canyon Formation, Williston basin.

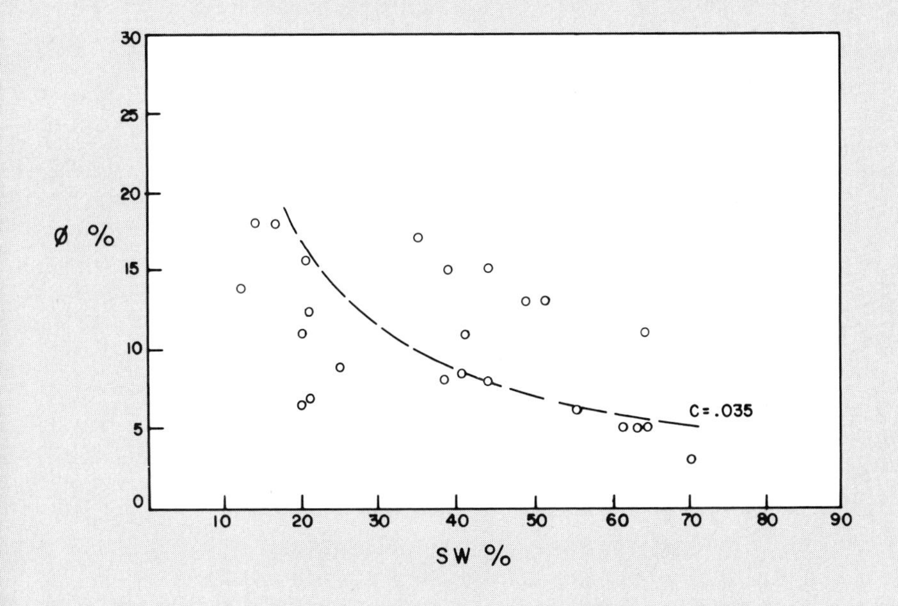

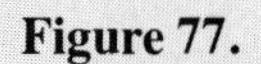

Figure 77.
Bulk volume water crossplot (ϕ vs. S_w), Mississippian Mission Canyon Formation, Williston basin.

C = bulk volume water.

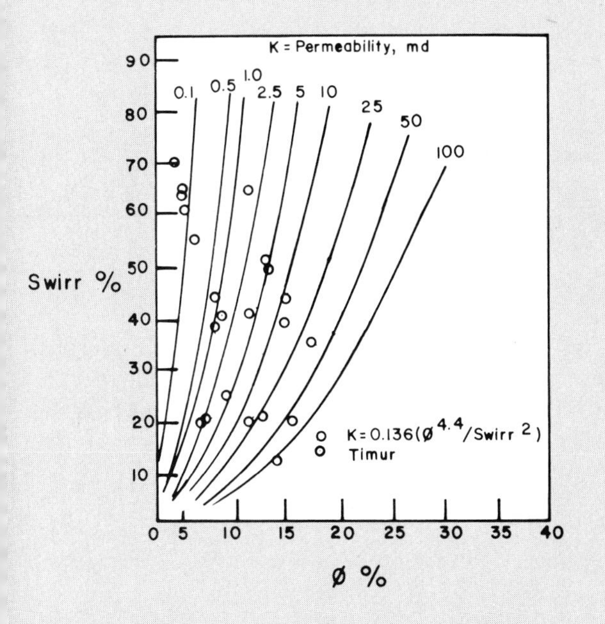

Figure 78.
Irreducible water saturation ($S_{w\,irr}$) versus porosity (ϕ) crossplot for determining permeability, Mississippian Mission Canyon Formation, Williston basin.

Case Study 3
Eocene Wilcox Sandstone
Gulf Coast

You are assigned the log evaluation of a lower Wilcox sandstone in South Texas. A fairly typical Gulf Coast log package was used in the well. It consists of an induction electric log with an SP and a R_{wa} quick look curve and a sonic log. The induction electric log has a deep induction log used to measure resistivities in the uninvaded zone (R_t) and a short normal to measure resistivities in the invaded zone (R_i). The log package is run on a single combination tool, and because it requires only one run in the well, it has saved your company valuable rig time.

Drilling operations, on your company's lower Wilcox wildcat well, halted at 10,936 ft after penetration of a sandstone. There was a sudden, large, gas increase in the drilling mud; gas increased by 3,200 units over background on the chromatograph. As a response to this, mud weight had to be increased from 14.8 lbs/gal to 15.4 lbs/gal to contain the gas within the formation. When drilling operations were resumed, gas continued to cut the mud with a weight of 15.4 lbs/gal going into the hole and 15.2 lbs/gal coming out of the hole. Also, the mud logger's chromatograph maintained about 100 units of gas, even when the well was deepened beyond the zone of initial gas show.

You select five depths (or points) from 10,930 to 10,970 ft within the Wilcox sand interval. Depths are picked on the basis of an even distribution through the interval being evaluated. Here, you use a distribution of points every 12 ft from 10,930 to 10,970 ft.

Because the only porosity tool at your disposal is the sonic log, you use one of the following equations to help you find porosity. The perferred equation is III because of your experience with its use in the Gulf Coast.

Equation Variables—$\Delta t = \mu sec/ft$ and is interval transit time read from the sonic log; $\Delta t_{sh} = \mu sec/ft$ and is interval transit time of shale read on a sonic log from a clean shale zone up the borehole from the Wilcox sand; $\Delta t_{ma} = \mu sec/ft$ and is interval transit time of the matrix for sandstone (known for Wilcox sands in Gulf Coast); and $\Delta t_f = \mu sec/ft$ and is interval transit time of freshwater-based muds (see Chapter IV).

Sonic Porosity Equations:

I. $$\text{sonic porosity } (\phi_s) = \frac{\Delta t - \Delta t_{ma}}{\Delta t_f - \Delta t_{ma}} \times \frac{100}{\Delta t_{sh}}$$

Where:

$\Delta t_f = 189$

$\Delta t_{ma} = 56.7$

II. $$\text{sonic porosity } (\phi_s) = \frac{\Delta t - 55}{1.4}$$

and

III. $$\text{sonic porosity } (\phi_s) = \frac{5(\Delta t - \Delta t_{ma})}{8 \times \Delta t}$$

Where:

$\Delta t_{ma} = 55.5\ \mu sec/ft$ for all sandstones.

Because the sonic log is strongly affected by gas, you need to use the equation for sonic porosity gas correction:

$$\phi_s = \phi_s \times 0.7$$

The recoverable reserves of gas are calculated from the gas volumetric equation and the following parameters are needed to complete the equation: drainage area = 240 acres; reservoir thickness = 15 ft; porosity = 18%; water saturation = 57%; gas gravity (estimated) = 0.62; recovery factor = 0.6; temperature (estimated) = 279°F; initial bottom hole pressure (IBHP, estimated) = 8,103 PSI; Z factor = 1.229; geothermal gradient = 0.0255 × formation depth; and pressure gradient = 0.74 × formation depth.

Your company has purchased a 25% working interest (WI) in the well, which has a net revenue interest (NRI) of 82.5%. Net revenue interest is the total interest (total interest equals 100%) minus any royalties such as an interest granted to a mineral rights owner.

The estimated cost of the well is 1.8 million dollars. You use a product price of $1.90 per mcf to find out the projected return your company can expect on its investment.

You have acquired certain information needed to help with an evaluation of the logs: $R_w = 0.022$ at T_f; $F = 0.62/\phi^{2.15}$ (formation factor for Gulf Coast sands); $R_{mf} = 0.222$ at T_f; $T_f = 260°F$; $\Delta t_{sh} = 116$; and surface temperature = 80°F.

The following Eocene Wilcox Sandstone Log Evaluation Table (work Table C) is designed to assist you with your work. The first three depths and their deep induction, short normal, and interval transit time (Δt) values have been determined for you. However, this information for the last two depths is left for you to complete.

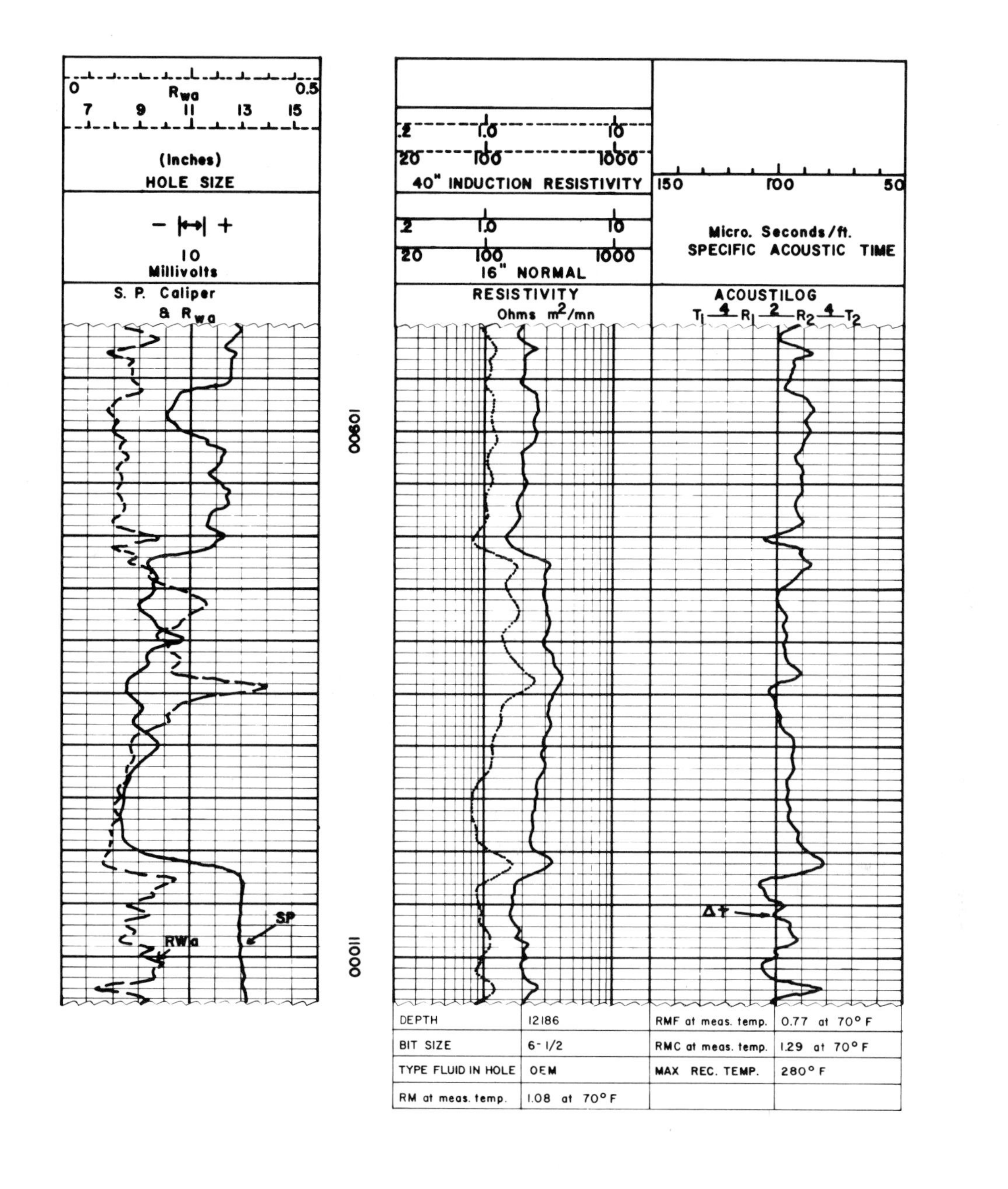

DEPTH	12186	RMF at meas. temp.	0.77 at 70° F
BIT SIZE	6-1/2	RMC at meas. temp.	1.29 at 70° F
TYPE FLUID IN HOLE	OEM	MAX REC. TEMP.	280° F
RM at meas. temp.	1.08 at 70° F		

Figure 79. Induction electric log with SP, R_{wa} curve, and sonic log, Eocene Wilcox Sandstone, Gulf Coast.

Note:

1. At top of Wilcox sand (10,935 to 10,958 ft) the deflection of R_{wa} quick look curve in track #1 to the right away from the SP curve. The deflection indicates the presence of hydrocarbons.
2. In track #2, the increase in resistivity in the upper part of the Wilcox sand (10,935 to 10,958 ft). The resistivity increase also indicates the presence of hydrocarbons.
3. The low resistivities and the deflection of the R_{wa} curve to the left at the base of the Wilcox sand (10,958 to 10,982 ft) indicating a probable water zone.

Work Table C:
Eocene Wilcox Sandstone Log Evaluation Table

Depth	ILd	S.N.	Δt	Φ_sIII	$\Phi_s \times .7$	S_{wa}	S_{WT}	S_{XO}	S_w/S_{xo}	MOS	BVW													
10930	1.5	3.0	99																					
10942	1.6	3.4	97																					
10954	1.3	3.1	99																					

10966
10978

Case Study 3 Answer

An evaluation of the logs begins with an assessment of sand development in the well through the Wilcox interval, because this interval is where a good gas show appeared. A porous and permeable sand is read on the SP log (Fig. 79), from 10,924 to 10,982 ft, by a leftward deflection of the curve from the shale baseline.

The R_{wa} quick look curve (Fig. 79) deflects to the right, away from the SP curve, in the upper part (10,935 to 10,958 ft) of the Wilcox sandstone. Such a curve deflection is evidence of the presence of hydrocarbons.

On the deep induction log (Fig. 79), reading R_t, resistivities increase in an upper zone covering approximately 10,924 to 10,958 ft. Increased resistivity also indicates hydrocarbons.

A fairly rapid decrease of resistivities into the lower zone and a deflection of the R_{wa} curve to the left at 10,958 to 10,982 ft means higher water saturations. These observations alert you that the lower zone may be water productive.

In order to establish whether or not the lower Wilcox zone will produce water, you decide that a rather detailed log evaluation is necessary. The evaluation will include a Pickett crossplot for water saturations, Archie equation calculated water saturations and crossplots for grain size, relative permeability, permeability, and bulk volume water.

A Pickett crossplot (Fig. 80) of deep induction log resistivities (R_t) versus sonic log interval transit (Δt) time, corrected for a sand matrix ($\Delta t - \Delta t_{ma}$), has water saturations ranging from 100% to less than 50%. High water saturations of greater than 71% on the Pickett crossplot are from the lower Wilcox zone. This supports your previous suspicions about its water productive nature.

Examination of the sonic log reveals porosities of 14 to 20% after correction for gas effect. These calculated sonic porosities are used in the Archie water saturation equation (S_{wa}) and in making crossplots.

A determination of Wilcox sandstone grain size as very fine-grained is made after you review a crossplot (Fig. 81) of water saturation Archie (S_{wa}) versus sonic porosity (ϕ). High water saturations of 70 to 80% in the lower zone (10,958 to 10,982 ft) probably invalidate grain size determinations, because the reservoir is *not at irreducible water saturation* ($S_{w\,irr}$).

Water saturations of the upper part (10,924 to 10,958 ft) of the Wilcox range from 43 to 62% (a relatively high level), and result from the very fine-grained size of the Wilcox sand. The relationship of grain size and water saturation is such that fine-grained sands have high irreducible water saturations (see Chapter VI), and consequently they also have high values for bulk volume water.

A crossplot (Fig. 82) of irreducible water saturation ($S_{w\,irr}$) versus porosity (ϕ) reveals a permeability range of 10 to 50 md. You conclude, therefore, that the reservoir has good permeabilities.

Data on a crossplot (Fig. 83) of irreducible water saturation ($S_{w\,irr}$) versus water saturation (S_{wa}), are plotted in a range from intermediate to low relative permeabilities to gas (K_{rg}). And, those data points which have low permeability to gas values (less than 10%) are from the lower Wilcox zone. As K_{rg} values decrease, relative permeability to water increases and the reservoir will produce some water. In the case of your company's Wilcox well, because of the position of the lower zone's data points, the amount of water produced may be appreciable. On the crossplot (Fig. 83) data points with values of relative permeability to gas which are greater than 10% should produce decreasing amounts of water. These data points are from the upper zone.

The moveable hydrocarbon index (S_w/S_{xo}) is less than 0.7 from 10,924 to 10,958 ft, indicating hydrocarbons have moveability. The high moveable oil saturation values (MOS) also suggest hydrocarbons will move.

The lower bulk volume water values (Fig. 84) and lower water saturations in the upper Wilcox sandstone, when compared to the lower, indicate only the upper Wilcox may be above a gas/water transition zone. The bulk volume water values which are much greater than 0.1 (Fig. 84) are from the lower part of the Wilcox from 10,958 to 10,982 ft. This interval is above irreducible water saturation.

All of the data produced by your evaluation support your early assessment of a reservoir with gas on top of water. You were immediately alerted to a potential problem after you examined the R_{wa} quick look curve and saw a deflection to the left through the lower Wilcox zone, and when you saw the fairly rapid decrease of resistivities into the lower zone.

An estimate of recoverable reserves by volumetric calculations of the Wilcox sandstone is 2.3 BCF using the following parameters: drainage area = 240 acres; reservoir thickness = 15 ft; porosity = 18%; water saturation = 57%; gas gravity (estimated) = 0.62; recovery factor = 0.6; temperature (estimated) = 279°F; initial bottom hole pressure (IBHP, estimated) = 8,103 PSI; Z factor = 1.229; geothermal gradient = 0.0255 × formation depth; and pressure gradient = 0.74 × depth.

You estimate the return on investment of the Wilcox sandstone as follows:

1. Total Well Cost × Working Interest = Working Interest Well Costs: \$1,800,000 × 0.25 = \$450,000.
2. Net Revenue Interest (lease) × Working Interest = Net Revenue Working Interest: 0.825 × 0.25 = 0.20625 (20.6%).

3. Reserves × Product Price = Gross Revenue: 2,300,000 MCF × \$1.90 = \$4,370,000.

4. Gross Revenue × Net Revenue Working Interest = Net Revenue (Working Interest): \$4,370,000 × 0.20625 = \$901,312.50.

5. Net Revenue (Working Interest) ÷ Working Interest Well Costs = Return on Investment (before taxes and operating expenses): \$901,312.50 ÷ \$450,000 = 2:1.

The relatively poor (2 to 1) return on your company's investment is weighed with your judgement about the reservoir's high water saturations. You think the reservoir will be able to produce gas at only fairly low levels so water production from the lower zone can be kept under control. However, a decision is made in conjunction with your company's management, to set pipe, because of a hope of future increases in gas prices. This pipe setting decision may be questioned by readers who find that the return rate, even with projected price increases, doesn't meet their economic criteria.

The Wilcox sandstone was perforated from 10,962 to 10,963 ft. The interval flowed 11 hours on a 10/64 inch choke at a rate of 1,584 mcfgpd, 5 barrels of condensate per day (BCPD) and 1,090 barrels of water per day (BWPD). The interval (10,962 to 10,963 ft) was squeezed (i.e. closed off) and an interval from 10,925 to 10,933 ft was then perforated. The results from this interval were as follows: Calculated absolute open flow (CAOF) was 7,000 mcfgpd and 7.8 BC/mmcf; SITP = 7,130 PSI: IBHP = 8,480 PSI: BHT = 283°F; gas gravity = 0.657; liquid gravity = 46.7°. The well produced 350 mmcf during the first ten months.

Answer Table C:
Eocene Wilcox Sandstone Log Evaluation Table

Depth	ILd	S.N.	Δt	$\Phi_s III$	$\Phi_s \times .7$	S_{wa}	S_{wr}	S_{xo}	S_w/S_{xo}	MOS	BVW	$\Phi_s II$	$\Phi_s \times .7$	$\Phi_s I$	$\Phi_s \times .7$	S_{wirr}
10924	1.6	2.7	89	24	17	64	33	100	.64	36	.106	24	17	21	15	12
10926	1.8	3.2	90	24	17	59	34	100	.59	42	.099	25	18	22	15	12
10928	1.5	3.0	93	25	18	62	36	100	.62	38	.109	27	19	24	17	12
10930	1.5	3.0	99	28	19	57	34	100	.57	43	.108	31	22	28	20	11
10932	1.7	2.9	100	28	20	51	33	100	.51	50	.101	32	22	29	20	10
10934	1.9	3.0	97	27	19	51	31	100	.51	50	.096	30	21	27	19	11
10936	1.7	3.3	97	27	19	53	36	100	.53	47	.101	30	21	27	19	11
10938	1.4	3.1	97	27	19	59	39	100	.59	41	.112	30	21	27	19	11
10942	1.6	3.4	97	27	19	56	39	100	.56	44	.106	30	21	27	19	11
10944	1.9	3.5	95	26	18	54	35	100	.54	46	.097	29	20	25	18	11
10946	2.3	4.0	91	24	17	52	33	100	.52	48	.088	26	18	23	16	12
10948	2.3	3.9	102	29	20	43	33	100	.43	57	.086	34	24	30	21	10
10950	1.8	3.5	103	29	20	49	36	100	.49	51	.098	34	24	31	21	10
10952	1.4	3.2	100	28	20	56	40	100	.56	44	.112	32	22	29	20	10
10954	1.3	3.1	99	28	20	61	41	100	.61	39	.116	31	22	28	20	11
10956	1.3	3.3	98	27	19	61	42	100	.61	39	.116	31	22	27	19	11
10958	1.2	2.9	94	26	18	69	42	100	.69	31	.124	28	20	24	17	11
10962	1.1	2.8	94	26	18	70	42	100	.70	30	.126	28	20	24	17	11
10964	1.1	3.0	93	25	18	73	44	100	.73	28	.127	27	19	24	17	12

10966	1.0	2.6	96	26	18	76	44	100	.76	24	.137	29	20	26	18	11								
10968	.8	2.5	96	26	18	81	47	100	.81	20	.145	29	20	26	18	11								
10970	.8	2.5	96	26	18	83	48	100	.83	18	.149	29	20	26	18	11								
10972	.8	2.4	95	26	18	83	47	100	.83	18	.149	29	20	25	18	11								
10974	.9	2.4	95	26	18	80	45	100	.80	20	.144	29	20	25	18	11								
10976	.9	2.4	92	25	18	80	44	100	.80	20	.140	26	18	23	16	12								
10978	.9	2.3	91	24	17	83	42	100	.83	17	.141	26	18	23	16	12								
10980	1.2	2.6	86	22	15	82	38	100	.82	18	.123	22	15	19	13	14								
10982	1.7	3.4	81	20	14	74	36	100	.74	26	.104	19	13	16	11	15								

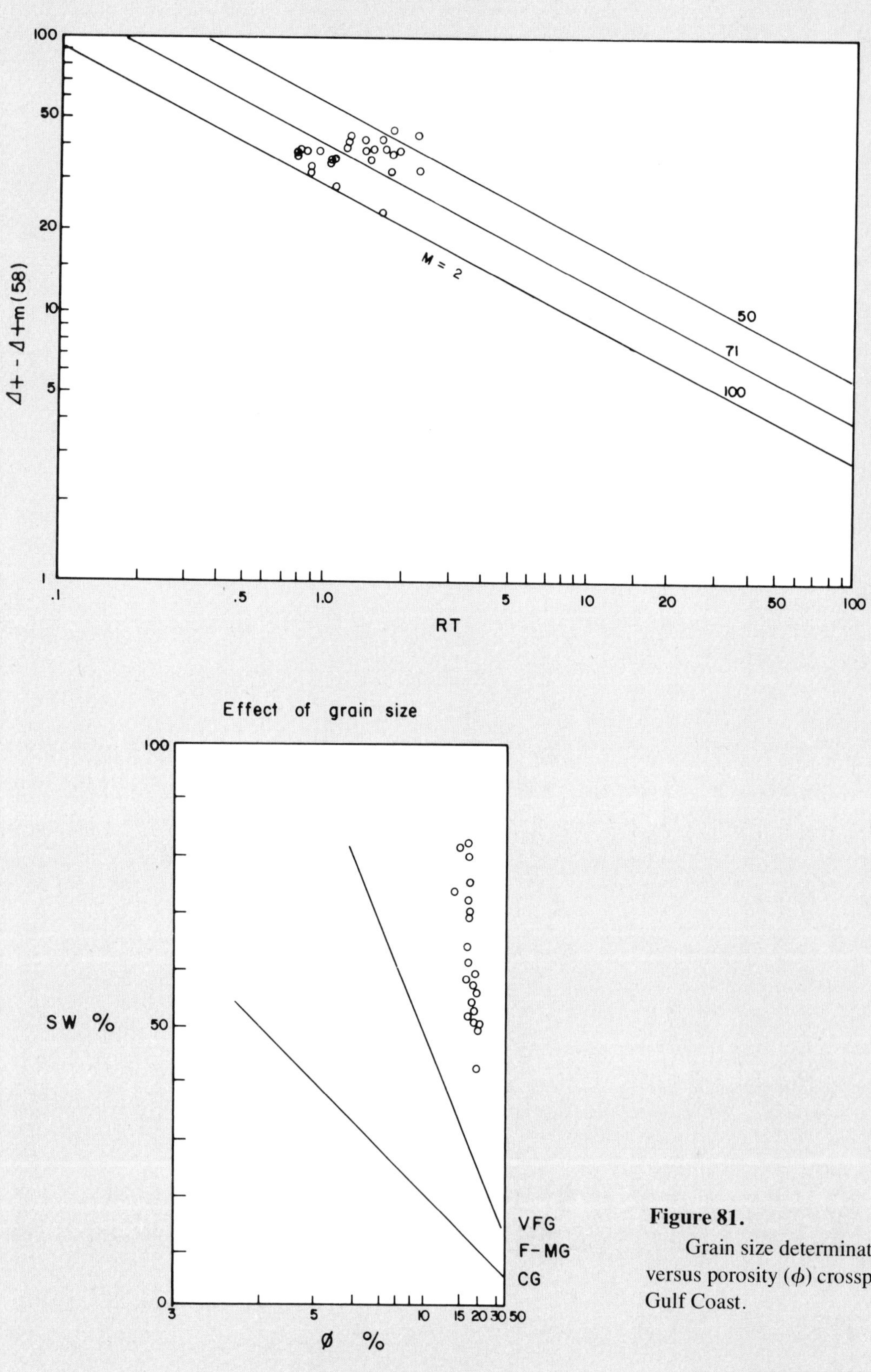

Figure 80.

Pickett crossplot of R_{ILd} versus $\Delta t - \Delta t_m$ (58), Eocene Wilcox Sandstone, Gulf Coast. Δt_m (58) is equal to the interval transit time (i.e. 58 μsec/ft) for the Wilcox Sandstone.

Solid dots are created by overlapping circles.

Figure 81.

Grain size determination by water saturation (S_w) versus porosity (ϕ) crossplot, Eocene Wilcox Sandstone, Gulf Coast.

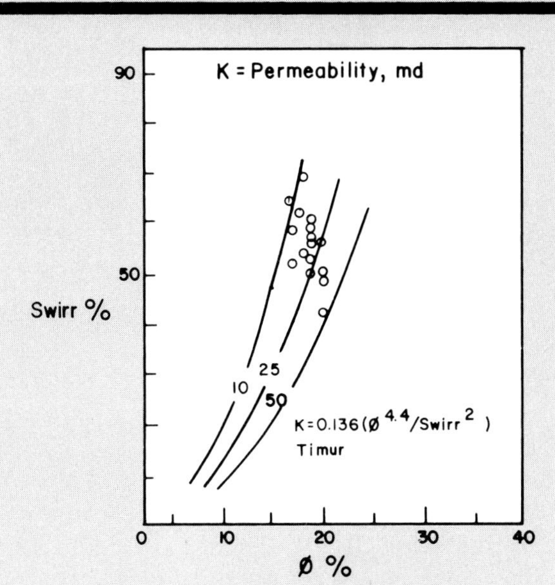

Figure 82.
Irreducible water saturation ($S_{w\,irr}$) versus porosity (ϕ) crossplot for determining permeability, Eocene Wilcox Sandstone, Gulf Coast.

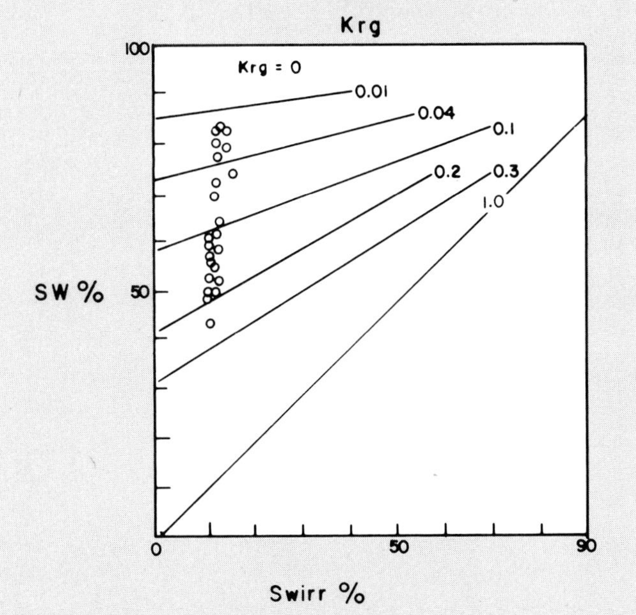

Figure 83.
Irreducible water saturation ($S_{w\,irr}$) versus water saturation (S_w) crossplot for determining relative permeability to gas (K_{rg}). Eocene Wilcox Sandstone, Gulf Coast.

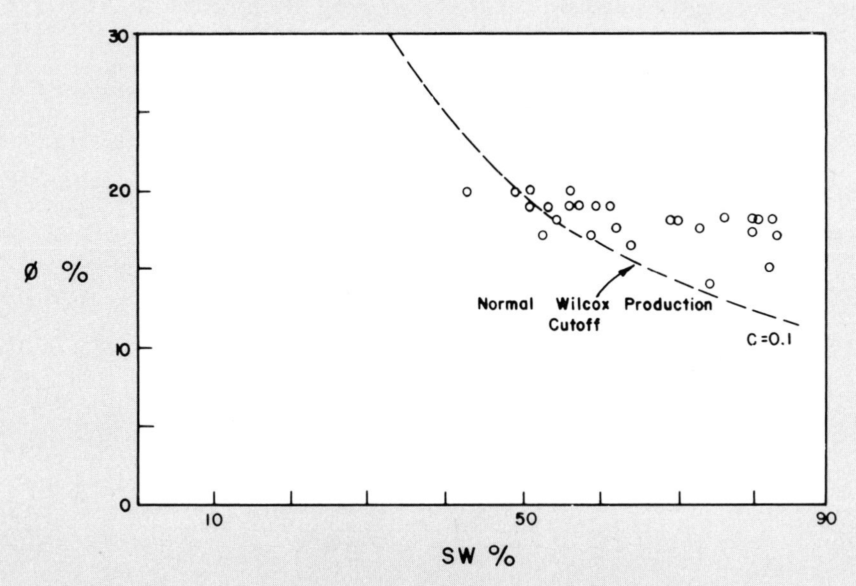

Figure 84.
Bulk volume water crossplot (ϕ vs. S_w), Eocene Wilcox Sandstone, Gulf Coast.

The bulk volume water values (BVW), plotted above the 0.1 hyperbolic line, are values from the lower part of the Wilcox Sandstone, which is *above* irreducible water saturation ($S_{w\,irr}$).

Case Study 4

Pennsylvanian Upper Morrow Sandstone

Anadarko Basin

An area covering several counties in the Anadarko basin is assigned to you. It is your responsibility to map and develop hydrocarbon prospects and oversee all company activity in these counties.

As you pursue the assignment, you review many logs from previously drilled wells; some are producing and some are dry and abandoned. One particular dry and abandoned well captures your attention, because its induction log has good resistivities and its sonic log has good porosities. The well wasn't tested before its abandonment. However, it was logged with an induction electric log (Fig. 85) to determine resistivities of the invaded (R_i) and uninvaded (R_t) zones, and a sonic log (Fig. 86) to determine sonic porosity.

Before you can calculate water saturations Archie, a value for R_w is needed. You decide to find a value for R_w by using the SP log and charts (see Fig. 11, Chapter II).

Because water production is a local problem but water samples are not available, you decide to check the amount of water the well may produce. To do this, irreducible water saturation ($S_{w\,irr}$) values are plotted versus water saturation Archie (S_{wa}) and a water-cut crossplot is constructed (charts for water-cut percent are in Appendix 4). Irreducible water saturation values calculated by the formula $S_{w\,irr} = \sqrt{F/2000}$ are: at a depth of 7,444 ft = 14; at 7,446 ft = 9; and at a depth of 7,448 ft = 11.

The sonic porosity formula is:

$$\phi_s = \frac{\Delta_t - \Delta t_{ma}}{\Delta t_f - \Delta t_{ma}}$$

Other important information you use to pursue your evaluation is: $R_{mf} = 0.527$ at T_f; $R_w = 0.11$ at T_f; $T_f = 130°F$; $F = 0.81/\phi^2$; and surface temperature = 70°F.

Volumetric recoverable oil reserves are calculated with the following parameters: drainage area = 160 acres; BOI = 1.3; recovery factor (RF) = 0.15; porosity = 16%; water saturation (S_w) = 57.5%; thickness (h) = 8 ft.

An analysis by the company's engineering department leads to a judgement that, if it appears the well will be productive, it can be re-entered. An estimated cost for re-entry and completion of the well is $275,000. With a gross product price of $32.00 per barrel and a lease which has a 3/16th royalty, what do you estimate as a return on investment?

Use the Pennsylvanian Upper Morrow Sandstone Log Evaluation Table (work Table D) to complete your evaluation.

Work Table D:
Pennsylvanian Upper Morrow Sandstone Log Evaluation Table

Depth	ILd	S.N.	Δt	Φ_{sI}	S_{wa}	S_{wr}	S_{xo}	S_w/S_{xo}	MOS	BVW
7444	12	34	70							
7446										
7448										

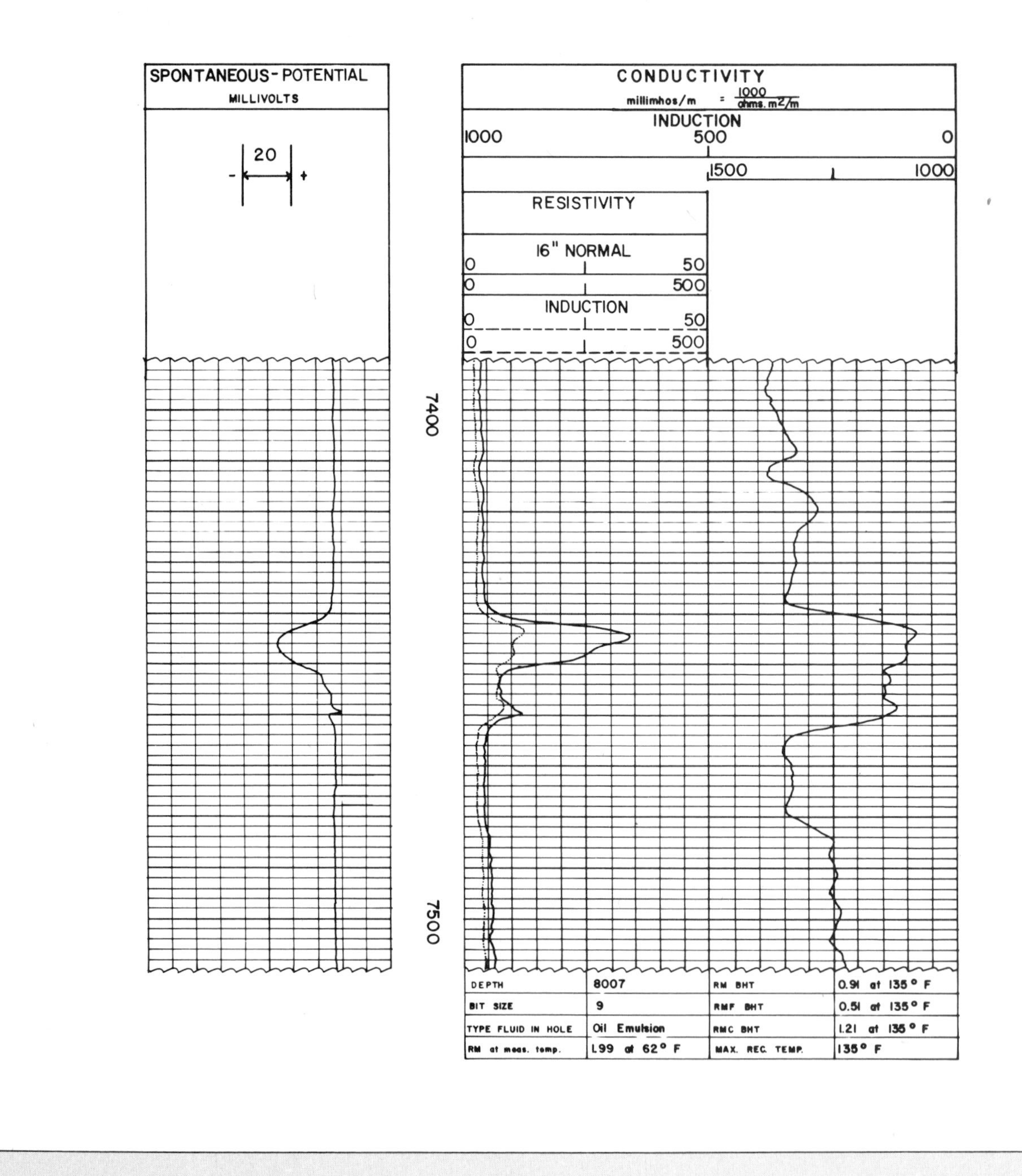

Figure 85. Induction electric log with spontaneous potential log, Pennsylvanian upper Morrow Sandstone, Anadarko basin.

Note:

1. The deflection of the SP curve in track #1 to the left away from the shale baseline (7,440 – 7,452 ft), opposite the porous and permeable upper Morrow Sandstone.
2. The separation of the short normal curve (R_i) from the induction curve (R_t) in track #2 indicates invasion has taken place, and that the upper Morrow Sandstone is permeable.

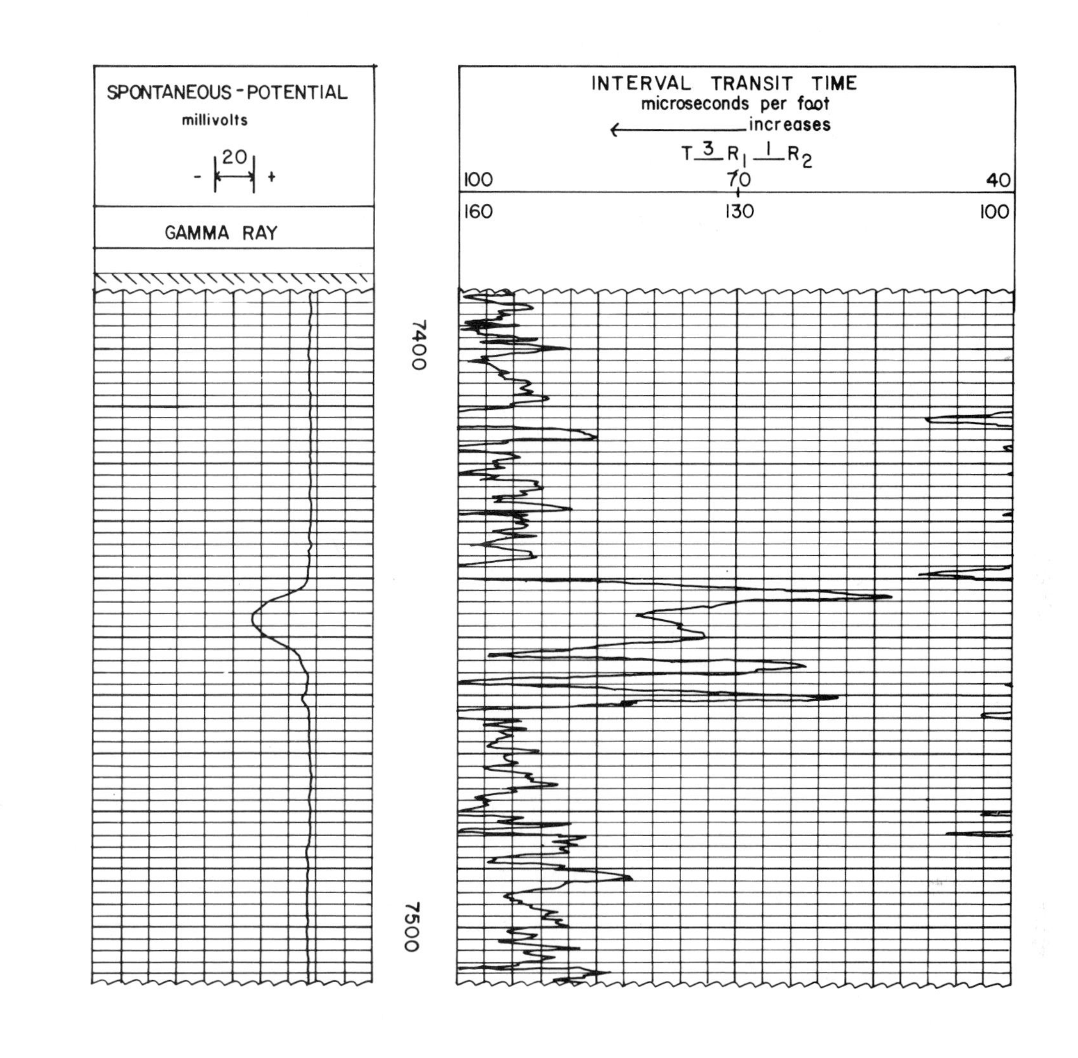

Figure 86. Sonic log with spontaneous potential log Pennsylvanian upper Morrow Sandstone, Anadarko basin.

Note:

The high interval transit time ($\Delta t = 81\ \mu$sec/ft. at 7,446 ft) which indicates high sonic porosity. See tracks #2 and #3 between 7,440 and 7,452 ft.

Case Study 4 Answer

The upper Morrow sandstone in the well occurs from 7,440 to 7,452 ft and is shown on the SP curve (Fig. 85) by its leftward deflection from the shale baseline. The permeable nature of the Morrow is seen by the separation between the short normal (R_i) and induction (R_t) resistivity curves. Separation between these curves indicates invasion has taken place. Porosity in the sand from 7,440 to 7,452 ft is apparent by the high (70 to 81 μsec/ft) interval transit time (Δt) on the sonic log (Fig. 86).

Grain size, determined from a plot (Fig. 87) of water saturation (S_w) versus porosity (ϕ), is very fine-grained. This means high water saturation and bulk volume water values. However, because of the very fine grain size, higher than normal values can be tolerated and the well may still not produce much water.

As suspected, water saturation (S_{wa}) values (work Table D) are high, ranging from 41 to 62%, and so are bulk volume water values (Fig. 88). Data points on the bulk volume water crossplot exhibit only minor scatter from the hyperbolic line, suggesting the reservoir may be at or near irreducible water saturation. But, the plot (Fig. 88) has only three points, not enough to firmly establish whether or not the well will produce some water.

Permeability (Fig. 89) varies from less than one to approximately 15 md. These are fairly low permeabilities, and probably occur here because the upper Morrow is very fine-grained.

Data points on the relative permeability to water crossplot (Fig. 90) increase in value from 0.04 to substantially over 0.1. Two of the three points, however, plot under 0.1. Because permeability relative to water is above zero, some water will certainly be produced, but probably not in any great amount.

A percent water-cut plot has two points between 20 and 40% water-cut. The third point plots around 70% and is from a depth of 7,444 ft at the end of the sand interval. The 70% point may not be reliable. Problems, caused by the resolution of logging tools along bed boundaries, cast doubts about the validity of the 70% water-cut data point. Therefore, you assume a water production amount of between 20 to 40%.

Finally, your calculations of the moveable hydrocarbon index (S_w/S_{xo}) and moveable oil saturations have favorable values.

You have completed your log assessment and are encouraged by evidence of good porosities on the sonic log and good indications of permeability on the resistivity log. Also, you are aware that the low cost of a re-entry well enhances the economics of the prospect.

However, you weigh the positive aspects of the prospect against the relatively thin zone and the few data points which you have used. The lack of points calls into question the statistical accuracy of the information, and in turn, your conclusions based on such limited data. Also, fine-grained reservoirs do not have the permeability of coarser grained reservoirs. And, you have determined a 20 to 40% water production from the well.

You estimate recoverable oil as 78,000 stock tank barrels (STB). An estimated investment return is: stock tank barrels (STB) $\times$ product price $\times$ net revenue interest (NRI) $\div$ total cost = investment return (78,000 $\times$ \$32.00 $\times$ 0.8125 $\div$ \$275,000 = 7.4:1 return). This 7.4:1 figure is an excellent return on investment and your company accepts your recommendation to re-enter the well.

The upper Morrow sandstone was perforated from 7,443 to 7,451 ft. Initial production flowing (IPF) was 100 barrels of oil per day (BOPD) and 6 barrels of salt water per day (BSWPD) with an oil gravity of 38.3°. During the first four months, production averaged 75 BOPD, and after six months, production stabilized at 50 BOPD. Two years after completion, an offset well located 1,320 feet away, was drilled. The new well had an initial production flowing (IPF) of 336 BOPD and no water. Production came from a 19 foot upper Morrow sandstone reservoir which was, structurally, 13 feet high to the original well.

Answer Table D:
Pennsylvanian Upper Morrow Sandstone Log Evaluation Table

Depth	ILd	S.N.	Δt	$\Phi_s I$	S_{wa}	S_{wr}	S_{xo}	S_w/S_{xo}	MOS	BVW	S_{wirr}
7444	12	34	70	14	62	72	80	.76	18	.087	14
7446	11	29	81	22	41	69	55	.74	14	.090	9
7448	11	25	76	18	50	63	73	.69	23	.090	11

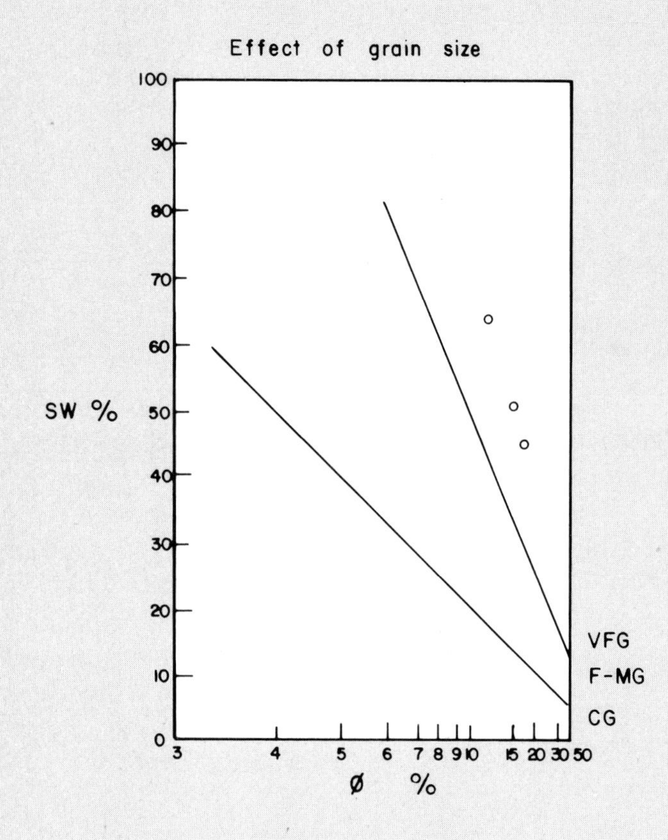

Figure 87.
Grain size determination by water saturation (S_w) versus porosity (ϕ) crossplot, Pennsylvanian upper Morrow Sandstone, Anadarko basin.

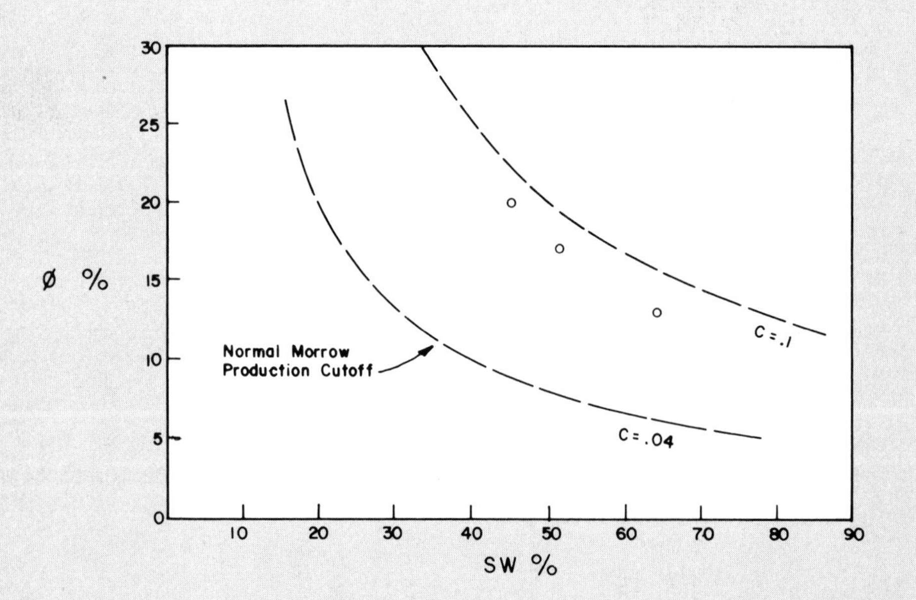

Figure 88.
Bulk volume water crossplot (ϕ vs. S_w), Pennsylvanian upper Morrow Sandstone, Anadarko basin.

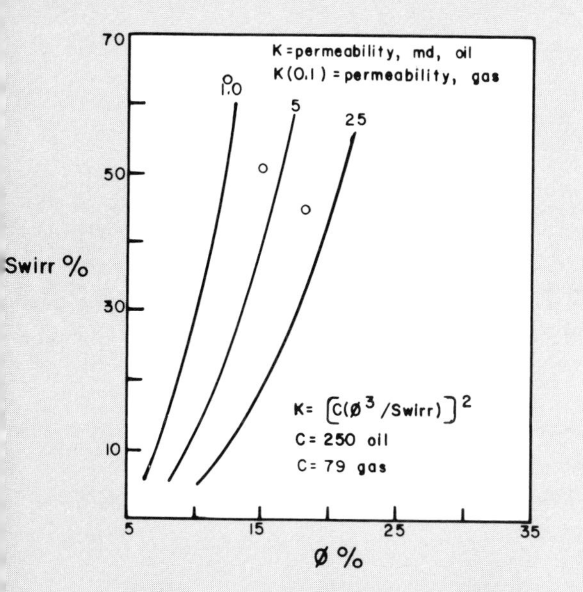

Figure 89.

Irreducible water saturation ($S_{w\,irr}$) versus porosity (ϕ) crossplot for determining permeability, Pennsylvanian upper Morrow Sandstone, Anadarko basin.

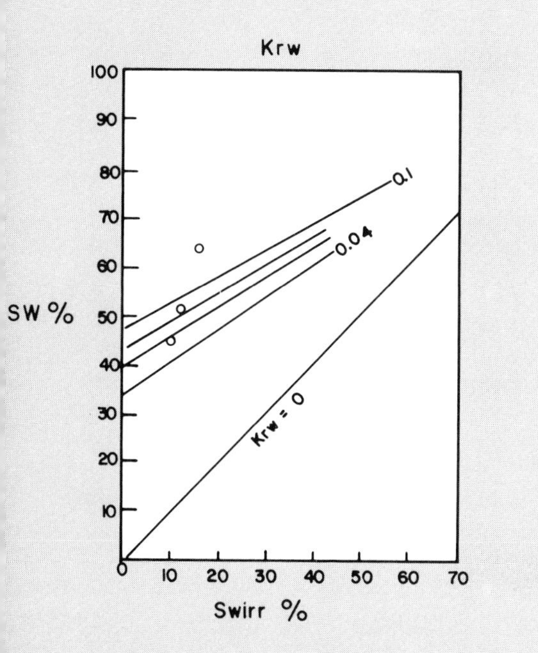

Figure 90.

Irreducible water saturation ($S_{w\,irr}$) versus water saturation (S_w) crossplot for determining relative permeability to water (K_{rw}), Pennsylvanian upper Morrow Sandstone, Anadarko basin.

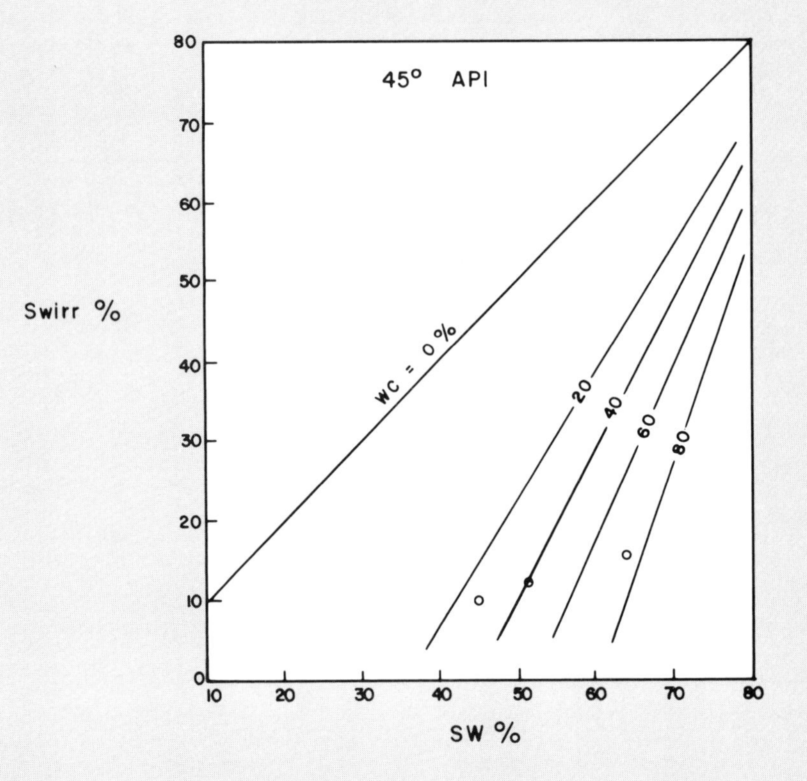

Figure 91.

Irreducible water saturation ($S_{w\,irr}$) versus water saturation (S_w) crossplot for determining percent water-cut, Pennsylvanian upper Morrow Sandstone, Anadarko basin.

Note:

WC = water-cut
API = oil gravity

Case Study 5
Cretaceous Pictured Cliffs Sandstone
San Juan Basin

A large block of acreage in the San Juan basin, much of it with gas production, has been purchased by the company you work for. The company is pursuing an active development drilling program of in-filling with new Cretaceous Pictured Cliffs Sandstone wells on 80 acre units, rather than on the 160 acre units, previously used.

You are presented with a log from a new in-fill well and are asked to evaluate it. Depth of the Pictured Cliffs Sandstone in the well is from 1,920 to 1,964 ft. The log package includes: an induction log with a Spherically Focused Log (SFL*) and an SP log (Fig. 92), and a Combination Neutron-Density log (recorded in sandstone porosity units†) with a gamma ray log (Fig. 93).

You already know from previous experience that shale in a zone can adversely affect logging measurements. Water saturations calculated by the Archie formula will have values which are too high (i.e. pessimistic values) if shale is present (see Chapter VI). Furthermore, shale in the reservoir will cause permeability problems.

A careful examination of the neutron porosity (ϕ_N), density porosity (ϕ_D), and gamma ray log convinces you that you are dealing with a shaly Cretaceous Pictured Cliffs sand (Fig. 93). A shaly sand analysis will be necessary. But, before following this investigative path, you decide to check the R_w value given for the area against a log calculated R_w. (See Chapter II, Table 3, for R_w formulas. Also see charts with Figures 12, 13, and 14).

R_w Calculation—Determine R_w using the following information: depth = 1,936; BHT = 89°F at 2,145 ft; surface temperature = 65°F; R_{mf} = 2.26 at 65°F; SSP = −57mv; T_f = 87°F; R_{mf} at 75° = 1.984; K = 71.526; R_{mfe}/R_{we} = 6.255; R_{mfe} = 1.687; R_{we} = 0.269. Therefore:

R_w at 75°F = 0.3026
R_w at T_f = 0.264

or

R_w at T_f = 0.26

To do a shaly sand analysis you use: a formula for calculating volume of shale, formulas for correcting both the neutron and density porosity for volume of shale, a formula for calculating neutron-density porosity, and finally, a formula which corrects water saturation for the effect of shale. The procedure is:

Shaly Sand Analysis—Shaly Sand Formulas; Schlumberger (1975):

$$V_{sh} = \frac{GR_{log} - GR_{min}}{GR_{max} - GR_{min}}$$

$$\phi_{Ncorr} = \phi_N - [(\phi_{Nclay}/0.45) \times 0.30 \times V_{sh}]$$

$$\phi_{Dcorr} = \phi_D - [(\phi_{Nclay}/0.45) \times 0.13 \times V_{sh}]$$

$$\phi_{N\text{-}D} = \sqrt{\frac{(\phi_{Ncorr})^2 + (\phi_{Dcorr})^2}{2}}$$

$$S_w = \frac{-\frac{V_{sh}}{R_{sh}} + \sqrt{\left(\frac{V_{sh}}{R_{sh}}\right)^2 + \frac{\phi^2}{0.2 \times R_w \times (1.0 - V_{sh}) \times R_t}}}{\frac{\phi^2}{0.4 \times R_w \times (1.0 - V_{sh})}}$$

Where (Formula Variables):

GR_{log} = Gamma ray reading from various depths in Pictured Cliffs Sandstone (for depths picked, see Log Evaluation Table; work Table E).
GR_{max} = 134 API gamma ray units; units are read on gamma ray log at a depth of 1,838 ft (shale).
GR_{min} = 64 API gamma ray units; units are read on gamma ray log at a depth of 1,921 ft (clean sand).
R_{sh} = 4 ohm-meters; resistivity of adjacent shale at a depth of 1,915 ft.
ϕ_{Nclay} = 0.53; neutron porosity of adjacent shale at a depth of 1,866 ft.
V_{sh} = volume of shale
$\phi_{N\,corr}$ = neutron porosity corrected for shale
$\phi_{D\,corr}$ = density porosity corrected for shale
ϕ = porosity corrected for shale

Other information you use to complete your log work is:

R_{mf} = 1.734 at T_f
T_f = 87°F

Because of the in-fill drilling program, your company is particularly interested in having you calculate recoverable reserves based on 80 acre units. Reservoir depletion will not be a problem, since Pictured Cliffs Sandstone wells normally don't drain 160 acres. The volumetric recoverable gas reserves are estimated from the following parameters: drainage area (DA) = 80 acres; reservoir thickness (h) = 30 ft; effective porosity (ϕ_e) = 17%; water saturation (S_w) = 53%; recovery factor (RF) = 0.75; gas gravity (estimated) = 0.51; temperature (estimated) = 99°F; initial bottom hole pressure (IBHP estimated) = 770 PSI: Z factor = 0.928; geothermal gradient = 0.051 × depth; and pressure gradient = 0.395 × depth.

A Cretaceous Pictured Cliffs Sandstone Log Evaluation Table (work Table E) assists you with your evaluation.

†See Chapter IV for a discussion of different matrix units used on the Combination Neutron-Density Log.

Work Table E:
Cretaceous Pictured Cliffs Sandstone Log Evaluation Table

Depth	ILd	SFL	Φ_N	Φ_D	GR	V_{sh}	$\Phi_{N\text{-}D}$	$\Phi_{N\text{-}D}^{\dagger}$	$\Phi_{N\text{-}D}^{\dagger\dagger}$	S_{wa}	$S_{wsh}^{\dagger\dagger\dagger}$	S_{wr}	S_{xo}	S_x/S_{xo}	MOS	BVW
1930	13	14	26.5	17	64	0	22	22	22	57	57					
1936	16	20	25	16.5	74	14	21	18	17	68	52					
1942	18	20	24	17.5	74	14	21	18	17	64	48					
1948	15	17	25.5	17	84											

$^{\dagger}\Phi_{N\text{-}D}$ **neutron-density porosity corrected for shale using the formula:** $\Phi_{N\text{-}D} = \Phi_{N\text{-}D} \times (1.0 - V_{sh})$

$^{\dagger\dagger}\Phi_{N\text{-}D}$ **neutron-density porosity corrected for shale using the following formulas:**

$$\Phi_{N\,corr} = \phi_N - \left[(\Phi_{N\,clay} / 0.45) \times 0.30 \times V_{sh}\right]$$

$$\Phi_{D\,corr} = \Phi_D - \left[(\Phi_{N\,clay} / 0.45) \times 0.13 \times V_{sh}\right]$$

$$\Phi_{N\text{-}D} = \sqrt{\frac{(\Phi_{N\,corr})^2 + (\Phi_{D\,corr})^2}{2}}$$

$^{\dagger\dagger\dagger}S_{wsh}$ **water saturation calculated by Schlumberger (1975) shaly sand equation.**

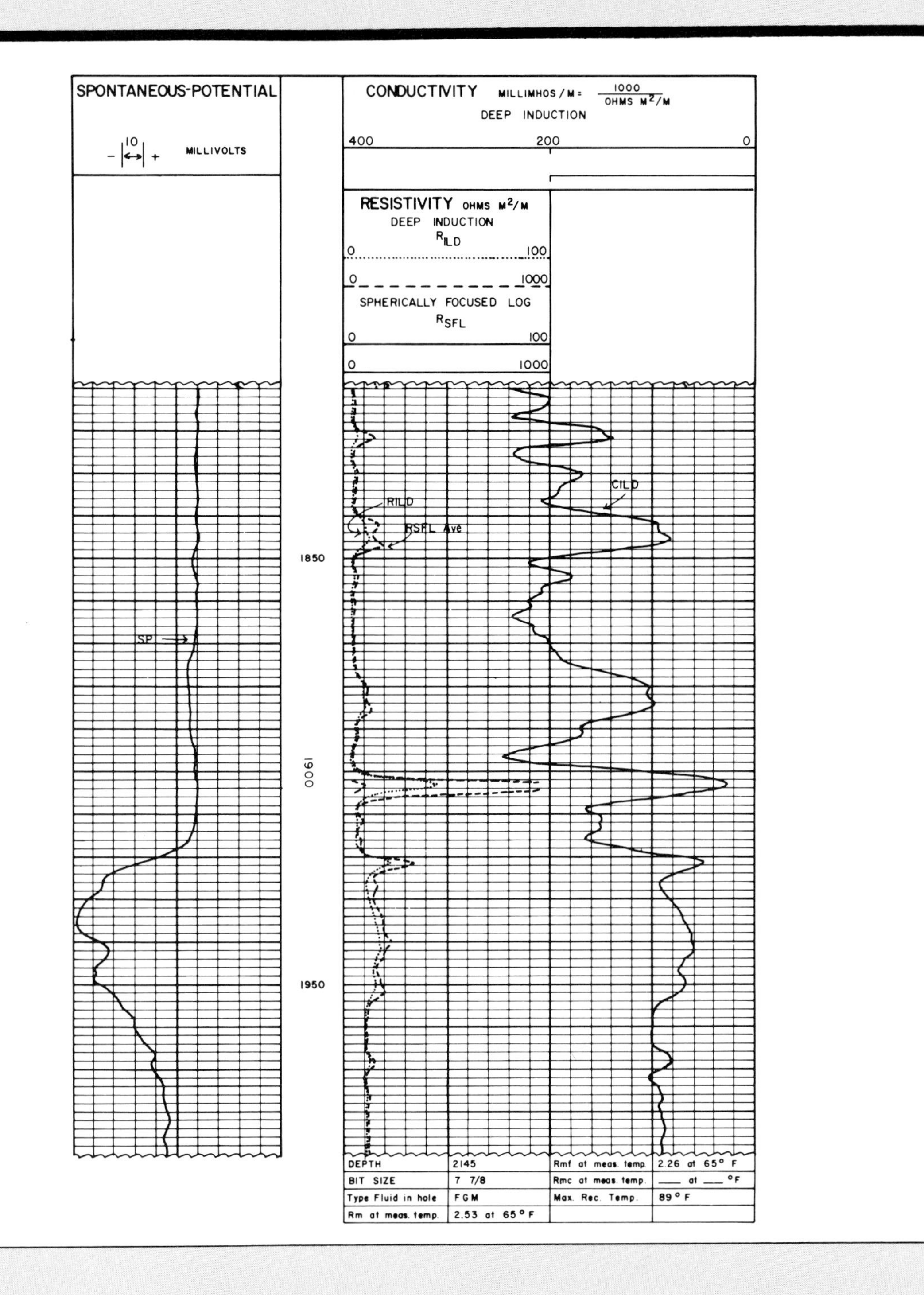

Figure 92. Induction-SFL* log with spontaneous potential log, Cretaceous Pictured Cliffs Sandstone, San Juan basin.

Note:

The increase in SP deflection (track #1) and increase in resistivity (tracks #2 and #3) upward in the Pictured Cliffs Sandstone (1,970 to 1,924 ft). These increases are the result of decreasing shale content upward in the Pictured Cliffs Sandstone.

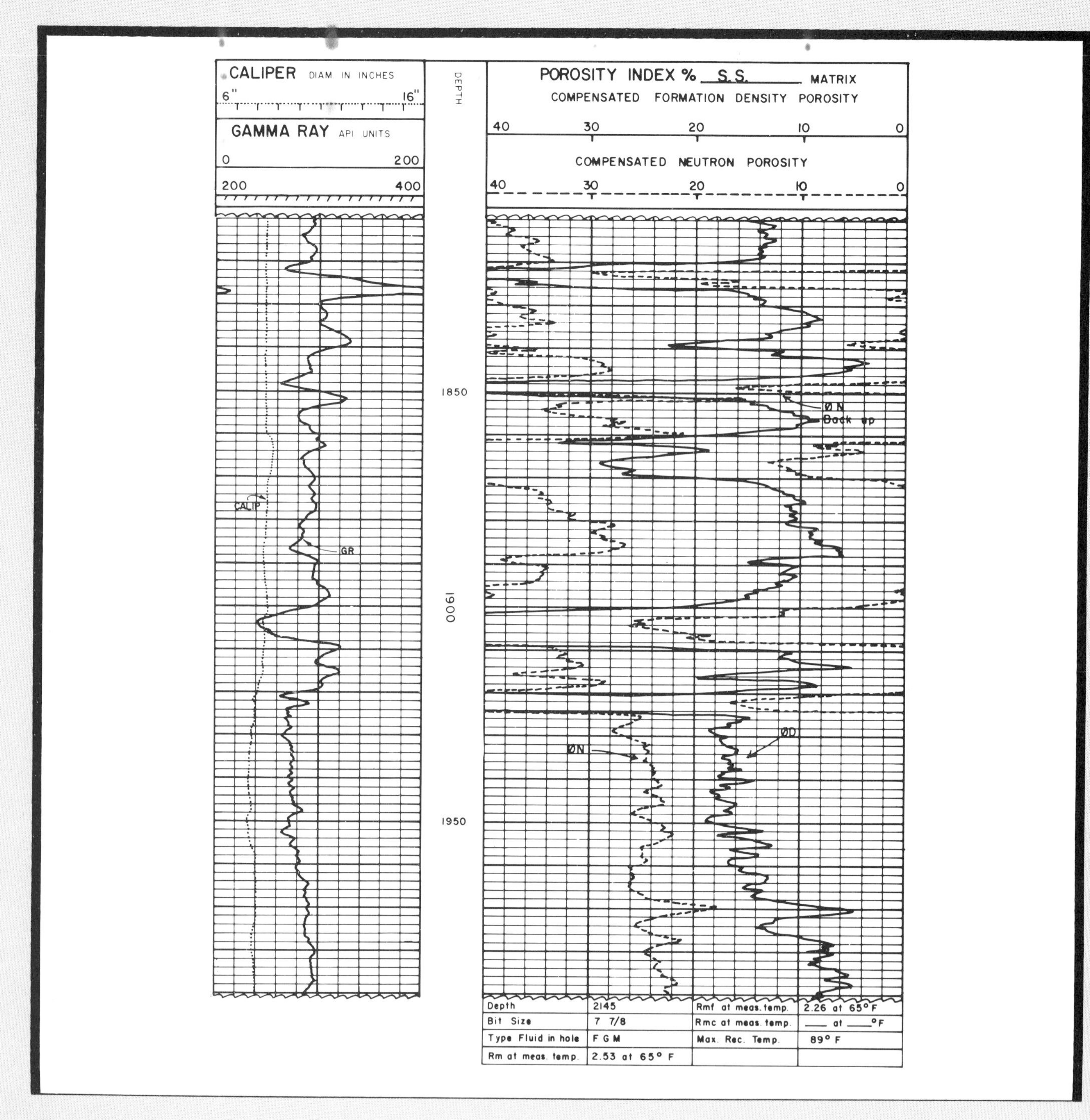

Figure 93. Combination Neutron-Density Log with gamma ray log and caliper, Cretaceous Pictured Cliffs Sandstone, San Juan basin.

Note:

1. Decrease in hole diameter (1,926 to 1,958 ft) on caliper log in track #1 indicating mudcake (permeability).
2. Neutron porosity is higher than density porosity in tracks #2 and #3. But in a gas-bearing sand, the neutron porosity should be less than density porosity (gas effect). It can be surmized that the reason for the high neutron porosity values in the Pictured Cliffs Sandstone is the high shale content of this sand. Shale has a high hydrogen concentration and therefore has high neutron porosity values.

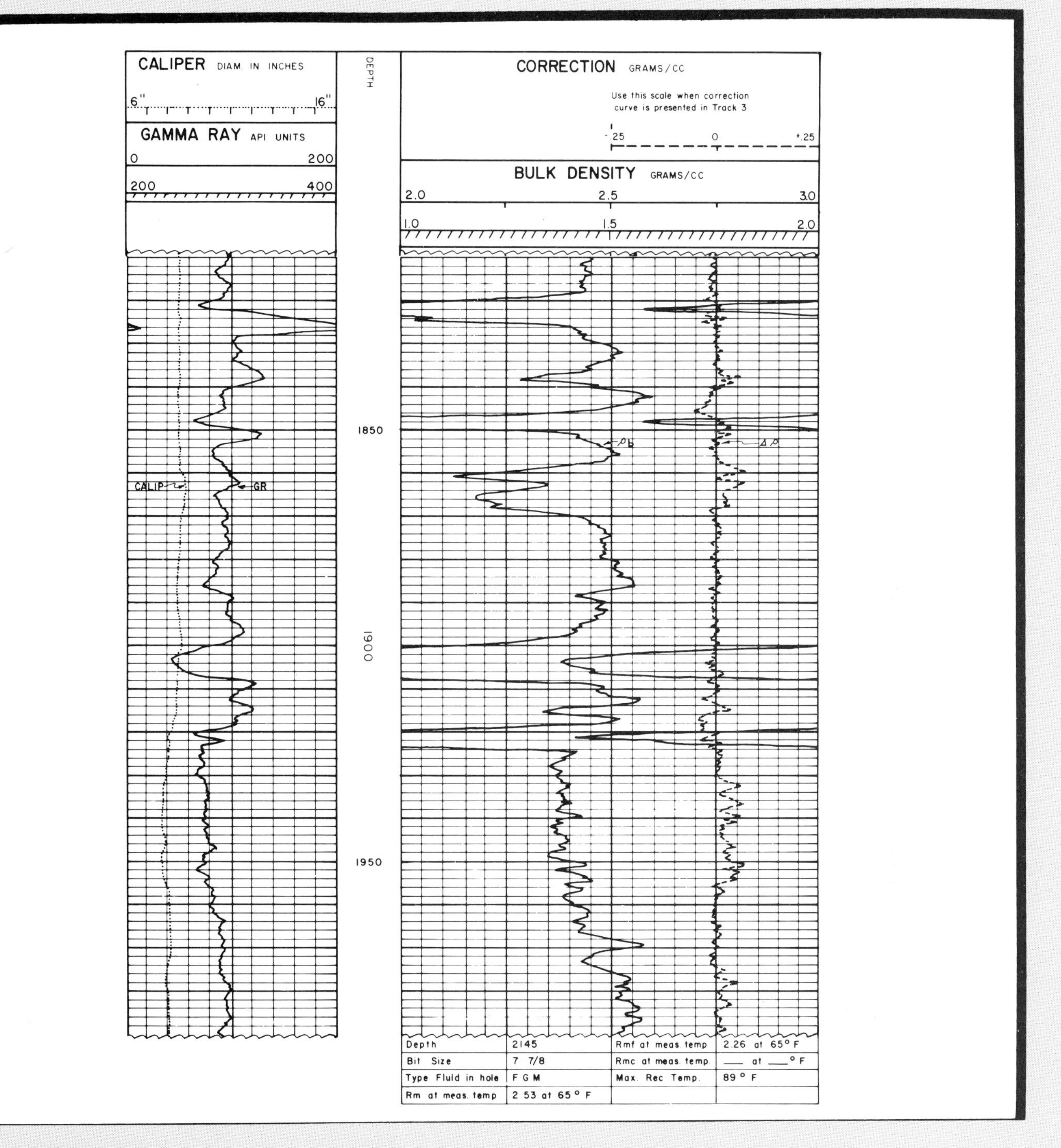

Figure 94. Density log with gamma ray log and caliper, Cretaceous Pictured Cliffs Sandstone, San Juan basin.

Note:

1. Decrease in hole diameter (1,926 to 1,958 ft) on caliper log in track #1 indicating mudcake (permeability).
2. Low bulk densities on the density log in tracks #2 and #3 indicating porosity.

Case Study 5 Answer

The behavior of the Combination Neutron-Density log, which was run on a sandstone matrix, alerts you to the presence of shale in the reservoir. You know that if a sand contains oil or water or both, the two curves—neutron and density—should have the same porosity values if the sand is logged on a sandstone matrix and if it is shale-free. However, if a sand contains gas (as in the case of the Pictured Cliffs) and is shale-free, the density log should have higher porosity values than the neutron log. (See Chapter IV about gas effect.) Your examination of the neutron-density log (Fig. 93) reveals neutron porosity (ϕ_N) is higher than density porosity (ϕ_D). The higher neutron porosity is recorded because shale has a high hydrogen ion concentration. The neutron log responds primarily to this concentration of hydrogen ions.

High values on the gamma ray log (Fig. 93) also alert you to the presence of shale in the Pictured Cliffs Sandstone. High gamma ray readings are caused by the greater levels of radioactivity found in shale.

Because shale can create permeability problems, as you analyze the well, you decide to pay careful attention to the various ways permeability can be detected or estimated.

You note a decrease in hole diameter occurring from 1,926 to 1,958 ft on the caliper log (Figs. 93 and 94). The decrease in hole diameter indicates the presence of mudcake, which means invasion has taken place.

A calculation of water saturation by the standard Archie equation gives values ranging from 62 to 83%. These high water saturation values result from the shale content of the reservoir, which is 6 to 29%. By using the Schlumberger (1975), shaly sand equations to correct for the effect of shale, the water saturations values can be lowered to 48 to 67%. Such a substantial adjustment to the water saturation values illustrates the importance of shaly sand analysis in the evaluation of certain reservoirs.

The high water saturations in the Pictured Cliffs Sandstone, even after using the shaly sand correction, are the result of the very fine grain size of the sand (Fig. 95). The reservoir's grain size, along with its shale content, is the reason for the low relative permeability to gas (Fig. 96) and the fairly low reservoir permeabilities (Fig. 97). The moveable hydrocarbon index is less than 0.7 and the moveable oil saturation is high. These parameters cause you to be reasonably certain gas will move out of the reservoir.

High bulk volume water values (0.11 to 0.14) also result from the very fine-grained shaly sand. But, the lack of data-point scatter from the hyperbolic line (Fig. 98) shows a reservoir at irreducible water saturation ($S_{w\,irr}$). And, at irreducible water saturation, the reservoir will not produce water, despite high water saturation values.

As you write your report on the Pictured Cliffs Sandstone in-fill well, you summarize all of the favorable indicators. First of all, the zone is approximately 30 ft thick and has permeability as shown by the presence of mudcake, indicated by the caliper log. Reservoir permeability is also indicated by the moveable hydrocarbon parameters (S_w/S_{xo} and MOS). Next, good porosities, ranging from 14 to 22% are determined from the Combination Neutron-Density log.

The bulk volume water plot is of special significance because it shows a reservoir at irreducible water saturation. Consequently, water-free production can be expected even though there are high water saturations. Volumetric calculations give a production estimate of approximately 0.322 BCF. All of the information you have developed, along with the shallow depth, low-cost, and very low-risk of the well, make a completion recommendation inevitable.

The Pictured Cliffs was perforated from 1,926 to 1,954 ft. After an acid treatment and a sand frac, the well initially produced 350 mcfgpd with no water on a 3/4″ choke. During the first year of production, the well produced approximately 60 million cubic feet (60,000 mcf) of gas.

Answer Table E:
Cretaceous Pictured Cliffs Sandstone Log Evaluation Table

Depth	ILd	SFL	Φ_N	Φ_D	GR	V_{sh}	Φ_{N-D}	$\Phi_{N-D}^{\dagger}$	$\Phi_{N-D}^{\dagger\dagger}$	S_{wa}	$S_{wsh}^{\dagger\dagger\dagger}$	S_{wr}	S_{xo}	S_x/S_{xo}	MOS	BVW	S_{wirr}
1926	11	16	25	14.5	70	9	20	18	18	77	67	39	100	.67	33	.139	11
1928	12	15	27.5	18	70	9	23	21	21	63	56	36	100	.56	44	.132	9
1930	13	14	26.5	17	64	0	22	22	22	57	57	32	100	.57	43	.125	9
1932	14	17	24.5	17	72	11	21	19	18	69	57	35	100	.57	44	.123	11
1934	15	19	24.5	16	74	14	21	18	17	70	55	36	100	.55	45	.119	12
1936	16	20	25	16.5	74	14	21	18	17	68	52	36	100	.52	48	.112	12
1938	17	20	24	17	72	11	21	19	18	62	51	34	100	.51	49	.112	11
1940	17	23	23.5	14.5	72	11	19.5	17	17	68	55	37	100	.55	45	.112	12
1942	18	20	24	17.5	74	14	21	18	17	64	48	33	100	.48	52	.109	12
1944	15	18	23	17.5	74	14	20	17	17	70	55	35	100	.55	45	.119	12
1946	14	15	23	16	76	17	20	17	15	83	59	32	100	.59	41	.125	14
1948	15	17	25.5	17	84	29	22	16	14	81	49	31	100	.49	51	.113	15
1950	15	18	23	18.5	70	9	21	19	19	62	55	35	100	.55	45	.118	11
1952	14	20	23	16	68	6	20	19	18	68	61	39	100	.61	39	.122	11

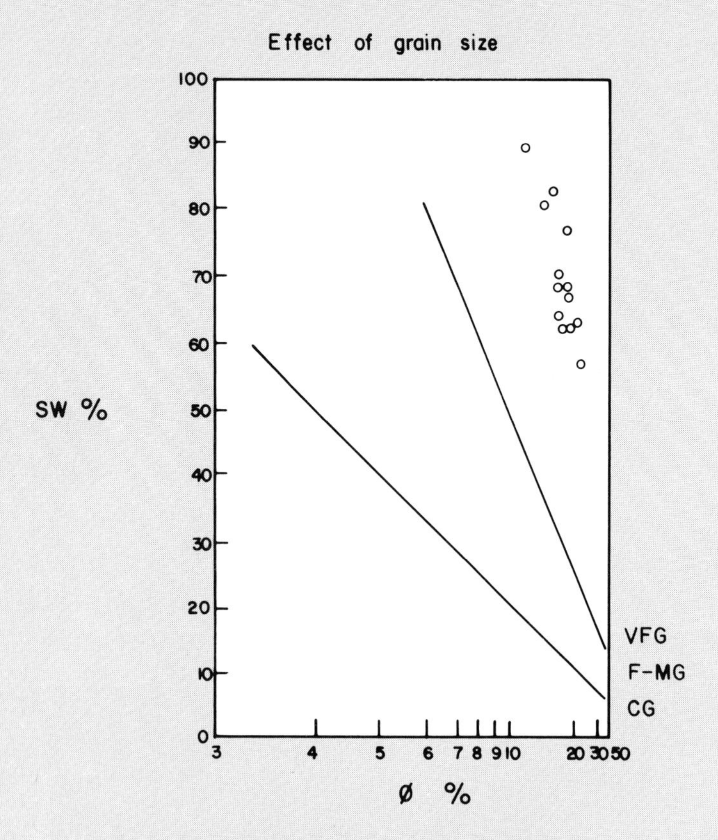

Figure 95.

Grain size determination by water saturation (S_w) versus porosity (ϕ) crossplot, Cretaceous Pictured Cliffs Sandstone, San Juan basin.

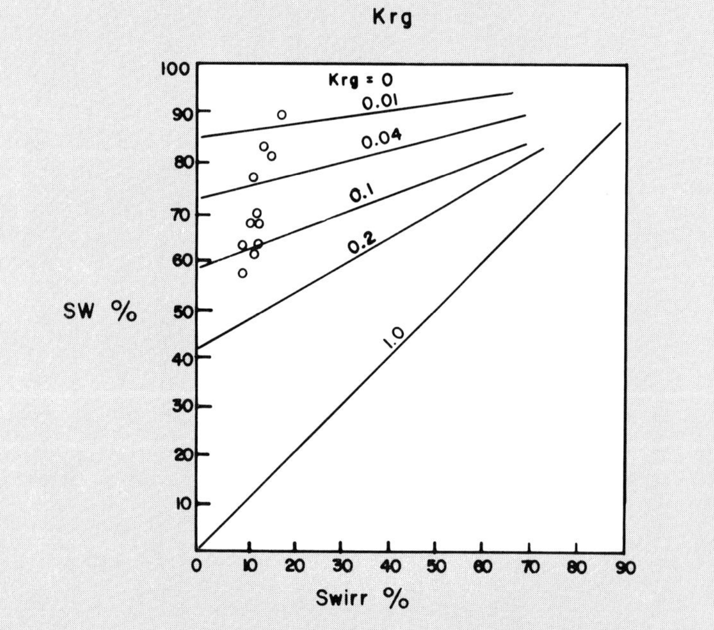

Figure 96.

Irreducible water saturation ($S_{w\,irr}$) versus water saturation (S_w) crossplot for determining relative permeability to gas (K_{rg}), Cretaceous Pictured Cliffs Sandstone, San Juan basin.

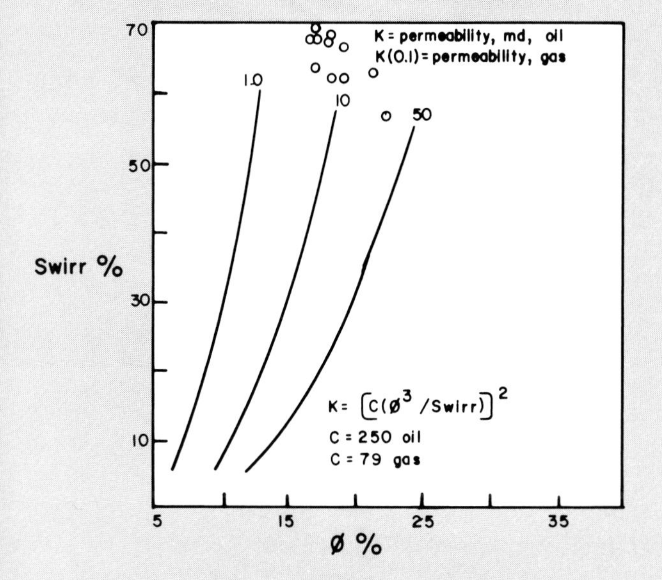

Figure 97.

Irreducible water saturation ($S_{w\,irr}$) versus porosity (ϕ) crossplot for determining permeability, Cretaceous Pictured Cliffs Sandstone, San Juan basin.

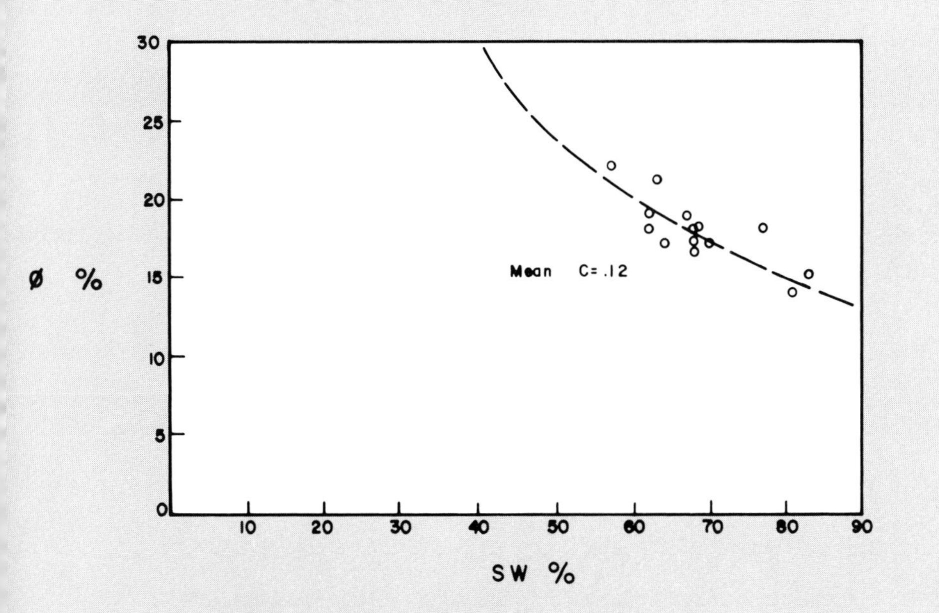

Figure 98.

Bulk volume water crossplot (ϕ vs. S_w), Cretaceous Pictured Cliffs Sandstone, San Juan basin.

Case Study 6
Devonian Hunton Formation
Anadarko Basin

A well owned by your company has just reached final depth after penetrating its target zone—the Hunton Formation in the Anadarko basin. The well was drilled because subsurface mapping indicated an updip fault and a structurally low well which produced gas and water. As you correlate logs from the newly completed well, you determine that it is some 40 ft structurally high to the well which produced gas and water.

When your company's well encountered the Chimney Hill Member of the Hunton Formation, a drilling break occurred. The penetration rate changed from 10 min/ft to 5 min/ft, and the mud logger's chromatograph increased 20 gas units.

By examining samples, you identify Hunton lithology as dolomite with some minor chert. It is gray to gray-brown, medium crystalline, and sucrosic with vuggy porosity. Samples collected through the drilling break do not exhibit fluorescence or cut.

Your usual procedure before beginning log analysis is a check of R_w values, generally known for the area, against a log-calculated R_w value. This time, you decide to determine a log value for R_w using charts (see Chapter II) and the SP log. The following information is assembled before you begin finding an R_w value:

R_w Determination Using SP Log and Charts—You determine SSP = −95mv where the SSP value is read from SP log; T_f = 221°F; and R_{mf} = 0.249 at T_f.

As you examine the log package on the new Hunton well, you identify values for: resistivity of the mud filtrate (R_{mf} = 0.249 at T_f), surface temperature (70°F) and formation temperature (221°F).

To correct neutron porosity for temperature in deeper wells (generally over 12,000 ft) such as this Hunton well, you need to add 1.5 porosity units (P.U.) to neutron porosity (ϕ_N) before you determine neutron-density porosity.

Recoverable volumetric gas reserves are calculated with these parameters: drainage area (DA) = 540 acres; gas gravity = 0.63 (estimated); temperature (estimated) = 228°F; BHP (estimated) = 7,260 psi; recovery factor (RF) = 0.75; Z factor = 1.165. In addition, a geothermal gradient of .014 × formation depth and a pressure gradient of 0.445 × formation depth are used in reserve calculations. You need to determine the following parameters: porosity (ϕ), water saturation (S_w), and reservoir thickness (h).

When the log evaluation is finished, you will make a completion decision. And, if you decide to set pipe you will be responsible for selecting perforations. A Devonian Hunton Log Evaluation Table (work Table F) helps organize the log data, but you must pick your own depths for calculation.

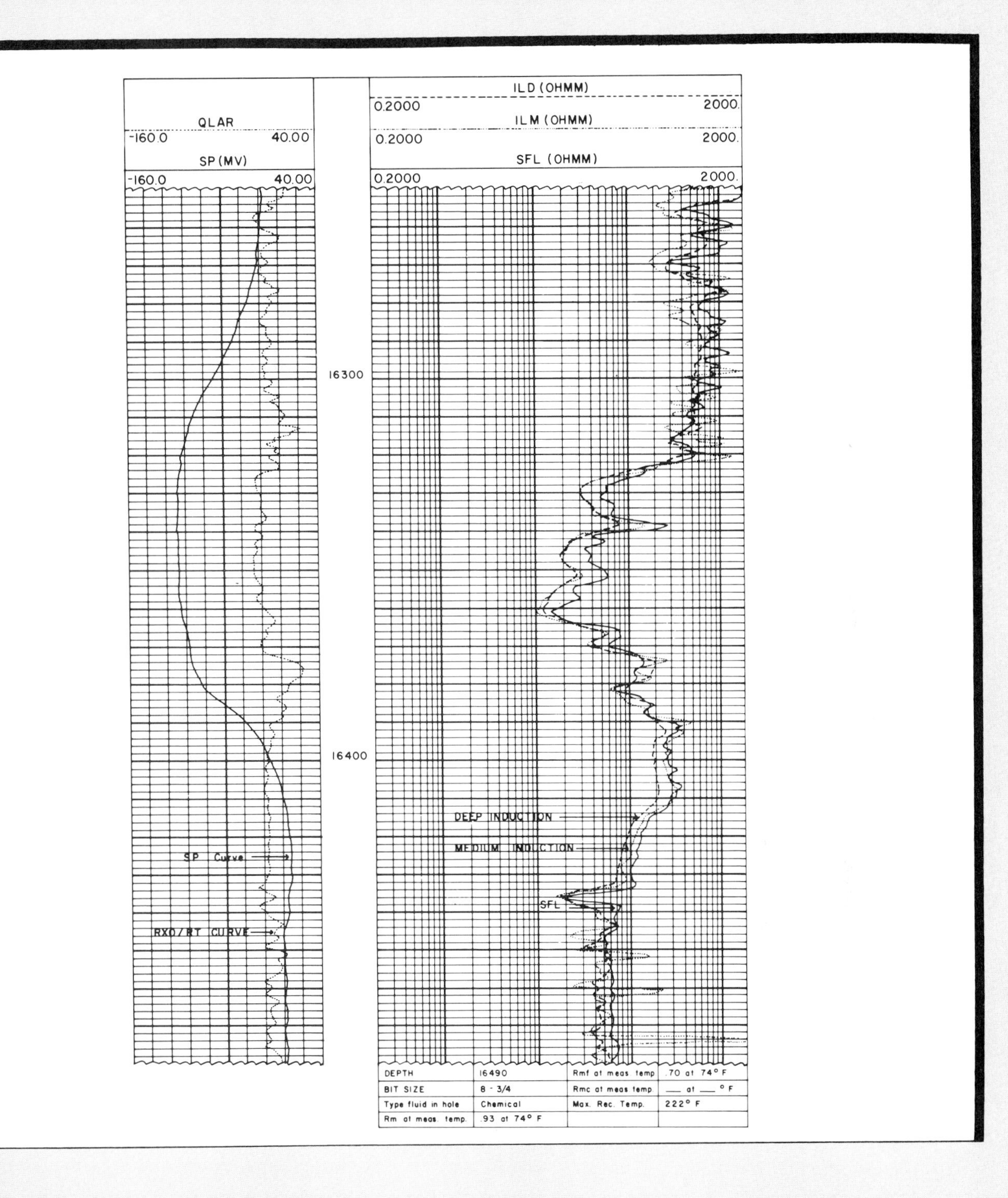

Figure 99. Dual induction-SFL* with spontaneous potential log and R_{xo}/R_t curve, Devonian Hunton Dolomite, Anadarko basin.

Note:

1. From 16,300 to 16,390 ft, the deflection of the R_{xo}/R_t curve to the right away from the SP curve in track #1 which indicates the presence of hydrocarbons (see Chapter VI).
2. From 16,325 to 16,370 ft, see the much lower resistivities in tracks #2 and #3. This is the result of either a water productive zone or a substantial change in the type of carbonate porosity (see Chapter VI, Table 8).

Work Table F:
Devonian Hunton Formation Log Evaluation Table

Depth	ILd	ILm	SFL	R_t	Φ_N	Φ_N†	Φ_D	$\Phi_{N\text{-}D}$	S_{wa}	S_{WT}	S_{XO}	S_w/S_{xo}	MOS	BVW										

$\dagger\Phi_N = \Phi_N + 1.5$ porosity units (P.U.)

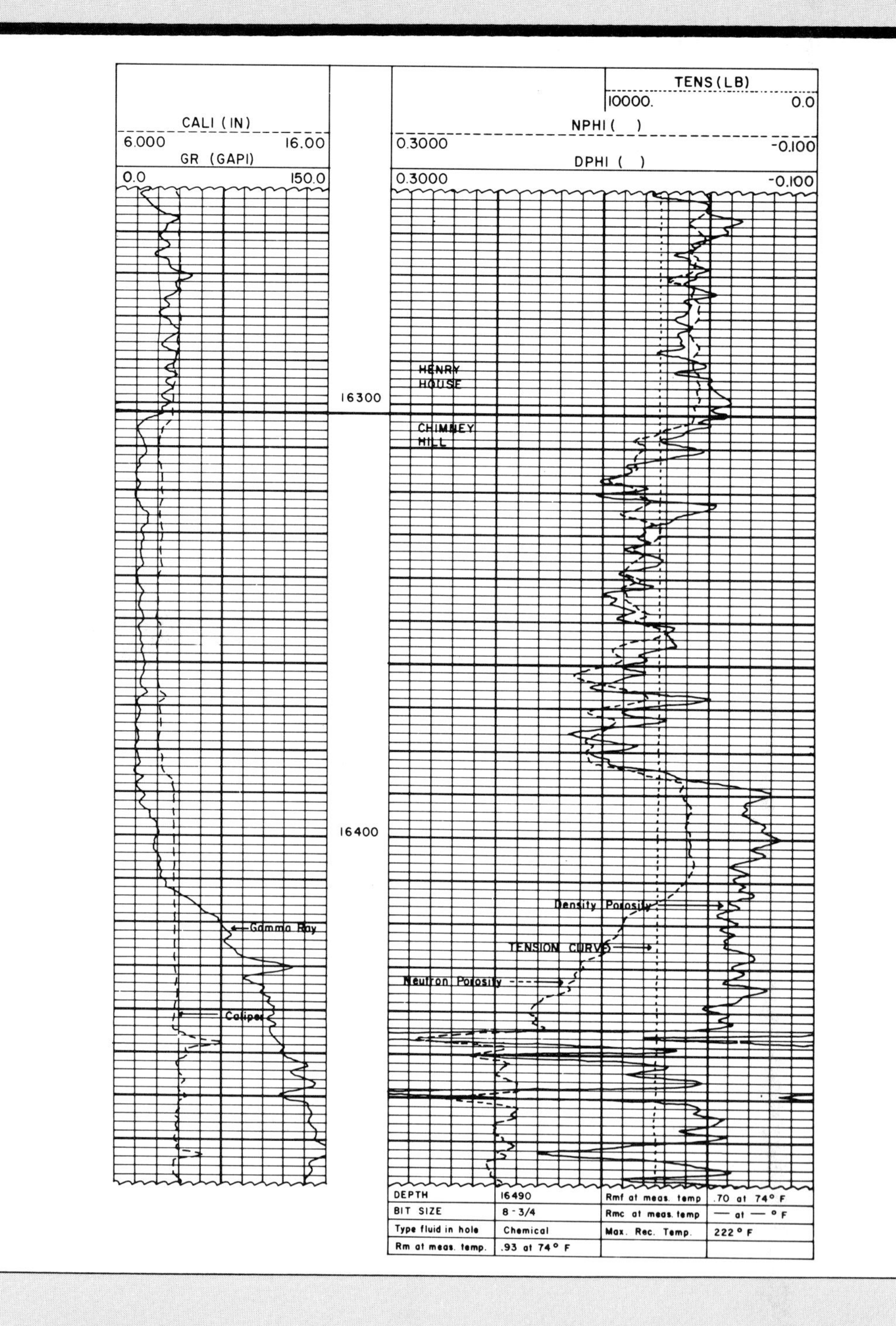

Figure 100. Combination Neutron-Density log with gamma ray log and caliper, Devonian Hunton Dolomite, Anadarko basin.

Note:

1. From 16,302 to 16,386 ft, the decrease in hole diameter on the caliper log in track #1 because of mud cake (permeability).
2. From 16,302 to 16,384 ft, see the increase in neutron and density porosity. Also, the neutron and density porosity values are approximately equal. Because the lithology of the Chimney Hill Member of the Hunton Formation is dolomite, if the Hunton was wet or an oil-bearing reservoir, the neutron porosity should be greater than the density porosity (lithology effect; Chapter IV). However, because the Chimney Hill porosity is gas-bearing, the neutron-density porosities are approximately equal (gas effect).

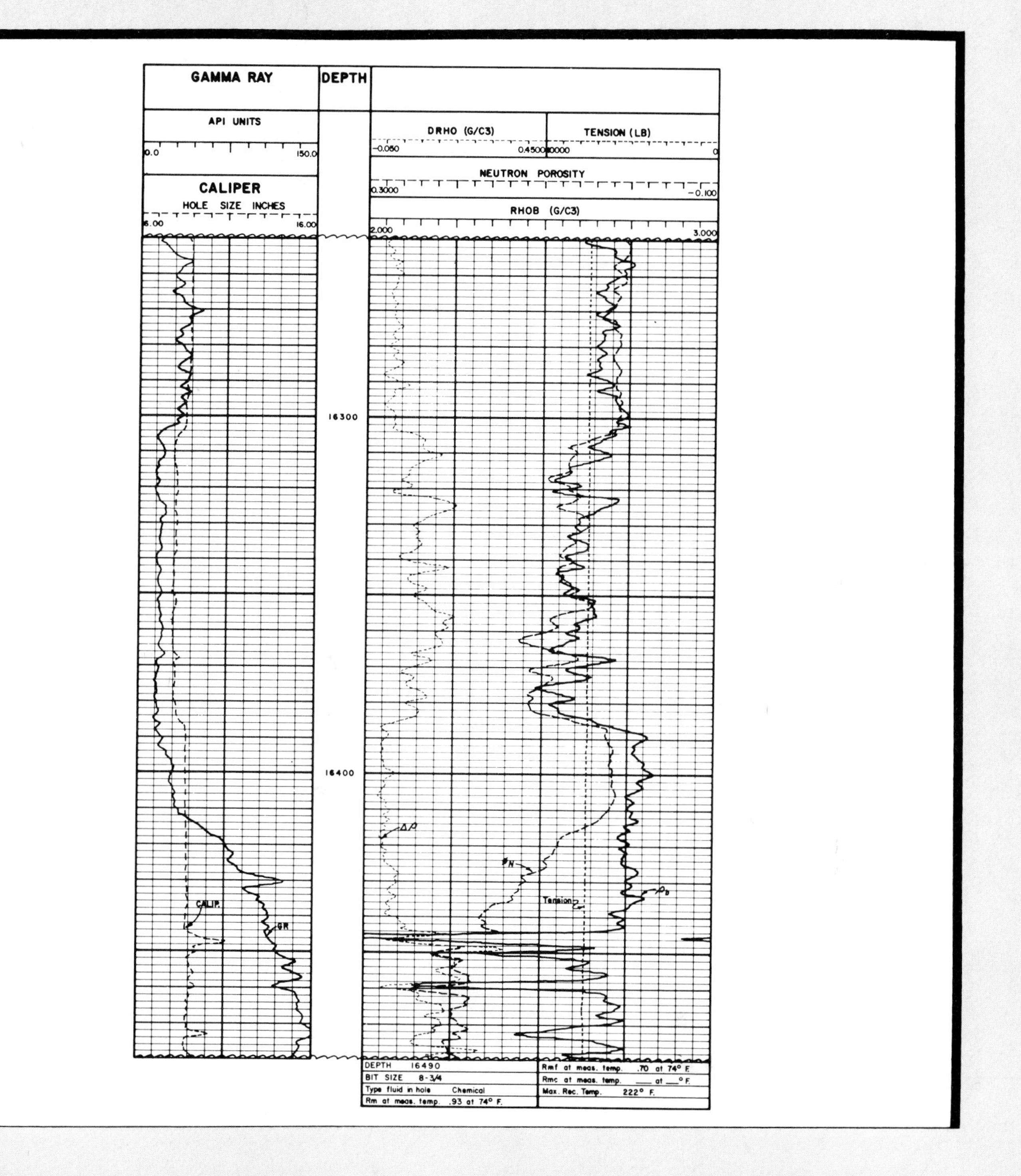

Figure 101. Density log with neutron porosity, gamma ray log, and caliper, Devonian Hunton Dolomite, Anadarko basin.

Note:

From 16,304 to 16,386 ft, the increase in correction (Δ_ρ) which has been applied to the bulk density log (ρ_b). This increase in correction is the result of the mudcake developed on the porous and permeable Chimney Hill Member of the Hunton Formation.

Case Study 6 Answer

The log package used in the well includes a Dual Induction Spherically Focused Log (SFL*) with an SP log and a R_{xo}/R_t curve (Fig. 99), and a Combination Neutron-Density log with a gamma ray log and caliper (Figs. 100 and 101).

On the Dual Induction Log, the R_{xo}/R_t quick look curve has a strong deflection to the right, away from the SP log, from 16,300 to 16,390 ft (Fig. 99). The rightward deflection of the R_{xo}/R_t curve indicates hydrocarbons are present. Further examination of the Dual Induction Log shows resistivities decline precipitously at an interval from 16,325 to 16,370 ft (Fig. 99). Such a rapid change to lower resistivities may be due to the presence of water or perhaps to a signficant change in lithology.

The SP curve in track #1 of the Dual Induction Log is used to find a value for R_w. The chart in Figure 13 helps you determine R_{mf}/R_{we} from SSP; the value located on the chart is 12. Next, calculate a value for R_{we} by the formula: $R_{we} = R_{mf}/(R_{mf}/R_{we})$. The resulting value for R_{we} is 0.021. Then, you determine by the chart (Fig. 14) that R_w at T_f is 0.025.

The caliper log on the Combination Neutron-Density Log alerts you to the occurence of invasion in the Hunton by mudcake development (Fig. 100). Another item of significance interpreted from the Combination Neutron-Density Log is gas effected dolomite (see Chapter IV).

On neutron-density porosity logs recorded over a non-gas-bearing dolomite, the neutron log will read a higher porosity than the density log when the logs are run on a limestone matrix. However, on this matrix through the porous Hunton dolomite, both the neutron and the density logs record essentially the same porosity, with porosity values ranging from 6 to 10% (Fig. 100). These equivalent porosity values can be explained by the presence of gas in the zone, because gas causes the neutron log to record too low a porosity, and the density log to record too high a porosity. The coming together of both the neutron and density curves, as noted in the Hunton dolomite, reflects a gas effected dolomite (Fig. 100).

The density log (Fig. 101) has a high correction on the Δ_ρ curve from 16,304 to 16,386 ft. This high amount of correction is the result of correcting the bulk density ρ_b for effect of mudcake.

Grain size, determined by a crossplot of water saturation versus porosity (Fig. 102), over the porous Chimney Hill Member of the Hunton from 16,306 to 16,384 ft is: coarse-grained from 16,306 to 16,324 ft; fine-grained to medium-grained from 16,326 to 16,372 ft; and coarse-grained from 16,372 to 16,384 ft. The coarser grain size data points on the crossplot are probably not the result of coarse, sucrosic dolomite porosity, but rather the presence of vuggy porosity. You reach this conclusion because you know that vuggy porosity has very low values for irreducible water saturation, and therefore, also has very low values for bulk volume water (see Chapter VI, Table 8). Conversely, because of higher bulk volume water values when only intercrystalline porosity is present, you conclude that the fine to medium grain sizes from 16,326 to 16,372 ft (Fig. 102) represent a zone in the Chimney Hill Member with predominantly intercrystalline (i.e. sucrosic) porosity and only minor vugs.

The low (less than 0.7) moveable hydrocarbon index (S_w/S_{xo}) together with high moveable oil saturation percentages suggests hydrocarbons were moved by invasion of the mud filtrate. Relative permeability to water is low and varies from 0.06 to 0 (Fig. 103) and relative permeability to gas (Fig. 104) is high with values ranging from greater than 20% to 100%. The relative permeability values for K_{rw} and K_{ro} are favorable indicators of commercial production.

Permeability (Fig. 105) averages 10 to 15 md and reaches a maximum permeability of 50 md. The bulk volume water crossplot (Fig. 106) confirms the presence of higher water saturations due to a finer grain size and a lack of vuggy porosity, or the higher water saturation values may be related to water saturation values above irreducible, where bulk volume water is greater than 0.015. Higher bulk volume water values are from the zone occurring over an interval from 16,325 to 16,370 ft. This is the interval which also happens to have the lower resistivities (Fig. 99). Bulk volume water values of 0.015 or less (Fig. 106) are from zones in the Hunton with both vuggy and intercrystalline porosity (see Chapter VI, Table 8).

The upper zone from approximately 16,300 to 16,322 ft of the Chimney Hill Member of the Hunton has several favorable indicators of a productive hydrocarbon zone. First, good permeability and porosity seem to be present. The porosity ranges from 6 to 10% and is intercrystalline and vuggy. There are low water saturations and low bulk volume water values; these, along with the high relative permeabilities to gas and the good reservoir permeabilities, support your decision to set pipe and to perforate the upper zone. The lower zone from 16,325 to 16,370 ft has higher water saturations and higher bulk volume water values, and therefore, it may be water productive. Because subsurface geology indicates the presence of downdip water, this zone might be avoided for perforating; however, the higher water saturation and bulk volume water values over the interval from 16,325 to 16,370 ft may occur because of changes in rock type. Evidence for a lithology change is the apparent loss of vuggy porosity as indicated by the increase in BVW from 16,325 to 16,370 ft. If, in fact, a loss of vuggy porosity has caused the abrupt change to lower resistivities, then the interval may not represent a zone above irreducible water saturation (i.e. "wet zone"). Unfortunately, without having a core to analyze, you can't determine with any certainty whether or not the 16,325 to 16,370 ft interval will be water or hydrocarbon productive, and so you decide to avoid perforating it in this well.

The estimated gas recovery is 2.76 BCF. Parameters used to arrive at this figure are: drainage area (DA) = 540 acres; porosity (ϕ) = 7%; water saturation (S_w) = 11%; reservoir thickness (h) = 8 ft; gas gravity = 0.63 (estimated); temperature (estimated) = 228°F; BHP (estimated) = 7,260 psi; recovery factor (RF) = 0.75; Z factor = 1.165. A geothermal gradient of 0.014 × depth and a pressure gradient of 0.445 × depth.

The Hunton was perforated from 16,306 to 16,314 ft with 27 holes. Completion was natural and the calculated open flow (CAOF) of the well was 67,023 mcfgpd and the shut-in tubing pressure was 4,639 psi. During the first six months of production, the Hunton produced 0.455 BCF plus 2,030 barrels of condensate.

Answer Table F:
Devonian Hunton Formation Log Evaluation Table

Depth	ILd	ILm	SFL	R_t	Φ_N	Φ_N†	Φ_D	$\Phi_{N\text{-}D}$	S_{wa}	S_{wr}	S_{xo}	S_w/S_{xo}	MOS	BVW	S_{wcorr}	S_{wirr}
16306	500	330	500	500	4.5	6	2.5	5	14	24	45	.32	31	.007	12	45
16308	500	500	600	500	8	9.5	4.5	7	10	27	29	.35	19	.007	8	32
16310	500	450	450	500	6.5	8	5.5	7	10	22	34	.30	24	.007	8	32
16312	500	500	500	500	7	8.5	4.5	7	10	24	32	.32	22	.007	8	32
16314	475	365	365	475	7	8.5	7.5	8	9	20	33	.28	24	.007	8	28
16316	300	600	300	300	8.5	10	8.5	9	10	24	32	.32	22	.009	7	25
16318	400	425	480	400	8	9.5	8	9	9	27	25	.35	17	.008	8	25
16320	450	1000	525	450	7	8.5	9.5	9	8	26	24	.34	16	.007	6	25
16322	250	300	300	250	2.5	4	2.5	3	33	27	96	.35	63	.010	6	75
16324	130	200	150	130	7.5	9	0.5	7	20	26	58	.34	39	.014	—	32
16326	55	60	110	55	6.5	8	3	6	35	37	79	.45	44	.021	18	37
16328	35	35	60	35	5	6.5	6	6	44	33	100	.44	56	.027	35	37
16330	30	31	55	30	5	6.5	6.5	7	44	35	100	.44	56	.029	—	34
16332	35	35	55	35	6	7.5	6	7	38	31	96	.40	58	.027	—	32
16334	40	41	70	40	6.5	8	6.5	7	36	34	85	.42	50	.025	—	32
16336	50	45	65	50	7	8.5	4.5	7	32	28	88	.36	57	.022	—	32
16338	75	120	210	43	8	9.5	7.5	9	27	64	38	.70	12	.024	21.5	25
16340	45	60	100	45	6.5	8	7.5	8	29	39	62	.47	33	.024	27	28
16342	28	28	45	28	8	9.5	5.5	8	37	32	93	.40	56	.030	—	28

16344	22	22	38	22	7.5	9	9	9	37	33	90	.415	53	.034	—	25
16346	19	20	39	19	7	8.5	9	9	40	37	89	.45	49	.036	—	25
16348	20	20	37	20	6	7.5	7.5	8	47	35	100	.47	53	.035	—	30
16350	25	28	52	25	7	8.5	7	8	39	37	86.5	.46	47	.032	—	28
16352	27	31	60	27	4	5.5	4	5	61	39	100	.61	39	.030	—	45
16354	20	20	30	20	6	7.5	3.5	6	59	31	100	.59	41	.035	—	37
16356	15	15	26	15	9	10.5	4	8	51	33	100	.51	49	.041	—	28
16358	14	12	28	14	8.5	10	6.5	8	53	37	100	.53	47	.042	—	28
16360	12	11	18	12	9.5	11	8	10	46	31	100	.455	54.5	.045	—	22
16362	13	12	16	13	12.5	14	6.5	11	40	27	100	.40	60	.044	—	20
16364	23	20	30	23	10.5	12	9.5	11	30	28	83	.36	53	.033	—	20
16366	50	50	78	50	6.5	8	8.5	8	28	31	71	.395	43	.022	27	28
16368	47	47	78	47	6	7.5	5	6	38	33	94	.41	56	.023	—	37
16370	42	35	60	42	10.5	12	7	10	'24	30	64	.38	40	.024	23	22
16372	90	90	115	90	9.5	11	5.5	9	19	28	52	.36	33	.017	17	25
16374	180	250	145	180	9	10.5	8	9	13	21	46	.28	33	.012	12	25
16376	190	120	120	190	10.5	12	13	12.5	9	18	36	.25	27	.011	8	18
16378	160	180	120	160	11	12.5	7.5	10	13	20	46	.27	33	.012	11	22
16380	110	100	100	110	11.5	13	11	12	13	22	42	.30	29	.015	11	19
16382	90	65	70	90	11	12.5	9.5	11	15	20	54	.28	39	.017	14	20
16384	120	88	140	120	7.5	9	5	7	21	26	60	.34	40	.014	19	32

†$\Phi_N = \Phi_N + 1.5$ porosity units (P.U.)

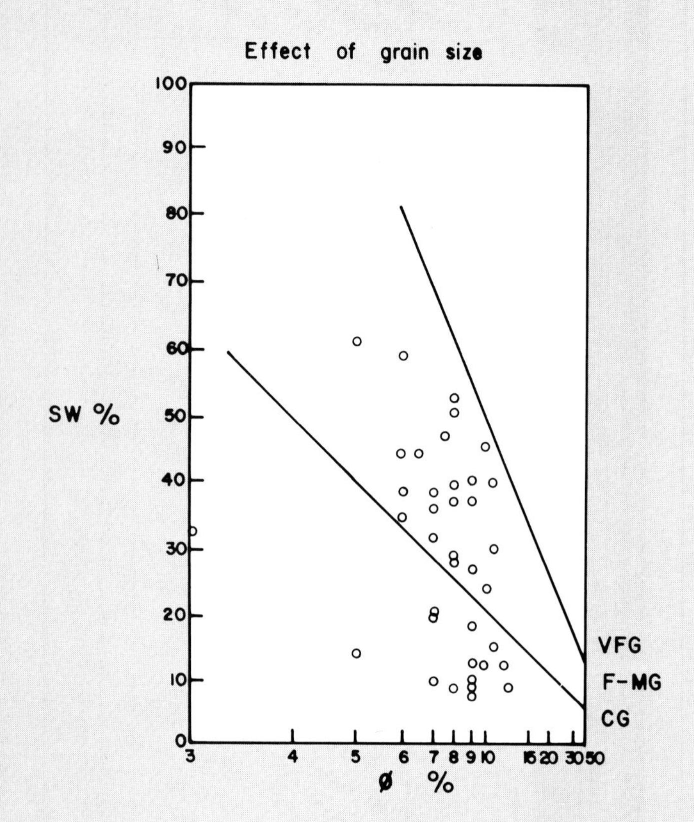

Figure 102.

Grain size determination by water saturation (S_w) versus porosity (ϕ) crossplot, Devonian Hunton Dolomite, Anadarko basin.

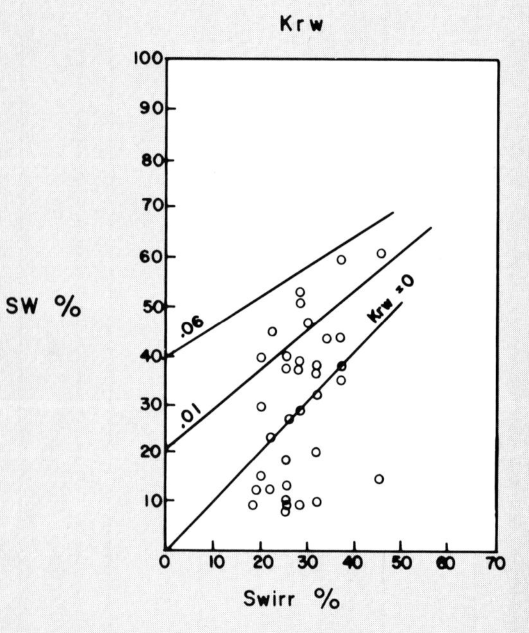

Figure 103.

Irreducible water saturation ($S_{w\ irr}$) versus water saturation (S_w) crossplot for determining relative permeability to water (K_{rw}), Devonian Hunton Dolomite, Anadarko basin.

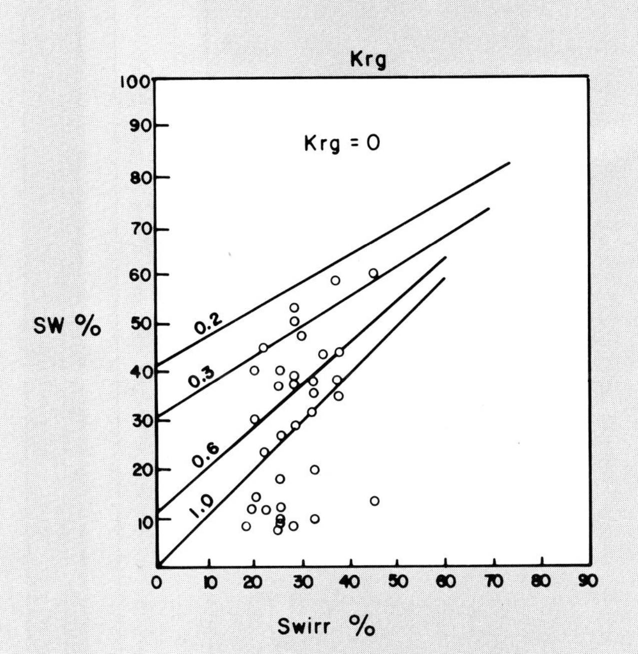

Figure 104.
Irreducible water saturation ($S_{w\ irr}$) versus water saturation (S_w) crossplot for determining relative permeability to gas (K_{rg}), Devonian Hunton Dolomite, Anadarko basin.

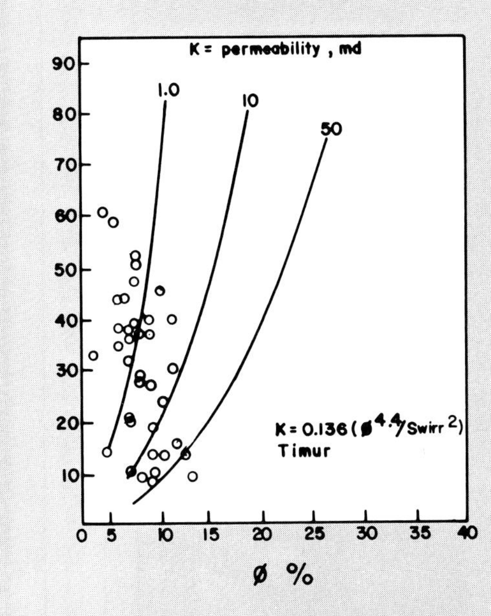

Figure 105.
Irreducible water saturation ($S_{w\ irr}$) versus porosity (ϕ) crossplot for determining permeability, Devonian Hunton Dolomite, Anadarko basin.

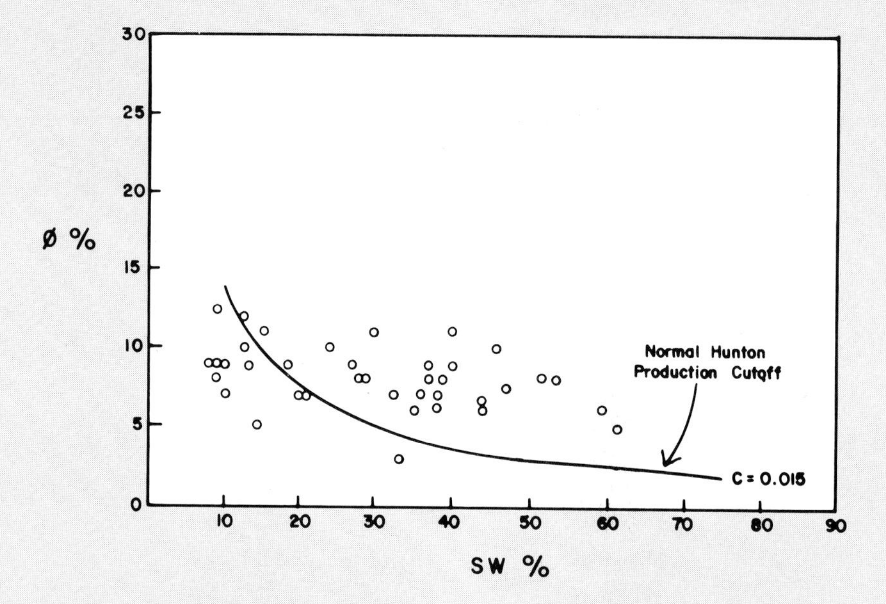

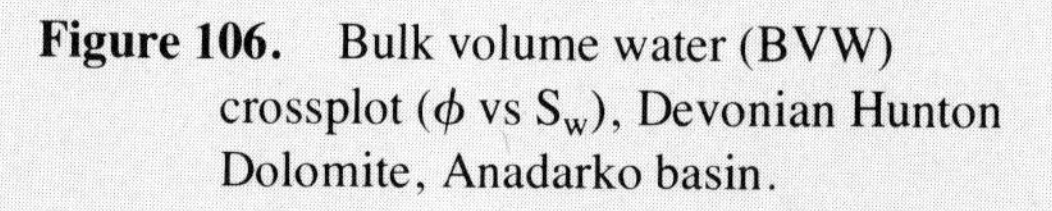

Figure 106. Bulk volume water (BVW) crossplot (ϕ vs S_w), Devonian Hunton Dolomite, Anadarko basin.

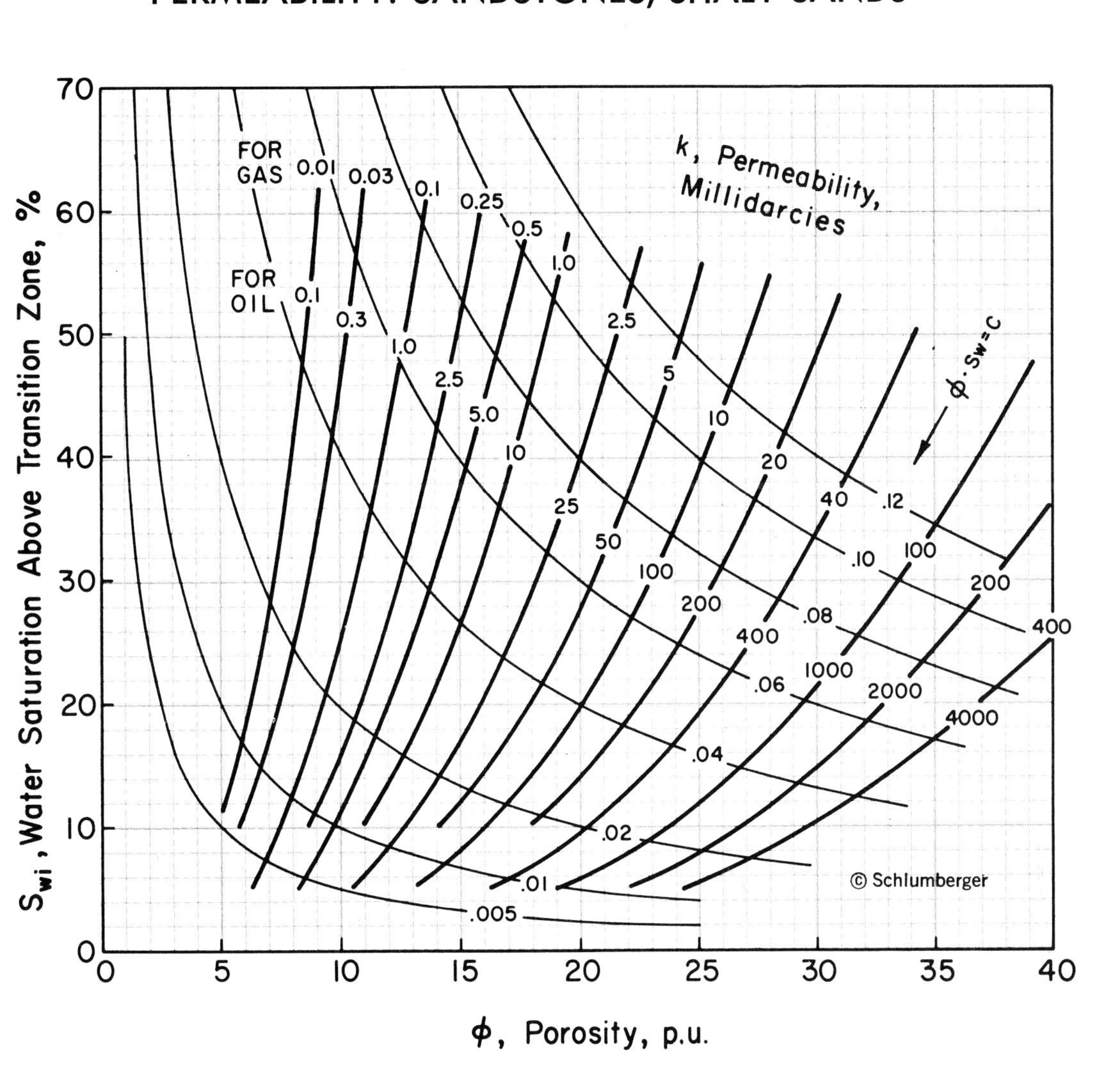

Appendix 1. Chart of porosity (ϕ) versus irreducible water saturation ($S_{w\ irr}$) for estimating permeability and determining bulk volume water ($C = S_w \times \phi$).

Courtesy Schlumberger Well Services.

Copyright 1969, Schlumberger.

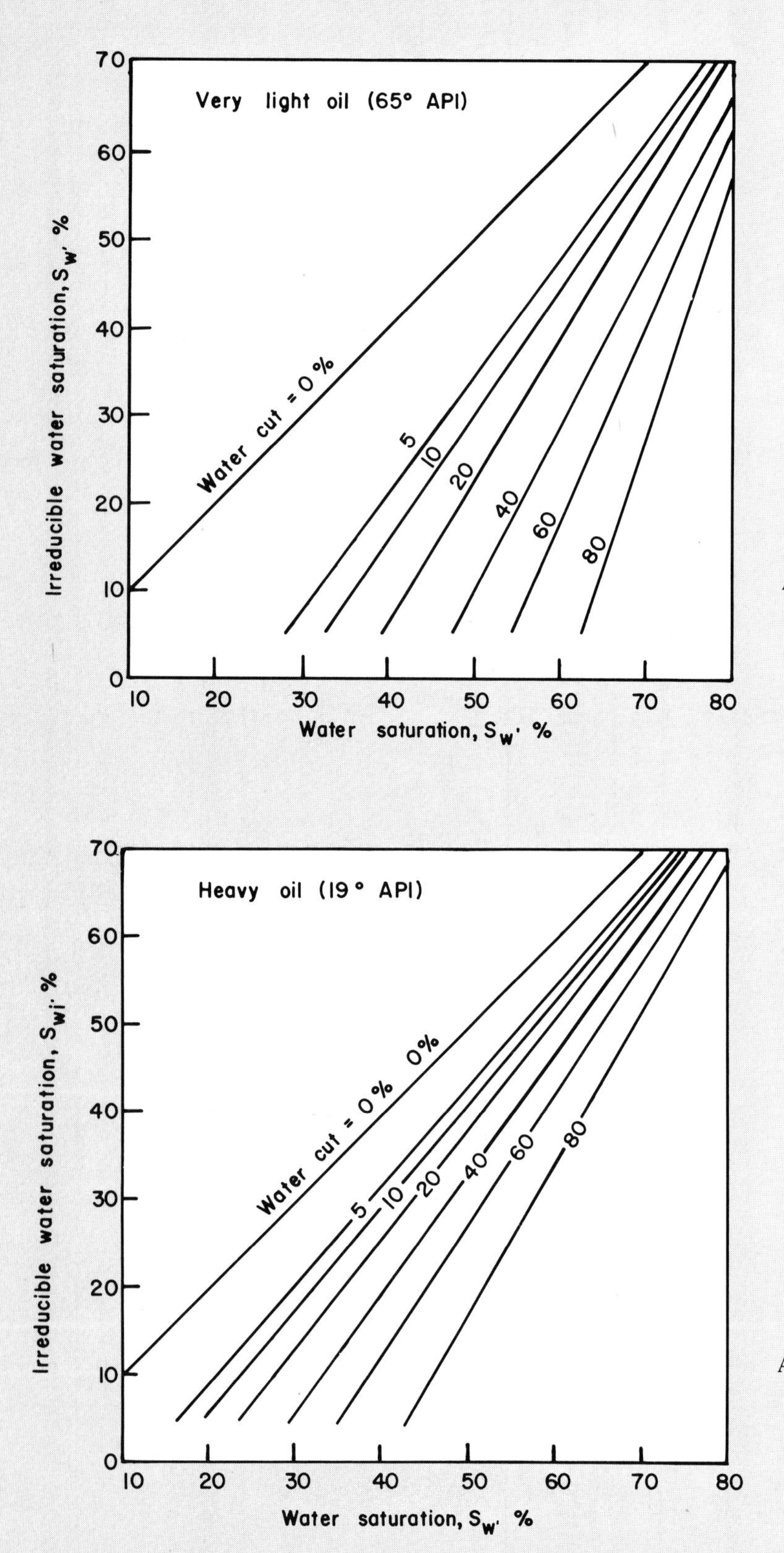

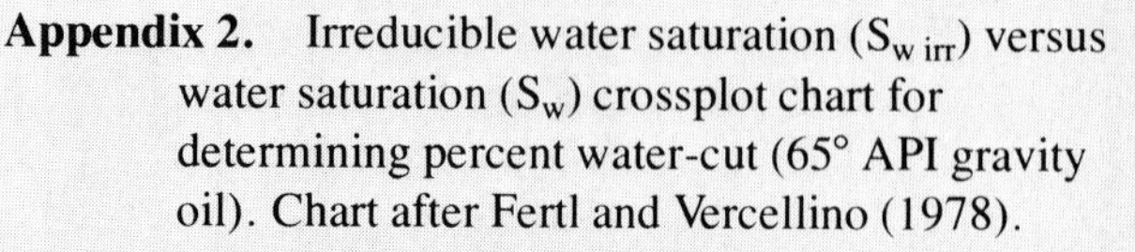

Appendix 2. Irreducible water saturation ($S_{w\,irr}$) versus water saturation (S_w) crossplot chart for determining percent water-cut (65° API gravity oil). Chart after Fertl and Vercellino (1978).

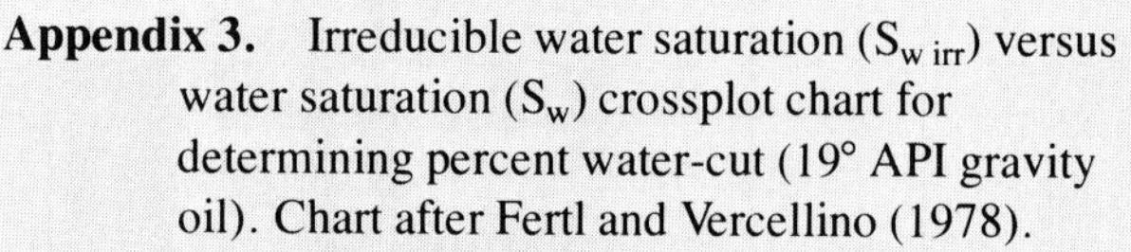

Appendix 3. Irreducible water saturation ($S_{w\,irr}$) versus water saturation (S_w) crossplot chart for determining percent water-cut (19° API gravity oil). Chart after Fertl and Vercellino (1978).

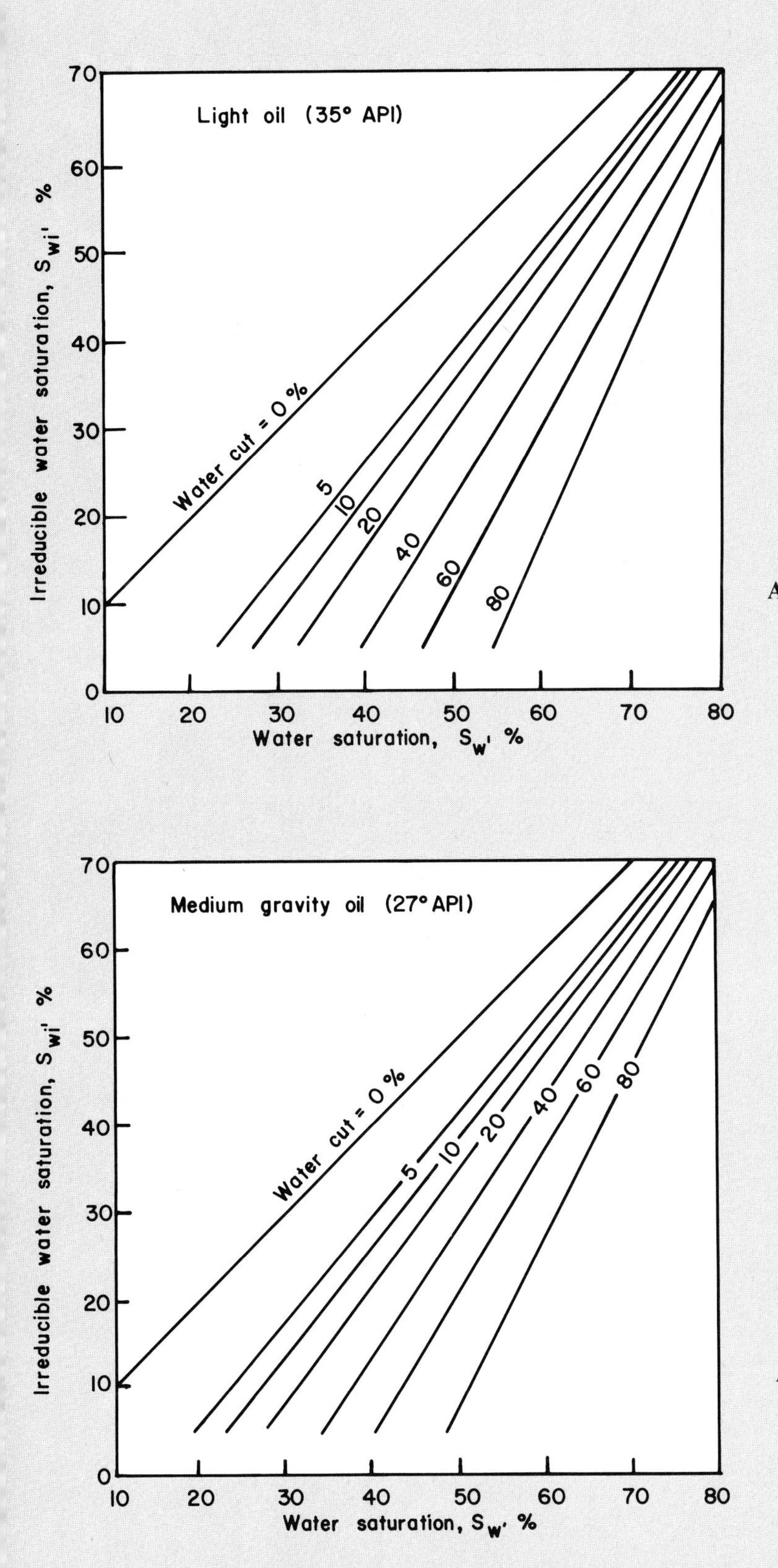

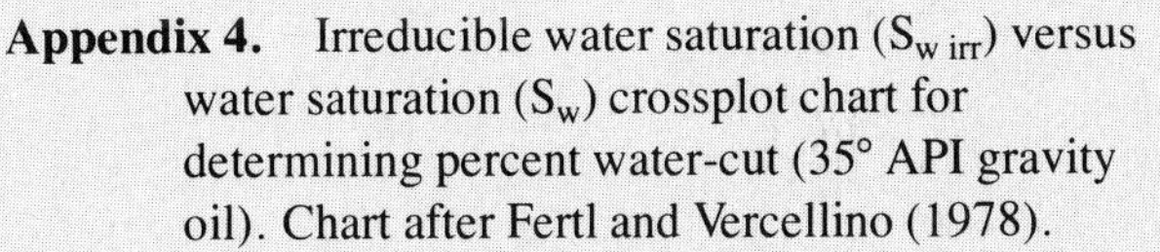

Appendix 4. Irreducible water saturation ($S_{w\,irr}$) versus water saturation (S_w) crossplot chart for determining percent water-cut (35° API gravity oil). Chart after Fertl and Vercellino (1978).

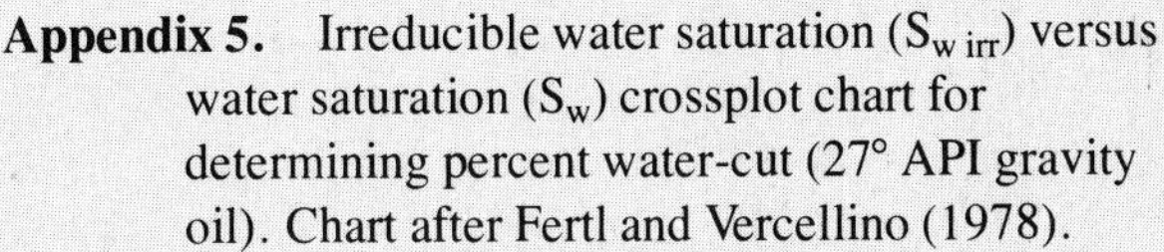

Appendix 5. Irreducible water saturation ($S_{w\,irr}$) versus water saturation (S_w) crossplot chart for determining percent water-cut (27° API gravity oil). Chart after Fertl and Vercellino (1978).

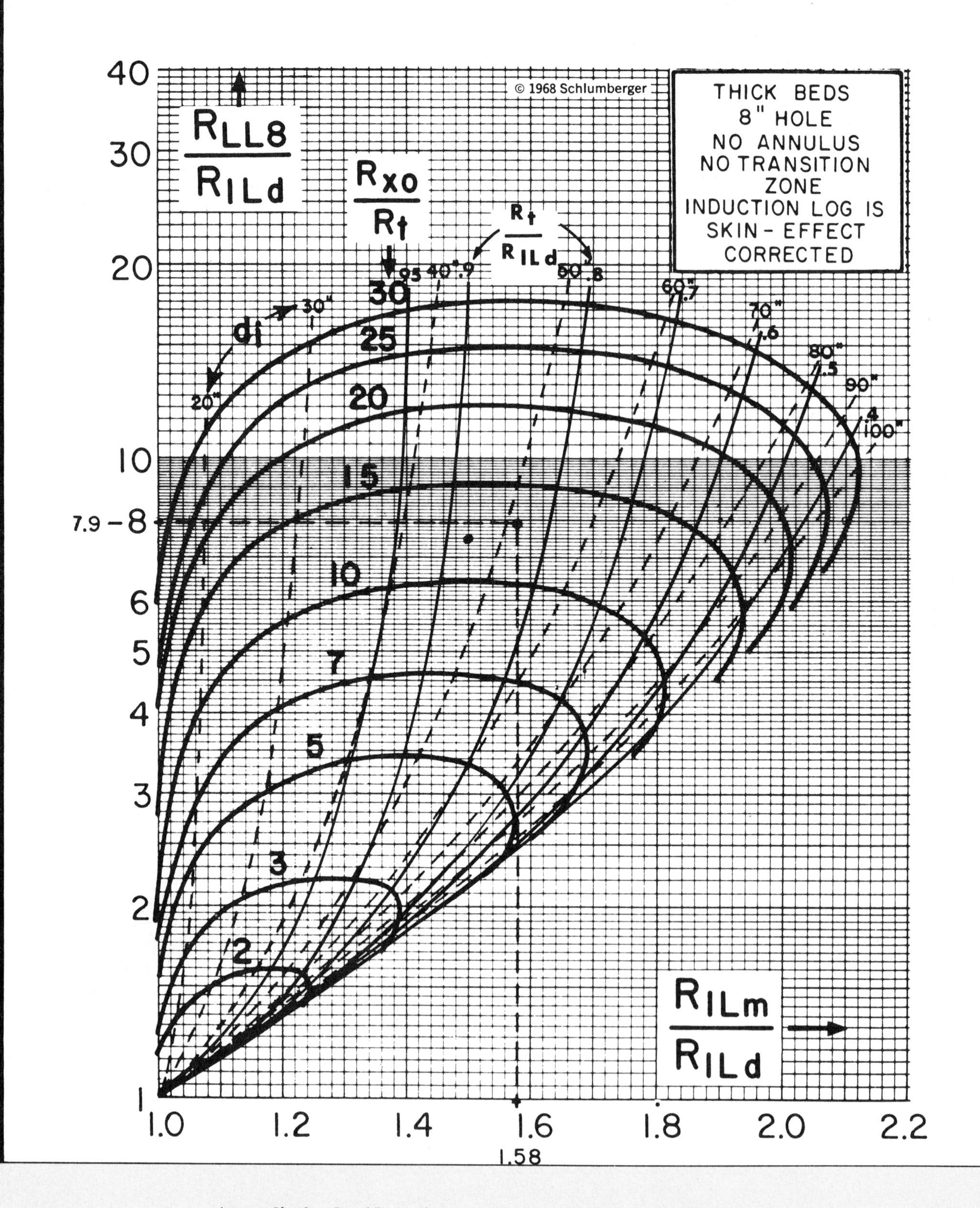

Appendix 6. Dual Induction-Laterolog-8* chart for correcting R_{ILd} to R_t.

Courtesy of Schlumberger Well Services.

Copyright 1968, Schlumberger.

REFERENCES

Alger, R. P., 1980, Geological use of wireline logs (p. 207-222) *in* G. D. Hobson, ed., Developments in petroleum—2: London, Applied Science Publishers, Ltd., 345 p.

Archie, G. E., 1942, The electrical resistivity log as an aid in determining some reservoir characteristics: Petroleum Technology, v. 5, p. 54-62.

Asquith, G. B., 1979, Subsurface carbonate depositional models—a concise review: Tulsa, PennWell, 121 p.

———1980, Log analysis by microcomputer: Tulsa, PennWell, 105 p.

Bateman, R. M., and C. E. Konen, 1977, The log analyst and the programmable pocket calculator: The Log Analyst, v. 18, no. 5, p. 3-11.

Carothers, J. E., 1968, A statistical study of the formation factor relation to porosity: The Log Analyst, v. 9, p. 38-52.

———and C. R. Porter, 1970, Formation factor-porosity relation from well log data: Soc. Professional Well Log Analysts, 11th Ann. Logging Symp., Trans., paper D.

Coates, G., and J. L. Dumanoir, 1973, A new approach to improve log-derived permeability: Soc. Professional Well Log Analysts, 14th Ann. Logging Symp., Trans., paper R.

Doll, H. G., 1948, The SP log, theoretical analysis and principles of interpretation: Trans., AIME, v. 179, p. 146-185.

Dresser Atlas, 1974, Log review—1: Houston, Dresser Industries, Inc.

———1975, Log interpretation fundamentals: Houston, Dresser Industries, Inc.

———1979, Log interpretation charts: Houston, Dresser Industries, Inc., 107 p.

Fertl, W. H., 1975, Shaly sand analysis in development wells: Soc. Professional Well Log Analysts, 16th Ann. Logging Symp., Trans., paper A.

———1978, R_{wa} Method—fast formation evaluation, *in* Practical log analysis—8: Oil and Gas Jour., (May 15, 1978 - Sept. 19, 1979).

———and W. C. Vercellino, 1978, Predict water cut from well logs, *in* Practical log analysis—4: Oil and Gas Jour., (May 15, 1978 - Sept. 19, 1979).

Hilchie, D. W., 1978, Applied openhole log interpretation: Golden, Colorado, D. W. Hilchie, Inc.

———1979, Old electric log interpretation: Golden Colorado, D. W. Hilchie, Inc., 161 p.

Hingle, A. T., 1959, The use of logs in exploration problems: Soc. Exploration Geophysicists, 29th Mtg. (Los Angeles).

Jaafar, I. B., 1980, Depositional and diagenetic history of the B-zone of the Red River Formation (Ordovician) of the Beaver Creek Field, Golden Valley County, North Dakota: M.S. thesis, West Texas State Univ. 68 p.

Johnson, H. M., 1958, The importance of accuracy in basic measurements for electric log analysis: 3rd Ann. Conf. on Well Logging Interpretation, McMurray College, Abilene, Texas.

Kobesh, F. P., and R. B. Blizard, 1959, Geometric factors in sonic logging: Geophysics, v. 24, p. 64-76.

Morris, R. L., and W. P. Biggs, 1967, Using log-derived values of water saturation and porosity: Soc. Professional Well Log Analysts, 8th Ann. Logging Symp. Trans., paper O.

Pickett, G. R., 1972, Practical formation evaluation: Golden Colorado, G. R. Pickett, Inc.

———1977, Recognition of environments and carbonate rock type identification: *in* Formation evaluation manual unit II, section exploration wells: Tulsa, Oil and Gas Consultants International, Inc., p. 4-25.

Sethi, D. K., 1979, Some considerations about the formation resistivity factor-porosity relationships: Soc. Professional Well Log Analysts, 20th Ann. Logging Symp. Trans., paper L.

Schlumberger, 1968, Log interpretation/charts: Houston, Schlumberger Well Services, Inc.

———1969, Log interpretation/charts: Houston, Schlumberger Well Services, Inc.

———1972, Log interpretation/charts: Houston, Schlumberger Well Services, Inc.

———1972, Log interpretation manual/principles, vol. I: Houston, Schlumberger Well Services, Inc.

———1974, Log interpretation manual/applications, vol. II: Houston, Schlumberger Well Services, Inc.

———1975, A guide to well site interpretation of the Gulf Coast: Houston, Schlumberger Well Services, Inc.

———1977, Log interpretation/charts: Houston, Schlumberger Well Services, Inc.

———1979, Log interpretation/charts: Houston, Schlumberger Well Services, Inc.

Society of Professional Well Log Analysists, 1975, Glossary of terms and expressions used in well logging: Houston, Soc. Professional Well Log Analysts, 74 p.

Simandoux, P., 1963, Mesures dielectriques en milieu poreux, application a mesure des saturations en eau: Etude du Comportement des Massifs Argileux, Revue de l'institut Francais du Petrole, Supplementary Issue.

Suau, J., P. Grimaldi, A. Poupon, and G. Souhaite, 1972, Dual Laterolog— R_{xo} tool: Soc. Petroleum Engineers - AIME, 47th Ann. Meeting (San Antonio), paper spe-4018.

Timur, A., 1968, An investigation of permeability, porosity, and residual water saturation relationships for sandstone reservoirs: The Log Analyst, v. 9, (July - August), p. 8-17.

Tittman, J., and J. S. Wahl, 1965, The physical foundations of formation density logging (Gamma-Gamma): Geophysics, v. 30, p. 284-294.

Tixier, M. P., R. P. Alger, W. P. Biggs, and B. N. Carpenter, 1963, Dual induction-laterolog—a new tool for resistivity analysis: Soc. Petroleum Engineers - AIME, 38th Ann. Meeting (New Orleans), paper no. spe-713.

Truman, R., R. P. Alger, J. Connell, and R. L. Smith, 1972, Progress report on interpretation of the dual spacing neutron log (CNL): Soc. of Professional Well Log Analysts, 13th Ann. Logging Symp. Trans., paper U.

Watney, W. L., 1979, Gamma ray-Neutron cross-plots as an aid in sedimentological analysis (p. 81-100), *in* D. Gill and D. F. Merriam, eds., Geomathematical and petrophysical studies in sedimentology: Pergamon Press, 266 p.

———1980, Cyclic sedimentation of the Lansing-Kansas City groups in northwestern Kansas and southwestern Nebraska: Kansas Geol. Survey Bull. 220, 72 p.

Wermund, E. G., 1975, Upper Pennsylvanian limestone banks, north central Texas: Univ. Texas, (Austin), Bur. Econ. Geol. circ. 75-3, 34 p.

Wylie, M. R. J., and W. D. Rose, 1950, Some theoretical considerations related to the quantitative evaluations of the physical characteristics of reservoir rock from electric log data: Jour. Petroleum Technology, v. 189, p. 105-110.

———A. R. Gregory, and G. H. F. Gardner, 1958, An experimental investigation of the factors affecting elastic wave velocities in porous media: Geophysics, v. 23, p. 459-493.

Explanation of Indexing:

A reference is indexed according to its important, or "key", words.

Three columns are to the left of a keyword entry. The first column, a letter entry, represents the AAPG book series from which the reference originated. In this case, R stands for methods in Geology Series. Every five years, AAPG will merge all its indexes together, and the letter R will differentiate this reference from those of the AAPG Studies in Geology Series (S), the AAPG Memoir Series (M), or from the AAPG Bulletin (B).

The following number is the series number. In this case, 3 represents a reference from Methods in Geology Series No. 3.

The last column entry is the page number in this volume where this reference will be found.

Note: This index is set up for single-line entry. Where entries exceed one line of type, the line is terminated. The reader must sometimes be able to realize keywords, although commonly taken out of context.

FLOW CHART FOR LOG INTERPRETATION

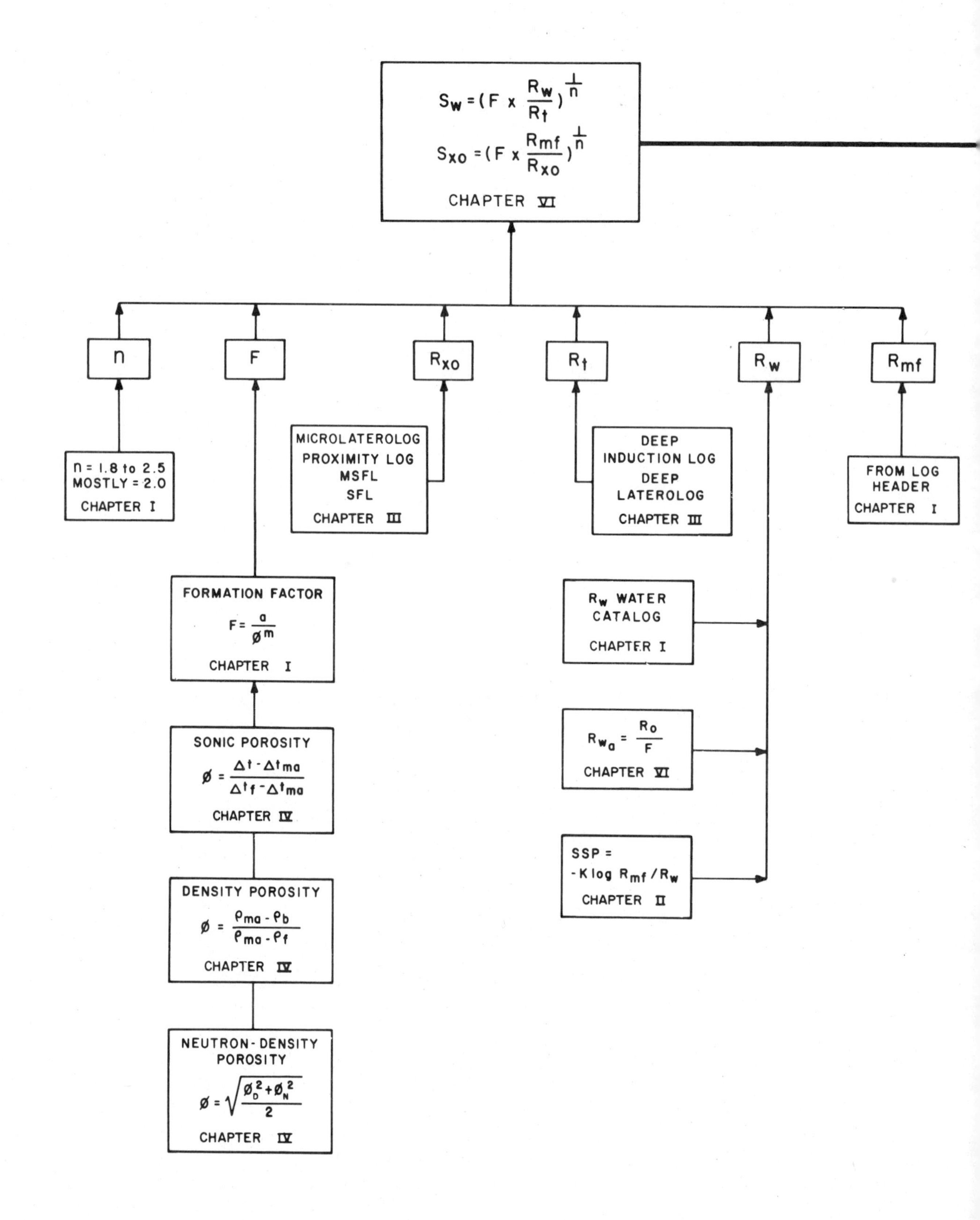
$S_w = (F \times \frac{R_w}{R_t})^{\frac{1}{n}}$
$S_{xo} = (F \times \frac{R_{mf}}{R_{xo}})^{\frac{1}{n}}$
CHAPTER VI
n
F
R_{xo}
R_t
R_w
R_{mf}
n = 1.8 to 2.5
MOSTLY = 2.0
CHAPTER I
MICROLATEROLOG
PROXIMITY LOG
MSFL
SFL
CHAPTER III
DEEP
INDUCTION LOG
DEEP
LATEROLOG
CHAPTER III
FROM LOG
HEADER
CHAPTER I
FORMATION FACTOR
$F = \frac{a}{\phi^m}$
CHAPTER I
R_w WATER
CATALOG
CHAPTER I
$R_{wa} = \frac{R_o}{F}$
CHAPTER VI
SONIC POROSITY
$\phi = \frac{\Delta t - \Delta t_{ma}}{\Delta t_f - \Delta t_{ma}}$
CHAPTER IV
SSP =
$-K \log R_{mf}/R_w$
CHAPTER II
DENSITY POROSITY
$\phi = \frac{\rho_{ma} - \rho_b}{\rho_{ma} - \rho_f}$
CHAPTER IV
NEUTRON-DENSITY
POROSITY
$\phi = \sqrt{\frac{\phi_D^2 + \phi_N^2}{2}}$
CHAPTER IV